CARBON AND CARBONACEOUS COMPOSITE MATERIALS

Structure-Property Relationships

CARBON AND CARBONACEOUS COMPOSITE MATERIALS

Structure-Property Relationships

Malenovice, Czech Republic 10–13 Oct 1995

Editors

K. R. Palmer, D. T. Marx & M. A. Wright

Southern Illinois University
Carbondale, Illinois

World Scientific
Singapore • New Jersey • London • Hong Kong

Published by

World Scientific Publishing Co. Pte. Ltd.

P O Box 128, Farrer Road, Singapore 912805

USA office: Suite 1B, 1060 Main Street, River Edge, NJ 07661

UK office: 57 Shelton Street, Covent Garden, London WC2H 9HE

British Library Cataloguing-in-Publication Data
A catalogue record for this book is available from the British Library.

CARBON AND CARBONACEOUS COMPOSITE MATERIALS
Structure-Property Relationships

ISBN 981-02-2801-5

This book is printed on acid-free paper.

Printed in Singapore by Uto-Print

Preface and Acknowledgments

The study of carbon and carbonaceous materials involves a number of different scientific disciplines varying from geologists, who usually bring their expertise in optical microscopy, in particular polarized light; to chemists, who deal with carbon reactivity; to physicists, who think of neutron absorption; to chemical engineers, who tend to be interested in formulations, carbonization and graphitization; and to materials scientists, who tend to relate microstructure to physical and mechanical properties. Obviously many other disciplines have interest in the subject and many other techniques can be identified as useful in the study of this rather remarkable material or set of materials. Indeed, interest in such materials rapidly accelerated when carbon fibers became a commercially important material, especially when Rolls-Royce planned to incorporate them into early versions of the RB-211 turbo jet engine. Unfortunately, reinforced epoxy fan blades never were a success; however, carbon fibers have become accepted in many structural materials, especially those used in the aerospace industry. Carbon fibers are now fabricated from PAN, petroleum and coal tar pitch and from a carbon containing gas phase. This latter material is quite novel and this conference proved to be a convenient and timely opportunity for a review of it.

Early in 1994, an international committee was put together to organize and coordinate sessions describing current research into carbon and carbonaceous products. The Technical University of Ostrava agreed to host the meeting at hotel Petr Bezruč which is situated in the village of Malenovice in the mountains Beskydy in the north eastern section of the Czech Republic. The Czech Academy of Sciences also agreed to act as local cosponsors. The United States Office of Naval Research, the College of Engineering College of Southern Illinois University at Carbondale and the Materials Technology Center provided additional technical support. The conference itself was conceived as a vehicle for world scientists to present their research without restriction. Thus, attendees were attracted from Belgium, China, England, India, Japan, Poland, Portugal, Spain, USA, as well as the host country, the Czech Republic.

The conference was organized into seven sessions: Applications of Carbon and Carbonaceous Composite Materials; Physical Properties of Carbon Materials; Carbon Fibers and Carbon Particulates; Fabrication of Carbon Artifacts, Fibers and C-C Composites; Interface Effects; Mechanical, Chemical and Tribological Properties; and a round table discussion of worldwide carbon research that completed the program. Two plenary lectures were given. The first, entitled "Clusters of Carbon," was presented by Professor Dr. R. Zahradnik of the Czech Academy of Sciences; the second was given by Max Lake of Applied Sciences of the USA and was entitled "Manufacturing Issues for Vapor Grown Carbon Fibers."

The high quality presentations made by the invited speakers set the standard for the papers that followed. Thus the organizers would like to express their appreciation for the effort put forth by; Dr. O.P. Bahl of the National Physical Laboratory, India; Mr. A. Bertram, Naval Surface Warfare Center, USA; Dr. P. Glogar, Czech

Academy of Sciences; Dr. J. P. Issi, Université Catholique de Louvan, Belgium; Dr. L. M. Manocha, Sadar Patel University, India; Dr. H. Marsh, Carbon Consultant, UK; Dr. B. McEnaney, University of Bath, UK; Dr. I. Mochida, Kyushu University, Japan; and Dr. S. Zengmin, Beijing Institute of Chemical Technology, China.

We are sure that we speak for all of the conference participants when we complement those members of the local planning committee for the magnificent job they did in making sure that the conference ran so smoothly. The hotel venue that they chose was most appropriate, the meals were first class, the outside trips ran very smoothly and the weather cooperated so wonderfully as to make the whole conference package a most enjoyable experience. Our thanks therefore to the following; Drs V. Dembovská, M. Kaloč, Z. Toman, P. Jelínek, V. Roubíček, K. Balík and J. Mead.

Finally, we would like to thank Dr. Peter Filip of the Technical University of Ostrava, who coordinated most of the day-to-day activities during the conference, and the graduate students who helped him.

Contents

Applications of Carbon

and

Carbonaceous Composite Materials

LOW-COST PITCH-BASED CFRC

M. HÁJEK
AGA Gas s.r.o., Molákova 576
180 00 Praha 8, Czech Republic

J. SVÍTILOVÁ and J. GREGOR
Institute of Rock Structure and Mechanics
Academy of Sciences of the Czech Republic
V Holešovičkách 41, 182 09 Praha 8, Czech Republic

ABSTRACT

A novel method of carbon-carbon composite manufacture was designed and the possibility of its realization was proved. It is based on the continual heat treatment of coal tar pitch which is soaked into a carbon fiber bundle and the subsequent hot-press molding. Furthermore, this method promises an easy realization at the industrial level and advantageous incorporation into the carbon fiber production process.

1. Introduction

Carbon fiber-reinforced carbon composites (CFRC), due to their outstanding properties, especially in the high-temperature region, are doubtless the material of the future. It is, for example, the only serious candidate for an inner material of the nuclear fusion reactor.

Unfortunately, to obtain dense CFRC materials with good mechanical properties, many impregnation-recarbonization cycles are required for conventional production. It is caused by low carbon yield of utilized matrix precursors, most commonly phenolic resins (first step) and coal tar pitch (next steps). These time- and energy-wasting steps make CFRCs extremely expensive so that their use is very limited. To make CFRCs more competitive, changes in fabrication are necessary.

New methods enabling the manufacturer to avoid standard multiple impregnation-recarbonization steps during manufacture of CFRC have begun to appear during recent years[1-5]. Their common feature is the enhanced carbon yield of the matrix precursor which can be achieved by, for example, non-specified "self-sintering" carbon[1] or by blending of coal tar pitch with finely powdered pitch-coke[2], special oligomers[3] or thermally treated "mesophase" pitch[4], etc. To form the composite successfully, all types of matrix precursors mentioned require hot molding to compensate for their high viscosity (note: only slight softening occurs and is achieved only within the high temperature range of about 440–500°C, depending on the heating rate). The coating of the carbon fiber bundle or the carbon cloth with such precursor is not simple, and it requires a very finely powdered consistency of matrix precursor. The coating can be carried out through either electrophoresis[1], dry electrostatic method, or in the fluidized bed[4].

The starting point of our attempt to improve the process is the idea of performing all the procedures of the thermal pretreatment of pitch *directly* and *continually* on the moving carbon fiber tow (the procedure can be incorporated advantageously into the carbon fiber production process). Impregnation of the bundle by pitch is simple, because it viscosity is low (at 200°C). Thus it avoids the problem of the costly powdering and coating technologies.

2. Experimental

Two types of coal tar pitch were employed in our experiments. The first was used in the raw state. The second was filtered at 200°C. Properties of both the raw and filtered pitches are reported in Table 1.

Table 1. Properties of Utilized Coal Tar Pitch

Parameter	Type 1	Type 2	Type 2, filtered
softening point (Mettler) (°C)	104.1	100.1	105.6
Alcan coke yield (%)	56.4	55.0	58.3
ash (%)	0.19	0.21	0.08
viscosity at 160°C (cP)	819	—	—
sulfur content (%)	0.46	0.46	0.31
TI (%)	31.6	25.3	—
QI (%)	11.95	7.2	1.2

The amount of pitch soaked into a Torayca T800H bundle of 6000 filaments at 200°C was controlled so as to reach the required matrix content in the final composite. But the control was very difficult under conditions of simple laboratory equipment so that fiber volume fraction did not exceed 50% while recommended values are about 60%. The heat treatment of the pitch-impregnated carbon bundle was achieved by pulling it through a tube furnace heated to a required temperature in a stream of nitrogen. The heat treatment temperature (HTT) is the key parameter of the process and its optimum range is narrow. The velocity of pulling was 1 m/min, and the length of the furnace was 2 m.

The heat-treated bundle (tow-preg) is a rod and is not very stiff. The bundle can be made flexible by "breaking" it by passing it over a pulley with a diameter of less than approximately 20 mm without significant matrix precursor loss, thus a cloth can be woven. Nevertheless, the present stage focuses on the unidirectional composite.

The tow-preg was wound on a drum with a diameter of 300 mm. Then it was cut into pieces of 0.5 m length and the linear density was checked by weighing. Further cutting to 68 mm rods followed, and the required number of rods were stacked into a steel mold with a cross-section of 36 x 72 mm (2 mm wide gaps were left at each end of the longer dimension of the mold). This set-up enabled one to directly determine a weight loss of the matrix precursor during the hot molding and subsequently the secondary carbonization operations. The mold was then heated at a rate of 2.5°C/min, and pressure of 100 MPa was applied simultaneously from 400°C onward. It was proved in the preliminary experiments that lower pressure is not sufficient and higher pressure caused oil leakage. After reaching 600°C, the temperature was kept constant for 30 min. The pressure was released after cooling the mold down to 500°C. Some additional experiments with variation of heating rate during hot molding were carried out for tow-pregs with a HTT of 470°C.

The secondary carbonization of the "green" composite was carried out to 1000°C at a heating rate of 5°C/min in a nitrogen environment at atmospheric pressure. CFRC plates 1.3–1.8 mm thick were manufactured under various conditions. Eight 35 x 6–8 mm specimens were cut from each plate; the remains were used for the microscopy observation and for real density (pycnometric in methyl alcohol) evaluation. The specimens were subjected first to density and then to flexural properties evaluation.

The density was calculated using the weights and dimensions of the specimens. A testing apparatus (MINIMAT from Polymer Science Ltd.) was used for the flexural strength tests in a center-loading three-point arrangement (support span 24 mm, cross-head velocity 0.5 mm/min, and load cell 1000 N). The static Young's modulus was determined from the linear part of the load versus displacement curve.

3. Results

The total weight loss of the matrix precursor achieved during the hot molding and secondary carbonization steps is very low and decreased from 4.2% at 450°C to 2.7% at 500°C as shown in Fig. 1. The plot of the dependence of flexure strength on HTT exhibits a maximum between 470 and 480°C as shown in Fig. 2. The plot of dependence of density of CFRC on HTT has a similar shape. The maximum is slightly wider (between 460 and 480°C) as shown in Fig. 3. Decreasing the heating rate in the hot molding step has a positive influence on both density and flexure strength of CFRC as reported in Figs. 4 and 5.

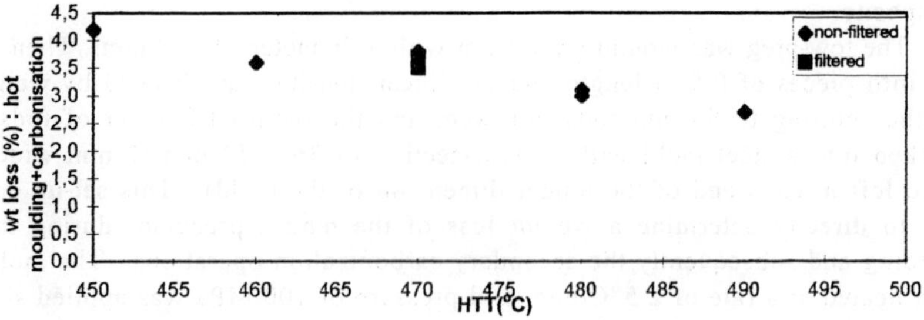

Fig. 1. Dependence of total weight loss during CFRC manufacture on HTT of tow-preg.

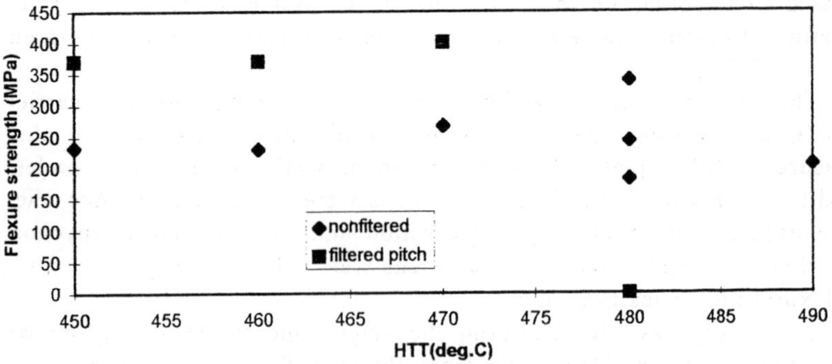

Fig. 2. Dependence of flexure strength of CFRC on HTT of tow-preg.

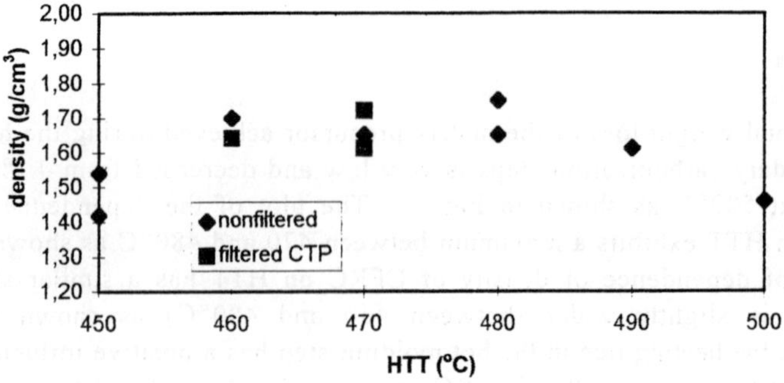

Fig. 3. Dependence of density of CFRC on HTT of tow-preg.

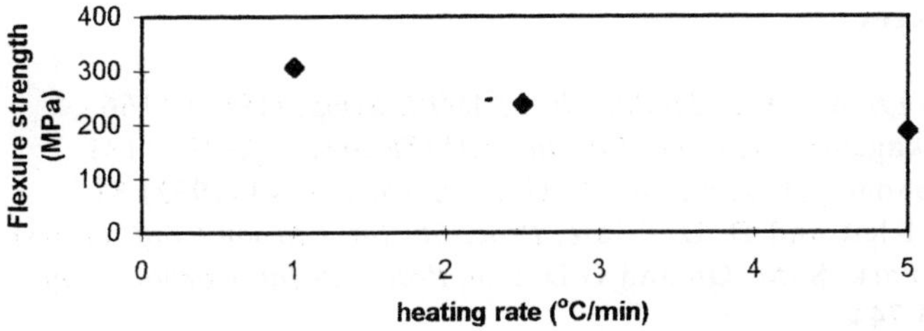

Fig. 4. Dependence of flexure strength of CFRC on hot mold heating rate.

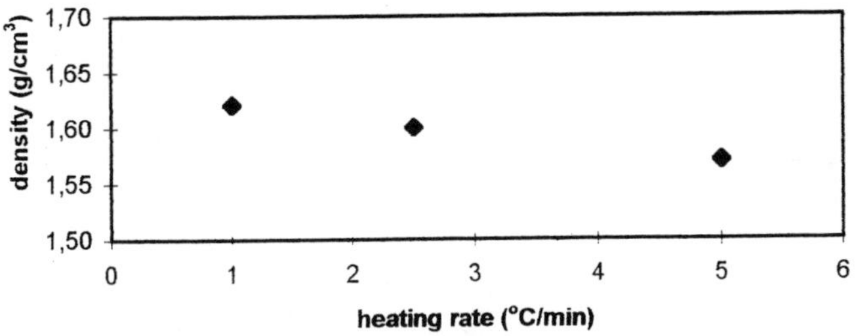

Fig. 5. Dependence of density of CFRC on hot mold heating rate.

4. Conclusions

Despite texture imperfections in which many cavities or cracks occur, the mechanical properties of CFRC prepared by our method are comparable with those referred to in references 3 and 4.

Considerable improvement of CFRC properties was achieved by utilization of filtered coal tar pitch.

The method is simple and utilizes cheap coal tar pitch as a raw material for carbon matrix. Furthermore, this method is easily realizable at the industrial level and can be incorporated advantageously into the carbon fiber production process.

Further optimization of the tow-preg manufacture, hot-press, and secondary carbonization conditions together with an optimal choice of the fiber/matrix ratio and fiber and/or pitch quality are expected to bring more uniformity to the texture and considerable improvement of the CFRC properties. Efficient tools for such improvement were demonstrated.

5. Acknowledgment

The financial support of this work by the Grant Agency of the Czech Republic, grant No. 106/93/0835, is gratefully acknowledged.

6. References

1. S. Sakagami, *Proc. 1st Jap. Int. SAMPE Symp.* (1989) 1166.
2. T. Nakagawa, *Proc. 1st Jap. Int. SAMPE Symp.* (1989) 1181.
3. J. Economy, H. Jung and T. Gogeva, *Carbon* **30** (1992) 81.
4. J. W. Klett and D. D. Eddie, *Proc. 6th Int. Carbon Conf.* (1993) 48.
5. Y.-D. Park, S.-M. Oh and G.D. Lee, *Proc. 5th Int. Carbon Conf., Carbon '92* (1992) 744.

C/TiNi COMPOSITES - A PROSPECTIVE MATERIAL FOR MEDICAL APPLICATIONS

P. FILIP, M. KALOČ, H. LORETROVÁ, E. LINZER,
B. HORÁK , and K. MAZANEC
Faculty of Metallurgy and Material Engineering
Technical University of Mining and Metallurgy
tř, 17, listopadu, 708 33 Ostrava - Poruba, Czech Republic

J. MUSIALEK
Municipal Hospital
Orthopaedic Department
Ostrava - Fifejdy, Czech Republic

M. A. WRIGHT
Materials Technology Center
Southern Illinois University at Carbondale
Carbondale, IL 62901, USA

1. Introduction

Carbon is the element most frequently found in bioorganic compounds. It is unique among the elements because of the variety and vast number of forms it can take. There are over a million carbon compounds, many thousands of which are vital to organic and life processes[1]. Some carbon forms have optimal chemical and mechanical structure (the benign biological reaction and the similarity in stiffness and strength between carbon and living tissues) and have been found suitable for use in partial or total replacement of damaged biological tissues such as bones, cartilages, ligaments, tendons and vascular structures. The acute and chronic responses to these special carbon forms are minimal and the said carbon structures are unaffected by the in-vivo environment and are incorporated in the living host. Three types of carbon are used extensively in biomedical applications: the low temperature isotropic (LTI) modification of pyrolytic carbon, vitreous (glassy) carbon and the ultralow temperature isotropic (ULTI) type of vapor deposited carbon[2]. Beside these three clinically significant materials, other forms of carbon are being considered for use in medicine. Carbon-carbon composite materials are discussed as possible bone substitutes or load carrying devices in joint replacement, and braided carbon-fiber systems have received attention as possible elements in the reconstruction of ligaments or tendons. Difficulties encountered with these two carbon materials are generally a result of their structure. The particles from composite particulate and fiber systems can migrate into blood and the lymphatic system[1]. On the contrary, the three above mentioned carbon materials, which are in use at present, are integral materials (LTI and vitreous carbon) or impermeable thin coatings (ULTI carbon). These materials (LTI, ULTI and glassy carbon) are subcrystalline forms with a lower degree of order between the layers contrary to graphite. The crystal structure of these carbons is called turbostratic and is two-dimensional. The mechanical properties are intimately related to the bonding characteristics since they are formed by covalent bonds between carbon atoms (bond energy 477 $kJmol^{-1}$) in combination with van der Waals forces (bonding energy 17 $kJmol^{-1}$). A controlled mixture of these two bonding mechanisms produces materials with low elastic moduli (20 GPa) and high flexural strength (250–600 MPa). The turbostratic carbon materials are extremely tough compared with bioceramics such as aluminum oxide. The energy to fracture Al_2O_3 is 0.18 MJm^{-3}, compared with LTI carbon with 5.5 MJm^{-3}. The strain to fracture for vapor deposited carbons is greater than 5%, making it possible to prepare highly flexible coatings (strain to failure of Al_2O_3 is ≈ 0.1%)[2].

The near equiatomic TiNi alloys exhibit shape memory behavior[3]. They are characterized by excellent corrosion resistance and biocompatibility which are attributed to the intermetallic character of this alloy[4,5]. This allows the TiNi material to be effectively used as a highly biofunctional and "smart" material in medicine[3–8].

The two most used shape memory phenomena in medicine are the so called shape memory effect (SME) and pseudoelasticity (PE). A shape memory material can be deformed up to strains of order 10% and this deformation is partially or totally reversible after heating the material above a critical temperature. If shape recovery occurs in constrained conditions, stresses higher than those necessary for original deformation are generated. The critical temperatures, extent of recovery, and the level of generated stresses depend on the chemical constitution of the TiNi alloy and on the previous mechanical and/or thermal treatment[3,9,10]. The SME is dominantly utilized for bone fracture fixation, cardiovascular purposes and for the anchoring of implants.

Pseudoelastic behavior is characterized by a reversible deformation of the order of 10%. In this case, contrary to the SME, the recovery of strain occurs without heating and occurs immediately after unloading of material[3]. This property is extensively used for surgical devices, in orthodontics and for tendon replacement treatment.

Composite materials, in which a combination of shape memory behavior with convenient properties of other material is used, are prospective materials from the point of view of enhanced biocompatibility, as well as biofunctionality[12,13,14]. It has been shown previously that plasma coated biomaterials based on TiNi and Al_2O_3 or $Ca_{10}(PO_4)_6(OH)_2$ have reasonably good mechanical properties and enhanced biofunctionality and seem to be very progressive in orthopaedics[15].

The aim of the present work is to prepare and to evaluate TiNi/C biomaterials which should be formulated to exhibit an appropriate elastic moduli with respect to interaction with tubular bones and osseointegration and to make possible the induction of growth of collagenous tissue. The design of prosthesis and implants was discussed previously[13,14]. Interest in this paper is focused on such topics as: change of transformation temperatures of

TiNi alloys as a consequence of processing technology, a study of the interface (interphase) of TiNi/C composite and measurement of the work of adhesion, and the damage induced by repeated transformation of TiNi as a consequence of repeated mechanical loading connected with phase transformations.

2. Experimental

In this study, near equiatomic titanium-nickel shape memory alloys were fabricated. Commercial grade TiNi alloys were supplied by FIBRA Ltd. after annealing in vacuum at 800°C in either completely austenitic (Ti-50.8 at%Ni, alloy A) or dominantly martensitic (Ti-49.5at%Ni, alloy B) state at room temperature. The alloys were available in plate form and in fine wires having a diameter of 70μm. The LTI carbon was vapor deposited on austenitic or martensitic TiNi substrates. The applied processing is described in detail in reference 17. Depending on the temperature and time parameters used during vapor deposition, different thicknesses of coatings between 2 and 6 μm were prepared. After carbon deposition, one part of each type (A and B) of specimen was annealed in a protective argon atmosphere at 600°C/3 hours/furnace. Such specimens are further called "coated and annealed specimens" (CAS). The other part represents "unannealed coated specimens" (UCS).

Differential scanning calorimetry (DSC) was performed using a Perkin-Elmer DSC apparatus operating at heating and cooling rates of 20 $Kmin^{-1}$. The thin wires (≈25 mg) were encapsulated directly into aluminum pans; the plates were first spark-cut to obtain the appropriate specimens weight (≈25 mg) for DSC.

The X-ray analyses were performed on the coated and uncoated TiNi plates by use of $Cu_{K\alpha}$ irradiation (Dron 2 diffractometer).

Scanning electron microscopy (SEM) was utilized to study the fracture surfaces of TiNi/C composites and the topology of the vapor deposited carbon on the TiNi substrates. The EDX and WDX analyses were performed using a Link 860 system. The characteristics of the interfaces of the carbon coated TiNi samples were studied on ground and mechanically polished surfaces by use of SEM and light microscopy (Neophot 2).

The wetting characteristics of the TiNi sessile drops on the carbon substrate were measured in a horizontal furnace with a protective argon atmosphere and camera-registration of the drops' shapes. The geometry of specimens and necessary dimensions were obtained by use of image analyses and developed software. The TiNi rods made from investigated materials B were cut (to weight ≈ 1g) and small rolls being placed on the carbon substrate were heated up to 1500°C. We suppose the behavior of both alloys A and B to be similar after melting because of the small concentration differences that exist between them. The wetting angle Θ and the surface energy γ_m of liquid metal TiNi were measured. Work of adhesion was calculated as $W_{ad}=\gamma_m(1+\cos\Theta)$.

The mechanical loading cycles were performed using wire specimens of alloy A, which behave pseudoelastically at room temperature. The specimens were repeatedly bent to the maximum calculated flexure strain ε = 4% (ε equals to wire diameter/2x bending radius) up to failure in a device developed for testing of dental wires (Fig. 1). The number of cycles to failure, N_f, changes in transformation temperatures after 2,000 cycles and properties of the interfaces were followed. Similar parameters (with exception of N_f) were investigated in case of wire specimens after 100 thermal cycles composed of cooling down to −196°C (liquid nitrogen) and heating up to 100°C (boiling water).

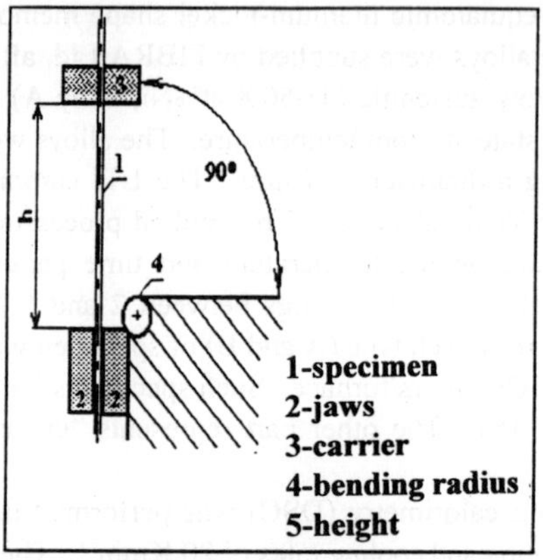

Fig. 1. Schematic of bending test device.

3. Results and Discussion

Typical DSC curves of a TiNi shape memory alloy A in as received condition (without carbon coating) are characterized with double peaks in each of the cooling cycles and a single peak in heating cycles (Fig. 2a). The peaks associated with the cooling cycle are exothermic and are associated with the transformations: B2 phase (austenite) to R-phase and subsequent B2 to martensite B19'. The transformation temperatures T_{Rs} and T_{Rf} of R-phase as well as the temperatures M_s and M_f of the martensite B19' formation, respectively, (subindices s and f represent start and finish of transformation process) can be estimated from the DSC curves. The peak associated with the heating cycle, on the other hand, is endothermic, and is characterized by the B19' to B2 transformation. Again, the start and finish temperatures of austenite B2, A_s and A_f can be estimated from this measurement. In all the cases, the area under the curve represents the enthalpy of transformation ΔH. In the case of alloy B, only one peak representing the B2 to B19' transformation is observed during cooling, conversely the B19' to B2 endothermic peak characterizes the heating process (Fig. 2b). There were no observed differences in the DSC results obtained from TiNi wires or TiNi plates. The different transformation temperatures and transformation sequences of alloy A or B are typical for TiNi alloys of different chemical composition (higher nickel content in alloy A)[3].

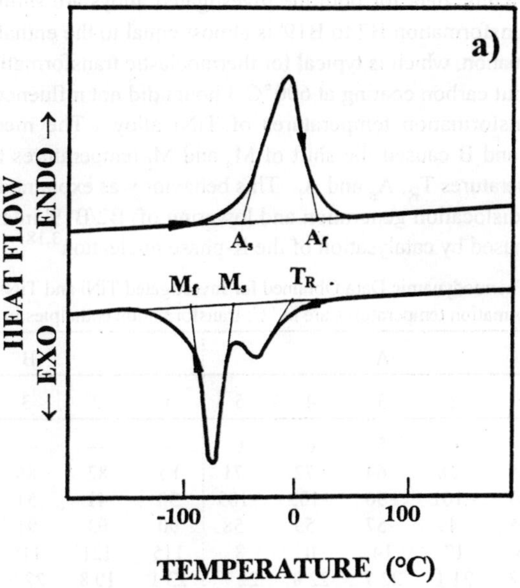
a)
HEAT FLOW
← EXO ENDO →
A_s
A_f
M_f
M_s
T_R
-100
0
100
TEMPERATURE (°C)

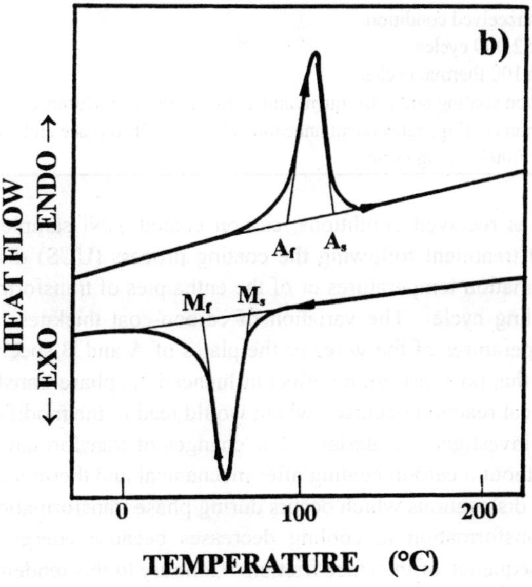
b)
HEAT FLOW
← EXO ENDO →
A_f
A_s
M_f
M_s
0
100
200
TEMPERATURE (°C)

The thermodynamic data for both the investigated alloys are summarized in Table 1. The enthalpy of transformation B2 to B19' is almost equal to the enthalpy of the reversible B19' to B2 transformation, which is typical for thermoelastic transformations. The annealing of TiNi alloys without carbon coating at 600°C/3 hours did not influence the transformation enthalpies and transformation temperatures of TiNi alloy. The mechanical cycling of uncoated alloys A and B caused the shift of M_s and M_f temperatures to lower values and increase of the temperatures T_R, A_s and A_f. This behavior was explained previously [3,18] and is connected with dislocation generation and lowering of B2/B19' interface mobility. The increase of T_R is caused by catalysation of the R-phase nucleation[3,18].

Table 1. Thermodynamic Data Obtained for Investigated TiNi and TiNi/C Systems (transformation temperatures are in °C; transformation enthalpies in $Jmol^{-1}$)

	A					B				
	1	2	3	4	5	1	2	3	4	5
T_R	5	6	5	6	6	—	—	—	71	70
M_s	-65	-68	-64	-72	-71	85	82	86	67	67
M_f	-95	-101	-96	-103	-105	50	41	51	33	34
A_s	-55	-49	-57	-59	-58	90	93	91	77	78
A_f	15	17	14	0	3	115	121	116	98	99
ΔH_{cool}	22.8	21.1	22.7	22.8	22.7	23.1	19.8	22.9	23.0	22.9
ΔH_{heat}	23.1	23.9	23.3	23.2	23.1	23.2	24.1	23.4	23.1	23.2

1 - TiNi in as received condition
2 - TiNi after 2,000 cycles
3 - TiNi after 100 thermal cycles
4 - After carbon coating and subsequent annealing at 600°C/3h/furnace
5 - After carbon coating, subsequent annealing at 600°C/3h/furnace and mechanical cycling (2,000 bending cycles)

Compared to as received conditions, carbon coated TiNi samples which were not subjected to a heat treatment following the coating process (UCS) did not undergo any changes of transformation temperatures or of the enthalpies of transformation in either the heating or the cooling cycle. The variation of carbon-coat thickness did not affect the transformation temperatures of the wires or the plates of A and B specimens. It could be concluded therefore that no constraining effect influenced the phase transformations in TiNi alloy nor any chemical reaction occurred which would lead to the modification of chemical composition of the investigated materials. The changes of transformation temperatures of UCS TiNi wires without a carbon coating after mechanical and thermal cycling correspond to the generation of dislocations which occurs during phase transformations. After cycling, the enthalpy of transformation at cooling decreases because energy is dissipated and consumed as a consequence of interface friction. Contrary to this tendency, on heating, the ΔH value increases which can be connected with stored elastic energy supporting the reversible thermoelastic transformation[3].

For the carbon-coated and annealed specimens (CAS), however, the decrease of M_s, M_f, A_s, and A_f temperatures (alloys A and B) as well as the increase of T_R in alloy A are typical. This is given by the diffusion processes leading to the change of chemical composition towards the higher nickel content in the TiNi matrix as well as to constraint. It is not possible to quantitatively distinguish between the two contributions (chemical and mechanical) influencing the transformation behavior. The higher nickel content in the TiNi matrix is responsible for the decrease of the transformation temperatures and under constraint, the rate of martensitic phase transformation is lower. After thermal and mechanical cycling of the CAS, the decrease of M_s and increase of the transformation temperatures seem to be repeatable. The true plastic deformation of CAS is probably less pronounced if compared to a specimen set with cycled UCS and uncoated wires, respectively. The transformation enthalpy is almost equal at starting the cycling and after 2000 mechanical or 100 thermal cycles.

The surface of annealed and unannealed TiNi/C specimens exhibited similar topologies (Fig. 3). The topologies did not change after annealing (compare Figs. 3a and 3b) and do not depend on the chemical composition of the TiNi alloy. The same is valid for the thickness of C coatings– no changes of the surface topology were observed for materials with different thicknesses of C coatings.

The differences between annealed (CAS) and unannealed (UCS) coated specimens were clearly demonstrated by light microscopy. Fig. 4a represents a microstructure of UCS specimen. The carbon coating is clearly visible on the TiNi wire and the corresponding interface was defectless. In the case of CAS, however, the clearly visible TiC interphase was detected (WDX) between TiNi wire and carbon-coating (Fig. 4b). Moreover, the linear EDX analysis has indicated that no changes of Ni content occurred across the wire thickness (Fig.5) during annealing. It is apparent from Fig. 4b that both interfaces between TiNi wire, interphase and carbon coating are compact without any defects.

The observed maximum, minimum and average number of cycles to failure for UCS and CAS are given in Table 2. The data for each group were obtained from seven measurements and show significant differences between annealed and unannealed carbon coated TiNi specimens. The lifetime is nearly two times higher in the case of the annealed specimens if compared with the unannealed ones. The results obtained for UCS correspond to the previously published results observed at TiNi alloys without coating. The higher values observed in CAS with TiC interphase should be attributed to the changes in the coating adhesion level.

Table 2. Observed Numbers of Cycles to Failure N^f_i (alloy A)

	N^f_{max}	N^f_{min}	N^f_{av}
UCS	1.62×10^5	0.8×10^5	1.3×10^5
CSA	3.31×10^5	2.3×10^5	2.8×10^5

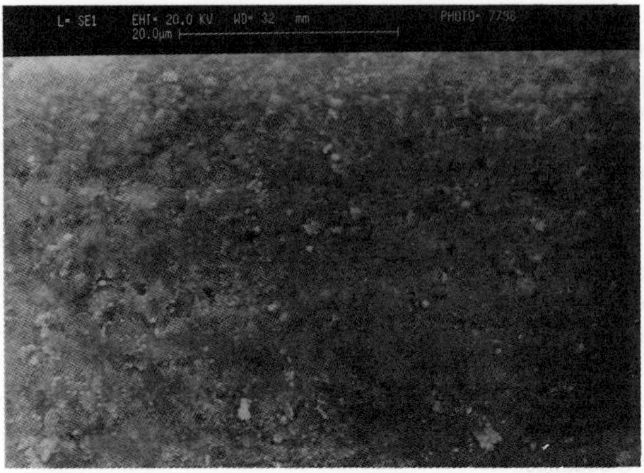

a

b

Fig. 3. Surface of carbon-coated TiNi specimens: a) unannealed coated specimens (UCS), b) coated annealed (600 °C/3 hours in a protective argon atmosphere).

a

b

Fig. 4. Microstructure of coated specimens: a) carbon coating on the TiNi wire surface of UCS, b) carbon coating and TiC interphase obtained after annealing (600 °C/3 hours).

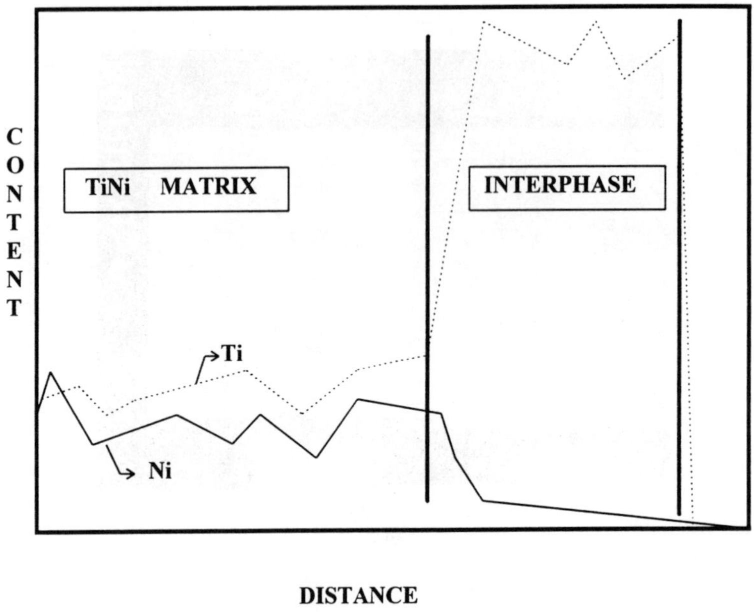

Fig. 5. Linear EDX analysis representing the changes of Ti and Ni content across the specimens thickness.

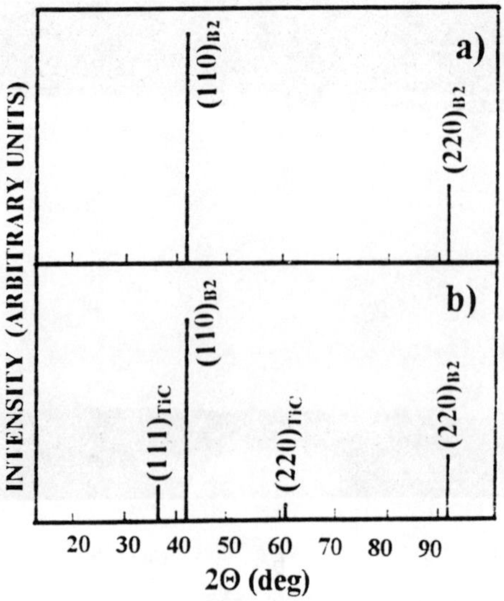

Fig. 6. X-ray diffractograms for a) UCB, and b) CAS plates of alloy A.

Figure 6a represents the X-ray diffractogram for UCS and Fig. 6b the diffractogram for the CAS. No intensive typical peaks were obtained from the carbon coatings. Thus the carbon probably has vitreous or crystalline structure with a minor turbostratic component, but the coating thicknesses may be too thin to observe stronger reflection with the X-ray apparatus used. As follows from Fig. 6b, the peaks of TiC appear in the X-ray diffraction pattern of CAS and this corresponds well with the EDX and WDX results. Moreover, the hard TiC interphase on the TiNi surface does not allow the dislocations to be generated on the surface and the formation of extrusion and intrusions is more difficult. This leads to the lower frequency of energy initiation.

Figure 7 represents the failure of specimens annealed after coating at 600°C/3 h, characterized by the interphase between TiNi and C. The high value of interfacial shear strength is typical for this specimen and the fracture is perpendicular to the wire axis. Contrary to this behavior the unannealed carbon coated specimens are typified by low interfacial shear strength, and frictional debonding and sliding of the carbon coating (Fig. 8). It is apparent that changes in the TiNi/C adhesion significantly modifies the final properties. Both the thermal and mechanical cyclings of UCS caused parts of the carbon coating to flake off. This behavior is not appropriate for implants because of the possible infiltration of debris to the blood and lymphatic system.

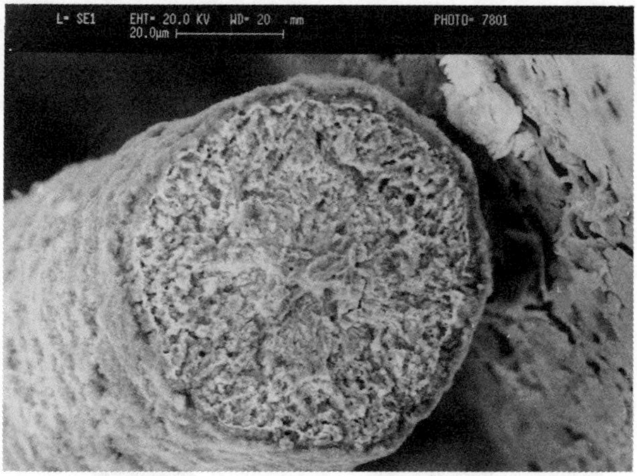

Fig. 7. Failure surface of CAS having the TiC interphase between TiNi and C.

The obtained results have demonstrated that the carbon coating without subsequent annealing is only weakly bonded to the TiNi substrate. After annealing of the specimens leading to the deformation of the TiC interlayer, the properties of TiNi/C composite materials are better. The same results were reported for other alloys. The bond strength of the ULTI carbon to stainless 316L steel and to Ti6Al4V alloy exceeds 70 MPa, due to the formation of interfacial carbides[1,2]. Differences between TiNi/C and 316 L/C or Ti6Al4V/C composites can be expected as a consequence of transformation and/or reorientation processes observed in TiNi alloys. It has been shown previously that significant contribution to the increase of the work of adhesion represents reorientation and stress induced martensite formation processes. During these processes, the energy dissipation takes place.

Fig. 8. Failure of UCS characterized by delamination of carbon coating.

Fig. 9. The geometry of "sessile TiNi drop" (1290 °C) on the carbon substrate with turbostratic structure.

The "sessile drop" geometry obtained during the wetting test at the temperature of 1290°C is presented in Fig. 9. Based on the measured values of wetting angle ($\Theta = 90.5°$) and surface energy ($\gamma_m = 0.9\ Jm^{-2}$), the work of adhesion $W_{ad} \approx 0.9\ Jm^{-2}$ was calculated for TiNi/C interface. This value of adhesive work is comparable to values estimated for the 316L/Al_2O_3 or Ni/Al_2O_3 systems and can be characterized as an intermediate one[19].

4. Conclusions

The chemical bonding represented by formation of TiC interphase after annealing of TiNi/C composite materials creates stronger interaction than the physical bonding which is typical for the unannealed TiNi/C system.

The failure mechanism in the TiNi/C system changes with increasing TiNi-C adhesion. The delamination and interfacial shear failure have to be expected in TiNi/C composites without TiC interphase formation. Cracking perpendicular to the wire (fiber) axis is observed in TiNi/TiC/C systems.

Excellent mechanical properties were found in coated TiNi wires. The number of cycles to failure during the bending test is higher in the TiNi/TiC/C composites when compared to the TiNi/C system without TiC interphase or TiNi.

Changes of transformation temperatures of TiNi alloy can be expected if annealing of the TiNi/C system should lead to the TiC interphase formation. This effect is caused by the increase of Ni-content in the coated alloy. Ti is bounded in TiC. Simultaneously the Ni diffusion from interface takes place.

5. References

1. D. Williams, *Concise Encyclopedia of Medical & Dental Materials*, Pergamon Press, Oxford (1990).
2. R. W. Cahn, P. Haasen, E. J. Kramer, Materials Science and Technology **14**, (Medical and Dental Materials), VCH, Weinheim (1992).
3. P. Filip, *Physical Metallurgy Parameters of Shape Memory Phenomena in TiNi Alloys and Their Practical Use*, Ph.D. Thesis, TU of Ostrava (1988), (in Czech).
4. P. Filip, V. Tomášek, K. Mazanec, *Kovové Materiály* **32** (1994) 193, (in Czech).
5. T. W. Duerig, K. N. Melton, D. Stoeckel and C. M. Wayman (eds.), *Engineering Aspects of Shape Memory Alloys*, Butterworth-Heinemann (1990).
6. P. Filip, *Progressive Biomaterials*, TU Ostrava (1994).
7. P. Filip, V. Bednář, *Orthodontics* **2(4)** (1993) 15 (in Czech).
8. P. Filip, J. Pech, K. Mazanec, *Berg-und Huettenmaenische Monatshefte* **139** (1994) 174.
9. P. Filip, K. Mazanec, *Materials Science and Engineering* **A143** (1994).
10. P. Filip, K. Mazanec, *Scripta Metall. et Mater.* **32** (1995) 1375.
11. P. Filip, K. Mazanec, *Scripta Metall. et Mater.* **30** (1994) 67.

12. P. Filip, K. Mazanec, J. Musialek, *Structure Optimization of TiNi Orthopaedic Implants*, ICOMAT'95, will be published in *Jnl. de Physique* (1996).
13. P. Filip, J. Musialek, *New Design of HIP and Small Joints Implants*, Orthopaedic Colloquium, Municipal Hospital Ostrava, February 1994 (in Czech).
14. P. Filip, J. Pech, *New Composite Materials for Dynamic Splinting*, Patent CZ-262 073, 1995.
15. P. Filip, K. Mazanec, *Microstructure and Properties of Hydroxyapatite Coatings on TiNi Shape Memory Alloys*, Proc. of 4 Internat. Metallurgical Symp., Ostrava, May 16-18., 1995.
16. P. Filip, A. Kneissl, K. Mazanec, *Microstructure and Properties of Hydroxyapatite Coatings on TiNi Shape Memory Alloys*, Internat. Metallography Conference, Colmar, May 10-12, 1995, France.
17. M. Kaloč, *Hutnickélisty*, **XXII**, (1967) 473.
18. P. Filip, J. Rusek, K. Mazanec, *Zeitschrift Metallkunde*, **82(H6)** (1991) 488.
19. E. D. Hodros, *Precipitation Processes in Solids*, Metall. Soc. of AIME, Warrendale, PA (1978).

ELABORATION AND CHARACTERIZATION OF C/C COMPOSITES FOR HIGH-ENERGY PHYSICS

S. BŁAŻEWICZ, J. CHŁOPEK, and J. PIEKARCZYK
University of Mining and Metallurgy,
al. Mickiewicza 30, 30-059 Kraków, Poland

J. BŁOCKI, J. GODLEWSKI, J. MICHAŁOWSKI,
K. PAKOŃSKI, and M. STODULSKI
Institute of Nuclear Physics,
ul. Kawiory 26A, 30-059, Kraków, Poland

1. Introduction

Over the past few years, there has been growing interest in the development and application of various composites for modern science and technology. Carbon-carbon composites, as the superior new material, offer many outstanding properties such as excellent mechanical properties to weight ratio, negligibly low thermal expansion, superior corrosion behavior, improved fatigue resistance, and high energy saving in fabrication and application[1,2]. Because of tailorable mechanical and thermal properties, they are considered as potential candidates for stable structure applications such as for supporting structure of silicon detectors in high-energy physics[3]. The tailoring for zero expansion is a well-known example[4]. On the other hand, the coefficient of thermal expansion can also be matched to silicon. It can be achieved either by the combination in stacking sequence of carbon fiber layers in the composite or by the choice of the operation conditions of C/C composite.

The most popular materials in this field of application are Be and BeO. Beryllium has a disadvantage due to potential health hazards during mechanical processing. Moreover, the thermal expansion coefficient of Be is different than that of silicon which is a serious disadvantage for a dimensionally stable structure[3,5]. Carbon fiber-reinforced organic and carbon composites are now considered as the base materials for dimensionally stable structures.

The aim of this study was to prepare thin-shell components for C/C composites as supporting structure for particle detectors operating in accelerators. Such components should have:

- high specific Young's modulus;
- long radiation length (a physical property of the material expressed in the first approximation as inversely proportional to density and the atomic number);
- radiation and water resistance;
- low coefficients of thermal and moisture expansion.

This paper presents the results of the studies on stability of the prepared C/C composites after irradiation tests and after their long-term exposure to water vapor.

2. Experimental Procedure

Two types of C/C composites were designed. One was formed by wet-winding of P-100 graphite fibers with a solution of organic matrix precursor (phenol-formaldehyde resin), followed by warm molding and final carbonization at 1000°C. Samples were then subjected to a single densification cycle by chemical vapor deposition of carbon. The isothermal infiltration process took place with methane as the carbon source. The second type was additionally heat treated at 2150°C. All specimens have been formed as thin plates (14 mm x 160 mm x 0.44 mm) consisting of three layers in 0°/90°/0° stacking sequence. To determine the radiation resistance, the samples were irradiated with gamma rays and neutrons. A Gammatron–S Siemens apparatus with a ^{60}Co gamma radiation source was used. The samples were irradiated gradually with the dose rate of 7130 rad/min; total dose was 10 Mrad. Other samples of C/C composites were irradiated with neutrons. The samples were placed in an aluminum container with AuCo,Ni activation foils used as indicators of neutron fluence. Carbon composites were irradiated by fast neutrons with a dose equal to 6.1×10^{16} n/cm^2. The changes in elasticity of C/C composites after irradiation were analyzed by measuring the flexural modulus and the velocity of ultrasonic wave propagation. Flexural modulus was measured in three-point bending mode at span-to-depth ratios greater than 16. The samples were deflected with a constant load up to 3N, without fracture and flexural modulus was estimated. The velocities of ultrasonic wave propagation parallel and perpendicular to the sample long axis were measured by an ultrasonic tester Mt-541. The detailed method of measuring is described elsewhere[6]. Moisture absorption of C/C composites was examined by measuring the water weight increment after immersion of samples in hot water. The effect of water vapor on elasticity of C/C composites was determined after exposing the samples to 100% humidity at 70°C.

3. Results and Discussion

Table 1 shows the mechanical and physical properties of C/C composites discussed in this work.

The densification of the carbonized samples carried out by the CVD method improved their mechanical properties expressed as the value of Young's modulus and flexural strength. This can be attributed to a decrease in porosity of the porous matrix. However, additional heat treatment of C/C samples up to 2150°C does not change these parameters. It is well known that in C/C composites, the matrix can contribute appreciably to the stiffness of the composite. The quality of the binder coke is one of the more important parameters that determines the final properties of C/C composite. In the case of anisotropic pitch precursor (mainly mesophase pitch), the preferred orientation of the binder in impregnation coke is initiated through a mesophase intermediate in a direction parallel to the fiber due to the thin slit faces enhancing the elastic properties of the composite. To prepare our C/C composite, phenol-formaldehyde resin was used as the carbon matrix precursor. The translation of fiber elasticity was achieved almost completely after carbonization. After graphitizing treatment, the yield in composite Young's modulus is almost unchanged as compared to additionally heat-treated samples.

Table 1. Properties of 2D C/C Composites

Composite Type	Flexural Strength (MPa)	Flexural Modulus (GPa)	Open Porosity (%)	Work to Failure (kJ/m^2)
C/C, 1000°C	498 ± 28	256 ± 19	14	6.7
C/C, 1000°C, CVD	650 ± 25	314 ± 11	8.94	11.1
C/C, 2150°C	639 ± 37	304 ± 18	8.45	9.8

Moisture absorption measurement showed that C/C composites obtained at 1000°C were considerably more sensitive against moisture (2.9 wt %) as compared to the graphitized samples (0.1 wt %). An appreciable increase in water absorption for the C/C composite obtained at 1000°C can be attributed to the higher value of porosity and is probably due to the specific adsorption of water on the oxygen-containing hydrophilic centers present after pyrolysis of organic matter.

The effect of water vapor on elasticity of C/C composites after exposing them to 100% relative humidity at 70° is shown in Fig. 1. This diagram indicates that C/C samples exposed to water vapor treatment retain their elasticity after about 50 hours of treatment with water vapor.

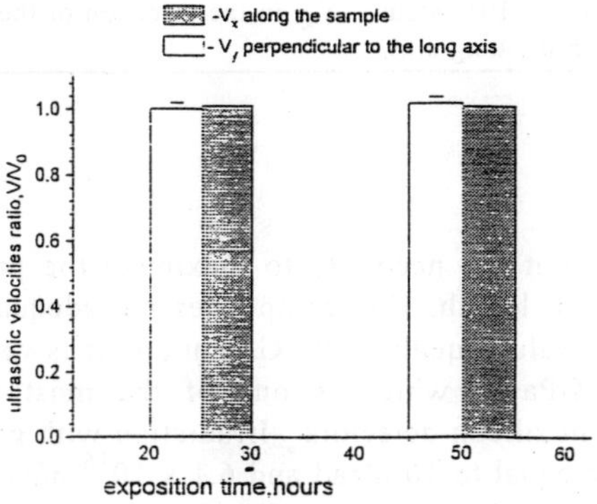

Fig. 1. Ultrasonic velocities ratio of C/C composite obtained at 2150°C on humidity (100%, 70°C).

Irradiation with gamma rays increases the Young's modulus (Table 2) of graphitized C/C composite by about 7% while for carbonized C/C samples a small lowering in elasticity (3%) is observed. Irradiation with neutrons (Table 3) causes only slight variation in elastic properties of examined composites.

Table 2. Effect of Gamma Irradiation* on Young's Modulus and Propagation Velocities of Longitudinal Ultrasonic Wave of C/C Composites

Composite	Treatment	Young's Modulus (GPa)	Velocity (m/s) V_x*	V_y*
C/C (1000)	unirradiated	314 ± 11	13980 ± 124	11132 ± 346
	irradiated	305 ± 14	13897 ± 145	11098 ± 298
C/C (2150)	unirradiated	304 ± 18	13971 ± 169	10645 ± 373
	irradiated	325 ± 18	14365 ± 167	10543 ± 221

*Gamma dose, 10 Mrad. V_x, in the direction of the long axis; V_y, perpendicular to the long axis.

Table 3. Effect of Neutron Irradiation* on Young's Modulus and Propagation Velocities of Longitudinal Ultrasonic Wave of C/C Composites

Composite	Treatment	Young's Modulus (GPa)	Velocity (m/s) V_x*	V_y*
C/C (1000)	unirradiated	314 ± 11	13980 ± 124	11132 ± 346
	irradiated	314 ± 14	14297 ± 175	11098 ± 298
C/C (2150)	unirradiated	304 ± 18	13971 ± 169	10645 ± 373
	irradiated	319 ± 25	14565 ± 177	10343 ± 245

*Neutron fluence, 6.1 x 10^{16} n/cm^2. V_x, in the direction of the long axis; V_y, perpendicular to the long axis.

4. Conclusion

Taking into account the necessity to maximize the product of the Young's modulus by a radiation length, C/C composites are adequate materials for high-energy physics. This value equals to 91 GPa·m and it is very similar to the value for beryllium (109 GPa·m) which is one of the most popular materials for supporting structures in silicon detectors. Irradiation with gamma rays or neutrons where the doses were equal to 10 Mrad and 6.1 x 10^{16} n/cm^2, respectively, causes slight variations of the elastic properties of composite. The changes do not exceed several percent of the respective values before irradiation.

5. Acknowledgment

The authors thank the Polish State Committee for financial support of this work, project no. 3-P4-07-006-04.

6. References

1. E. Fitzer, A. Glogkidis and M. Heine, Carbon Fibers and Their Composites, *Carbon '84, France*, 1984.
2. E. Fitzer, *Carbon* **25** (1987) 163.
3. G. A. Feofilov, L. Ricatti and L. Schukraft, *Proc. Int. Workshop on Advanced Materials for High Precision Detectors*, Geneva (1994) 77.
4. D. Bashford, D. Eaton and A. Pradier, *Proc. Int. Workshop on Advanced Materials for High Precision Detectors*, Geneva (1994) 9.
5. J. B. Donnet and R. C. Bansal, *Carbon Fibers* (Marcel Dekker, Inc., New York, 1990).
6. J. Piekarczyk and R. Pampuch, *Brittle Matrix Composite 2*, ed. A. M. Brandt and I. H. Marshall (Elsevier Applied Science, London, New York, 1989) p. 312.

CARBON FIBER REINFORCED THERMOPLASTIC COMPOSITES

Y. LU
Institute of Carbon Fiber & Composites (Box 101)
Beijing University of Chemical Technology
Beijing 100029, P. R. China

ABSTRACT

A relationship between mechanical properties and resin content of continuous carbon fiber reinforced polyether ketone with cardo (PEK-C), polyether sulfone (PES) and phenolphthalein polyether sulfone copolymer (PPES) was studied and the morphology of both impact and delaminated broken sections and the interface of the composites were observed by SEM. A fuzzy evaluation method for characterization of adhesion between carbon fiber and resins was proposed and the role of resins and carbon fiber contributed to the interlaminar shear strength of the composites was discussed.

1. Introduction

Continuous carbon fiber reinforced composites offer the potential for simple manufacturing techniques. Some high temperature resistant thermoplastics have been used as the matrix for advanced composites[1-3]. Most of the efforts in the development of thermoplastic resins for high performance composites have involved the use of new specialty plastics[4-6].

In this study three amorphous specialty plastics were used as matrix materials for continuous carbon fiber composites. They are polyether ether ketone with cardo (PEK-C), polyether sulfone (PES) and phenolphthalein polyether sulfone copolymer (PPES) with the chemical composition as follows:

PEK-C (Tg=221°C)

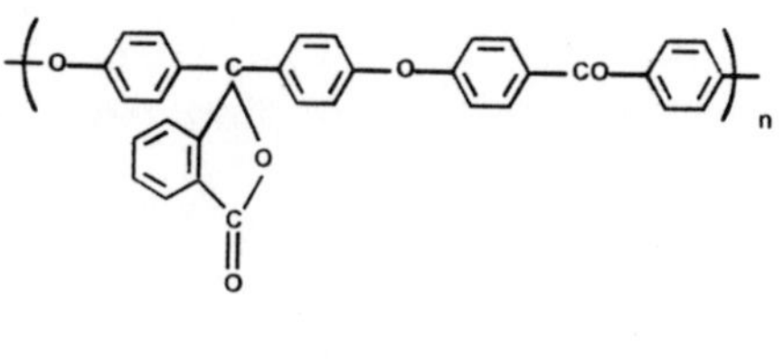

PES (Tg=224°C)

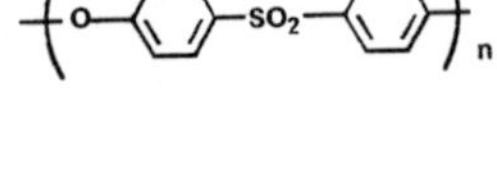

PPES (Tg=201°C)

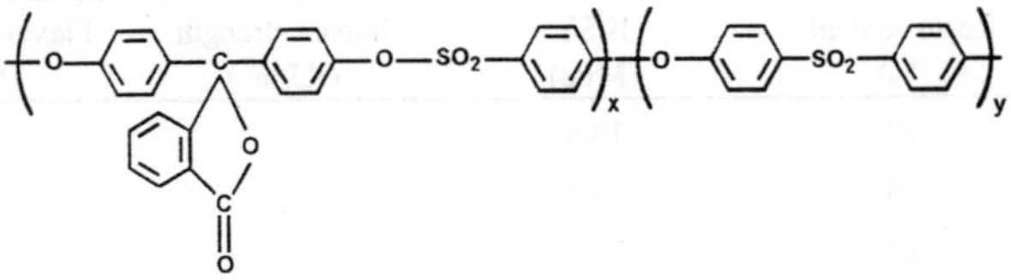

2. Experimental

2.1. *Materials*

PEK-C and PPES were produced by Xuzhou Paint Factory of China, PES was produced by Jilin University of China and carbon fibers were supplied by Liaoyuan Carbon Factory of China.

2.2. *Prepreg Preparation*

The prepreg was made by impregnating continuous carbon fibers in three kinds of resins and dichloride methane solution with concentration of 15% (wt./v.). Since these resins are thermoplastics, the prepreg has no shelf life restriction and no special storage requirements.

2.3. *Laminate Fabrication*

The laminate was fabricated from the prepreg and laid up 10 plies unidirectionally in a mold which was pressed at 210°C for 10 min with pressure of 1 MPa before preheating at 350°C for 20 min.

2.4. *Mechanical Tests*

The interlaminar shear strength (ILSS) and flexural resistance tests were performed on an Instron 1121 universal materials testing machine according to ASTM standards GC3357-82 and GB3356-82, respectively. The impact resistance tests were performed on a Charpy XCJ-40 machine. The dimensions of the unnotched specimens used were 100 mm x 6 mm x 3 mm.

2.5. *Fractography Analysis*

Scanning electron microscopy, using a Cambridge Instruments S250 MK3, was carried out on the tested specimens to give information about fracture morphology and fiber distribution.

3. Results and Discussion

3.1. *Mechanical Properties*

The results of ILSS, impact and flexural tests from carbon fiber reinforced PEK-C, PES and PPES are shown in tables 1, 2 and 3.

Table 1. Mechanical Properties of Carbon Fiber Reinforced PEK-C Composites

Resin content (%)	ILSS (MPa)	Impact strength (KJ/m^2)	Flexural strength (MPa)
10	13.6		
20	23.4		
27	39.5		
29	41.3	114	877
38	45.9	98	1047
44	52.8	83	1042

Table 2. Mechanical Properties of Carbon Fiber Reinforced PES Composites

Resin content (%)	ILSS (MPa)	Impact strength (KJ/m^2)	Flexural strength (MPa)
20	38.9		
30	50.5	70	850
40	51.1	55	971
50	66.2	56	980
58	68.0		

Table 3. Mechanical Properties of Carbon Fiber Reinforced PPES Composites

Resin content (%)	ILSS (MPa)	Impact strength (KJ/m^2)	Flexural strength (MPa)
30	59	108	493
40	66	109	699
50	66	55	1000

3.2. Morphological Characteristics

Microscopic evaluation of three resins and carbon fiber composites indicates that the fiber packings are of the uniform random kind distribution[7].

For PEK-C/carbon fiber composites, micrographs of impact broken section did not detect the clean fibers pulled out of the resin. It has found that the composites have a calabash morphological feature. For PES/carbon fiber composites more resin surrounded the carbon fibers than for PPES/carbon fiber composites.

3.3. Evaluation of Interfacial Adhesion

In order to describe quantitatively the interfacial adhesion of the composites by means of SEM micrographs, a fuzzy evaluation method is proposed. We assume a function:

$$F = A/A_0$$

in which A_0 is the surface area of carbon fiber and A is the resin covered area around the carbon fiber. F = 0 means that there is no resin around the surface of the carbon fiber and F = 1 means that there is resin fully around the surface of the carbon fiber so that $0 < F < 1$ indicates the amount of resin coverage on the surface of the carbon fiber.

If F = 0–0.2 adhesion is very poor
> 0.2–0.4 adhesion is poor
> 0.4–0.6 adhesion is fair
> 0.6–0.8 adhesion is good
> 0.8–1.0 adhesion is very good

According to the results of SEM, the adhesion between fiber and resins can be quantitatively determined. For PEK-C/carbon fiber composites, F = 0.45, meaning the adhesion is fair. For PES/carbon fiber composites, F = 0.6, also meaning that the adhesion is fair. But the adhesion of PES/carbon fiber is better than that of PEK-C/carbon fiber. For PPES/carbon fiber composites, F = 0.1, meaning that the adhesion is very poor.

3.4. Contribution of Resin, Carbon Fiber to Mechanical Properties of Composites

Mechanical properties of composites were contributed from resin, fiber and interphase. Through PEK-C, PES and PPES, which have different interfacial adhesion with carbon fiber, the contribution of resin and fiber to mechanical properties of composites can be discussed.

By comparing ILSS and interfacial adhesion of PEK-C, PES and PPES/carbon fiber composites, we can find that their F values are different. However, their ILSS values are in the same level. This implies that they may have different mechanism of ILSS fracture. ILSS value involves both interplies shear failure and intraplies shear failure. Because the adhesion between PPES and carbon fiber is lesser, the ILSS value is determined by the intraplies shear failure of PPES. But ILSS values of PEK-C and PES composites with carbon fiber are contributed from both interplies and intraplies shear failure because of their good interfacial adhesion.

From the results of studying these three resins and carbon fiber composites, the flexural strength increases with resin content and the impact strength decreases with resin content. This contrary effect of resin content reflects the role of resins to flexural strength and fiber to impact strength and a proper resin content is required.

4. References

1. E. Fitzer and P. A. Staniland, *ACHEMASIA '89* (1989) 186.
2. F. N. Cogswell, G. Cuff and C. R. Gore, *Polym. Rubb. Compos. Process. Appl.* **15** (1991) 1.
3. G. R. Belbin and P. A. Staniland, *Phil. Trans. R. Soc. Lond.* **A32** (1987) 451.
4. L. Zhang and T. Chen, *CNGK* 85108751.
5. Zh. Wu, Eng. *Plast. Appl.* **3** (1986) 23.
6. X. Jin and K. Liu, *Acta Polym. Sinica* **3** (1988) 237.
7. F. N. Cogswell, *SAMPE Q.* **3** (1983) 33.

THE APPLICATION OF THE METHOD "MAXIMAL PRESSURE IN THE GAS POCKET" BY THE MEASUREMENT OF THE SURFACE TENSION OF COAL TAR PITCHES

E. LINZER and M. KALOČ
VSB-Technical University of Ostrava
708 33 Ostrava-Poruba, Czech Republic

ABSTRACT

The surface tension of the binding pitches is a very important parameter in the carbon anodes and graphite electrodes industry. There are many methods for measuring the surface tension of pitches. The method named "maximal pressure in the gas pocket" was tested in this work. The experimental apparatus has simple construction and guarantees the reproducibility and accuracy of the measurement. It is now possible to determine the density of the liquid pitch. The calibration of the apparatus was performed with glycerol and provided adequate precision for these purposes. One of the most important factors in this method is the kinetics of the formation of the gas pocket.

1. Introduction

When using coal-tar pitch as the binding agent in the manufacture of industrial carbon, reasonable knowledge of its surface properties is of considerable significance. The equilibrium form of a droplet of melt at the surface of a solid substance is governed above all by cohesion and adhesion. The adhesive work can be expressed by Dupré's equation:

$$A_{adh} = \sigma_{(l)-(g)} + \sigma_{(s)-(g)} - \sigma_{(s)-(l)} \qquad [\mathrm{N.m^{-1}}] \quad (1)$$

where: $\sigma_{(l)-(g)}$ is the surface tension of molten pitch [N.m^{-1}]
$\sigma_{(s)-(g)}$ is the surface tension of the subsurface mass (s) [N.m^{-1}]
$\sigma_{(s)-(l)}$ is the interphase tension at the interface of the subsurface mass-molten pitch [N.m^{-1}]

In the case of a homogeneous liquid, the cohesive work is written as follows:

$$A_{koh} = 2\sigma_{(l)-(g)} \qquad [\mathrm{N.m^{-1}}] \quad (2)$$

To determine the parameters cited above, knowledge of the surface tension $\sigma_{(l)-(g)}$ of molten pitch is indispensable. (The application methods of experimental investigation into the surface tension of molten pitches are dealt with in another paper of this conference[1].)

2. Experimental Procedure and Results

The method of maximum pressure in a bubble was applied here. Experimental equipment was proposed and completed for measurement making use of two capillary tubes with different depths of immersion. The principal part of the equipment is the mechanism of displacement of the capillary tubes, enabling continuous alteration of the immersion depth, having a scale calibrated in steps of 0.125 mm. Measurements were carried out with calibrated steel capillary tubes with an inside-diameter of 1.48 mm. In addition, this equipment consists of a crucible furnace with controlled sample placement. A Hg-thermometer is used for measurement of the temperature. The gas phase is air, and the pressure in the capillary tubes is measured with a micromanometer.

The applied method and design of equipment provide simultaneous determination of density and surface tension.

The accuracy of the equipment was verified by means of glycerol (the surface tension of σ = 66 $mN.m^{-1}$ was taken from a standard table). The surface tension, determined experimentally with different immersion depths, was 67.3 ± 1.33 $mN.m^{-1}$. A comparison with the tabular value proves reasonable accuracy of the measuring instrumentation.

Measurements were run with four samples designated ŽIAR 80, SIGRI 80, DEZA 90, and LINZ 90 within the temperature interval of 190 to 260°C. These temperatures were governed, in the case of the lower temperature, by the high viscosity of the solidifying pitch, while at 260°C, by the beginning of outgassing.

Calculation of the surface tension $\sigma_{(l)-(g)}$ of molten pitch was made by using Eq. 3:

$$\sigma_{(l)-(g)} = \frac{r}{2}(p_{max} - p_h) = \frac{r}{2}[\rho_l gh - (\rho_t - \rho_g) g h_t] \tag{3}$$

where:
- p_{max} is the gas pressure for releasing a bubble [$N.m^{-2}$]
- p_h is the hydrostatic pressure of a column of molten pitch [$N.m^{-2}$]
- h_t is the immersion depth of the capillary tubes [m]
- ρ_t is the density of molten pitch [$kg.m^{-3}$]
- ρ_l is the density of the liquid in manometer [$kg.m^{-3}$]
- ρ_g is the density of the applied gas [$kg.m^{-3}$]
- g is the acceleration of gravity [$m.s^{-2}$]

The density of molten pitch ρ_t can be determined by using Eq. 4:

$$P_{\max} = P_h + P_\sigma = (\rho_t - \rho_g)gh_t + P_\sigma \qquad [\mathrm{N.m^{-2}}] \quad \mathbf{(4)}$$

from the slope of the straight line of the function $P_{max} = f(h_t)$.

The surface tension values of the individual pitch samples, as determined experimentally, are shown in Figs. 1 and 2 indicated with index (m).

Evaluation of the surface tension of molten pitches was made by means of experimental data of two measurements of each sample, and these data were processed into a functional dependence by the following equation:

$$\sigma_{(l)-(g)} = a_0 + a_1 t + a_2 t^2 + a_3 t^3 \qquad \mathbf{(5)}$$

The acquired dependences of $\sigma_{(l)-(g)}$ = f(t) are shown in Figs. 1 and 2, indicated with index (r).

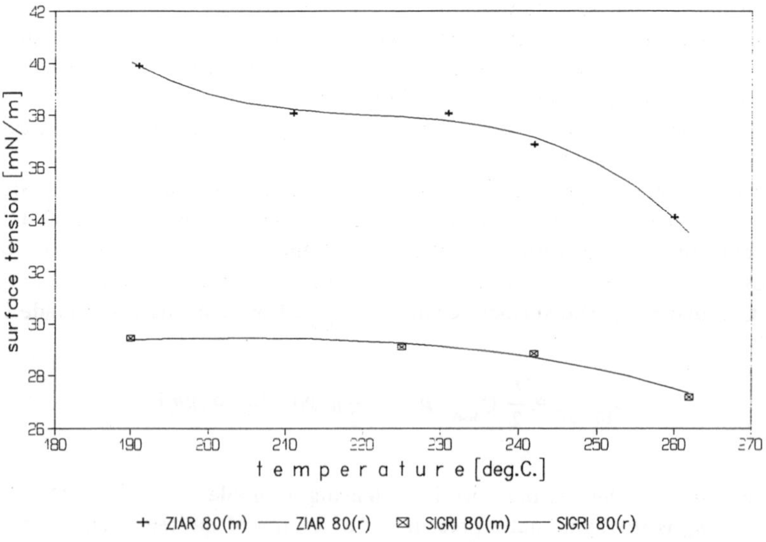

Fig. 1. Temperature dependence of the surface tension of the molten pitches ŽIAR 80 and SIGRI 80.

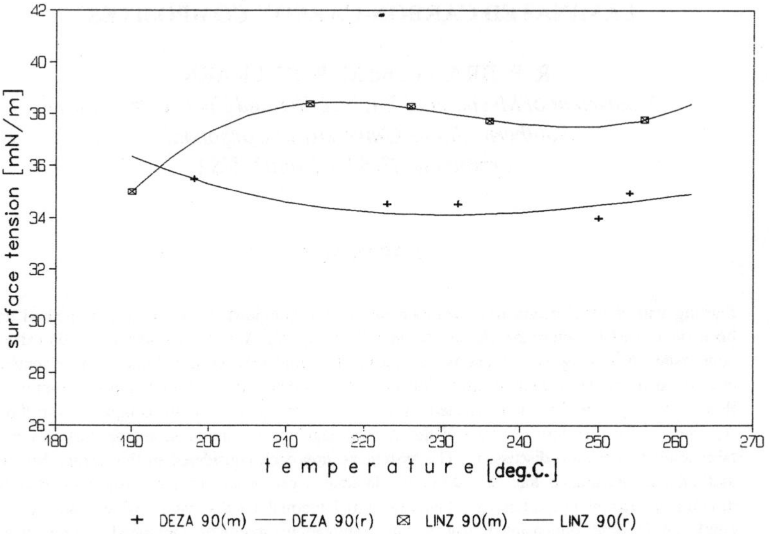

Fig. 2. Temperature dependence of the surface tension of the molten pitches DEZA 90 and LINZ 90.

3. Conclusion

The obtained dependences $\sigma_{(l)-(g)} = f(t)$ reveal different surface tension values for the individual pitch samples at the temperature interval in question. From the series of pitches investigated, the ŽIAR 80 sample reveals the lowest values. From the viewpoint of temperature dependence, the drop of $\sigma_{(l)-(g)}$ is of significance with samples SIGRI 80 and ŽIAR 80 at a temperature of 230°C, while on the contrary, a slight increase of tension was recorded with the samples of DEZA 90 and LINZ 90. In the case of sample LINZ 90, this phenomenon starts from a temperature of 240°C.

4. References

1. M. Kaloč, E. Linzer and D. Košťal. The Surface Tension of Binders and their Measurement Methods, in *Carbon and Carbonaceous Composite Materials: Structure Property Relationships*.

GLOBAL THERMAL CONDUCTIVITIES IN LAMINATED CARBON-CARBON COMPOSITES

R. P. BRADY and M. R. KULKARNI
Department of Mechanical Engineering and Energy Processes
Southern Illinois University at Carbondale
Carbondale, IL 62901-6603, USA

ABSTRACT

Starting with thermal conductivity components for the individual lamina or ply, a method has been developed to determine global conductivities (K_x, K_y, K_z) in a laminated carbon-carbon composite. Assuming that all laminae are identical in thickness and in volumetric fiber content, effective thermal conductivity in each global direction is determined for the laminated composite. Results are reported for a few selected standard laminate lay-ups and are compared with those reported in the literature. Details of the methodology and its limitation and applicability in a fabrication process are discussed. The systematic approach considered in this comprehensive method can be utilized for: a) fabricating laminated carbon-carbon composite with specific directional thermal properties; b) obtaining global thermal conductivities, which can then be employed in multi-dimensional heat transfer simulations involving laminated carbon-carbon composites; and c) carrying out *"what-if"* types of simulations to assess the effect of volumetric fiber content and the fiber orientation in individual lamina on the global conductivities of the composite. The techniques presented here are applicable for non-carbon types of laminated composites as well.

Notation

k_1 Thermal conductivity of the lamina in the longitudinal (fiber) direction
k_2 Thermal conductivity of the lamina in the transverse direction
k_f Thermal conductivity of the fiber
k_m Thermal conductivity of the matrix material
k_x Global conductivity component for the lamina in the x-direction
K_x Global conductivity component for the laminate in the x-direction
$\overline{K}_x$ Global conductivity component for the laminate in the x-direction w.r.t. k_m
k_y Global conductivity component for the lamina in the y-direction
K_y Global conductivity component for the laminate in the y-direction
$\overline{K}_y$ Global conductivity component for the laminate in the y-direction w.r.t. k_m
k_z Global conductivity component for the lamina in the z-direction
K_z Global conductivity component for the laminate in the z-direction
$\overline{K}_z$ Global conductivity component for the laminate in the z-direction w.r.t. k_m
R Thermal conductivity ratio, k_f / k_m
V_f Volumetric fiber content in a lamina
θ Fiber orientation angle with respect to the global x-axis

1. Introduction

Nowadays, the use of fiber-reinforced composite materials is very common, and the use is on the rise. These high performance composite materials are lightweight, durable and can be made non-corrosive. Most composite materials developed in the earlier years have been fabricated to improve mechanical properties such as strength, stiffness and toughness; but, more and more materials are needed for high thermal performance including their ability to withstand high temperatures and for rapid heat removal from thermally sensitive devices or heat exchangers[1]. The greatest percentage of fiber-reinforced materials is used in the aircraft/aerospace industry. In particular, graphite fibers are most commonly used[2]. Carbon, in its various forms, is one of the most versatile thermal management materials, because its thermal conduction properties may be tailored to meet the needs of a given engineering application[3].

Fiber-reinforced composites can be classified as single-layer composites and multi-layered (angle-ply) composites. The single-layer composites include composites having the same orientation and properties in each layer. The multi-layered composites can be broken down into laminates and hybrids. In this paper, laminated carbon-carbon composites have been considered. However, the work is equally applicable to unidirectionally reinforced continuous-fiber single-layer composites including composites having the same orientation and properties in each layer. A global thermal conductivity model in a laminated carbon-carbon composite is reported.

Although there exists in the literature a large knowledge base for the mechanical properties of composite materials, only limited information is available on the thermal properties. The limited works include those by Springer et al.[4], Demain et al.[1], Grove[5], Pilling et al.[6], and Hasselman et al.[7] Springer et al. considered composite thermal conductivities of unidirectional composites and obtained expressions for predicting conductivities in the principal directions, along and normal to the fibers. The parallel conductivity is obtained using mixture rule, while the normal conductivity is obtained using an analogy between thermal lead-ins and the constitudinal shear loading of a unidirectional composite[4].

Demain et al. reported the measurements of thermal and electrical conductivities of polycarbonate specimens as a function of chopped pitch-based carbon fiber concentration[1]. Grove modeled the transverse thermal conductivity in continuous unidirectional fiber composite materials by combining finite element analysis and spatial statistical techniques[5]. Pilling et al. reported measurements of thermal conductivity between 80 and 270 K of epoxy resin specimens reinforced with carbon fibers[6]. They measured both the in-plane and out-of-plane thermal conductivity. Lastly, Hasselman et al. modeled the effective thermal conductivity of uniaxial composite with cylindrically orthotropic carbon fibers and interfacial thermal barrier[7].

While the published literature is quite extensive, what is lacking is a systematic and comprehensive method that could be utilized either for fabricating laminated carbon-carbon composites with specific directional thermal conductivities or for carrying out *"what if"* types of simulation to assess the effect of volumetric fiber content and the fiber orientation in individual lamina on the global conductivities of the composite. Once the global conductivities are known, they can be employed in multi-dimensional heat transfer simulations involving laminated carbon-carbon composites.

2. Approach

In order to determine the global conductivity components for laminated composites, a number of well-established concepts are brought to bear in this methodology. The concepts are: in-plane and out-of-plane conductivities for the *individual* lamina, the one-dimensional heat transfer through a composite wall, and electrical analog for the heat transfer. A computer program is written to evaluate global conductivities for various volumetric fiber contents and for a number of different fiber orientations in the individual lamina.

As stated earlier, composite materials offer the possibility of achieving specialized mechanical, thermal and electrical properties that are quite different or possible in conventional "monolithic" materials. However, designing with these materials involves numerous challenges stemming from difficulties in accurately predicting composite material response from knowledge of the individual constituents[8]. In traditional modeling, the central concern is the role of microstructure vs. macrostructure with the goal being consolidation of individual constituent responses into a constitutive description that accounts for their cumulative effect in a manner that is suitable for computational analysis and prediction. In this work, the starting material is individual lamina or ply. At least for now, all laminae are assumed to be identical in thickness and also in volumetric fiber content. What is varied is the fiber orientation in each individual lamina. The global conductivities obtained in this work are compared with those reported in the literature[5].

3. Methodology Details

As stated earlier, the methodology is based on the following well-established concepts: in-plane and out-of-plane conductivities for the individual lamina, one-dimensional heat transfer through a composite wall, and electrical analog for the heat transfer.

3.1. Lamina: In-Plane and Out-of-Plane Conductivity

The thermal properties are a function of the properties of each of the constituents. If the individual lamina is considered to be made of fiber and matrix material, then for each ply, the thermal conductivity in the longitudinal (fiber) direction (k_1) and transverse directions (k_2) may be calculated[9]:

$$k_1 = k_f V_f + k_m(1 - V_f) \tag{1}$$

$$k_2 = k_m \left[\frac{k_f(1 + V_f) + k_m(1 - V_f)}{k_f(1 - V_f) + k_m(1 + V_f)} \right] \tag{2}$$

where V_f is fiber volume fraction; k_f is fiber conductivity and k_m is matrix material conductivity.

These principle conductivities (k_1 and k_2) can then be used to obtain global (K_x, K_y, K_z) conductivity components for *individual* lamina by

$$k_x = |k_1 \cos\theta| + |k_2 \sin\theta| \tag{3}$$
$$k_y = |k_1 \sin\theta| + |k_2 \cos\theta| \tag{4}$$
$$k_z = k_2 \tag{5}$$

where θ represents the fiber orientation angle with respect to the global x-axis. These components for the individual lamina are used further to derive global thermal conductivity components for a laminated composite. It is assumed that the contact resistance between the lamina and between the individual components is negligible.

3.2. *One-Dimensional Heat Transfer through a Composite Wall*

Heat transfer through a composite wall has been studied quite extensively. Selected references include the textbooks by Holman[10], Incropera et al.[11] and Gebhart[12]. Although the heat flow through the composite material can be two- or three-dimensional, it is often reasonable to assume one-dimensional conditions. Further, in order to experimentally determine the thermal conductivity of materials a condition of one-dimensional heat transfer is usually imposed. Since the main objective of this work is to determine the thermal conductivity values for a laminated composite, it is reasonable to employ one-dimensional heat transfer analysis.

It should be noted that once these global thermal conductivity components are determined, they can be employed in multi-dimensional heat transfer simulations involving laminated carbon-carbon composites. Also, once a direct link has been established between the individual lamina and the laminated composite, it would be possible to fabricate laminated carbon-carbon composite with specific *directional* thermal properties. One could also carry out "what if" types of simulations to assess the effects of volumetric fiber content and the fiber orientation in individual lamina on the global conductivities of the composite. Conversely, depending on the expected thermal loading, the thermal performance of composite material can be tailored for specific target locations and also for specific target direction(s). An example where this would be important is aircraft braking materials[13].

3.3. *Electrical Analogy for the Heat Transfer*

One of the widely utilized methods of studying one-dimensional heat transfer through composite wall is the method of electrical analogy. In electrical circuits, we have a resulting current, when a voltage differential is applied across a resistance. In heat transfer analysis, we have a resulting flow of heat, when a temperature difference exists across a material. If the material is a good conductor, then a greater heat flux is achieved, while for insulating materials, the heat flow is greatly hindered. To make the analogy complete, a thermal resistance is defined as

$$R_{th} = \frac{\delta}{kA} \tag{6}$$

where δ is the material thickness, k is the thermal conductivity of the material and A is the cross-sectional area perpendicular to the heat flow. So, the electrical analogy can be illustrated as

$$\textit{Electrical:} \quad \textit{Current} = \frac{\textit{volage}}{\textit{resistance}}, \quad I = \frac{V}{R} \tag{7}$$

$$\textit{Thermal:} \quad \textit{Heat flow} = \frac{\textit{temperature difference}}{\textit{resistance}}, \quad q = \frac{\Delta T}{\delta / kA} \tag{8}$$

In the discussion thus far, we have considered only one material to establish the electrical analogy. A laminated composite (laminate) material consists of a number of individual lamina (ply). So, the equivalent electrical circuit will comprise a number of individual lamina resistances in series for the heat flow in the out-of-plane direction and a number of individual lamina resistances in parallel for the heat flow in each of the two in-plane directions. Now, if we assume that each individual lamina has the identical thickness, then using the series and parallel circuit analysis, the following expressions can be obtained for the global thermal conductivity components for the composite laminate:

$$K_x = \frac{1}{N}\sum_{i=1}^{N} k_{x,i} \tag{9}$$

$$K_y = \frac{1}{N}\sum_{i=1}^{N} k_{y,i} \tag{10}$$

$$K_z = \frac{N}{\sum_{i=1}^{N}\left(\frac{1}{k_{z,i}}\right)} \tag{11}$$

where N is the number of plies in a laminate, z is the out-of-plane direction and x and y are the in-plane directions. The summation is carried out using the individual lamina conductivity components discussed earlier in Eq. (3) through Eq. (5).

It should be noted that while the above expressions for laminate global conductivity components are valid when all individual laminae have identical thicknesses, the methodology can be easily extended to include those situations where the laminate is comprised of lamina of varying thicknesses. Though it is not a common practice to employ variable lamina thickness, the technique may prove to be very valuable for tailoring thermal properties for specific target locations and specific target directions.

4. Results

In order to compare and contrast our results with those from the literature[5,14], we assumed that all the lamina in a laminate had identical volumetric fiber content. The quantities that were varied include fiber orientation in individual lamina and the total number of lamina in a laminate. Parametric results were obtained for laminates under varying volumetric fiber content and also under varying thermal conductivity ratio, R. This ratio is defined as a ratio of fiber conductivity (k_f) to matrix conductivity (k_m). In fact, the entire conductivity analysis was carried out in relation to the matrix conductivity,

$$\overline{K}_x = K_x / k_m \tag{12}$$

$$\overline{K}_y = K_y / k_m \tag{13}$$

$$\overline{K}_z = K_z / k_m \tag{14}$$

The true global conductivity components can be determined by multiplying the relative components obtained via modeling by the matrix conductivity. During this study, the volumetric fiber content was varied from 0.0 to 0.5, while the conductivity ratio was varied from 2 to 1000.

Figs. 1a through 1c and 2a through 2c provide the global conductivity results for a pseudo-random fiber distributed laminate. The fiber distribution is called pseudo-random for two reasons: even though there are 45 lamina in this laminate, the fiber orientation is assigned a value 1 degree for the first lamina, 2 degrees for the second and so on. Secondly, since all the assigned fiber orientation angles are between 1 and 45 degrees as opposed to 1 and 90 degrees, there is definitely a preferred direction, namely the x-direction. The plots indicate that to be the case. Among the two in-plane conductivity components, the x-component is higher than the y-direction counterpart, at times 100% higher.

When the results from Figs. 1c and 2c are compared with those from Grove[5], there is an excellent agreement. Also, as reported before[5,14], at a given conductivity ratio, R, the global conductivity component seems to increase almost linearly with the volumetric fiber content. Further, as expected, at a given volumetric fiber content, the global conductivity component increases with increasing value of R. This is evident from Figs. 1a through 1c.

In Figs. 2a through 2c, a wider range of conductivity ratio, R, is explored. There seems to be a marked difference between the trends for in-plane and out-of-plane global conductivity components. While the z direction components seem to reach a plateau at higher values of R, the x and y components seem to increase monotonically with R. The apparent independence of z conductivity component at higher R values was expected based on the earlier works[5].

Results for a few selected laminates are tabulated in Table 1. These were obtained for a conductivity ratio, R, of 100 and for a volumetric fiber content, V_f, of 0.5. Since for a chosen value of R and V_f, the out-of-plane conductivity is independent of the fiber orientations in the individual lamina, the $\overline{K}_z$ was expected to be a constant. Also, since the global

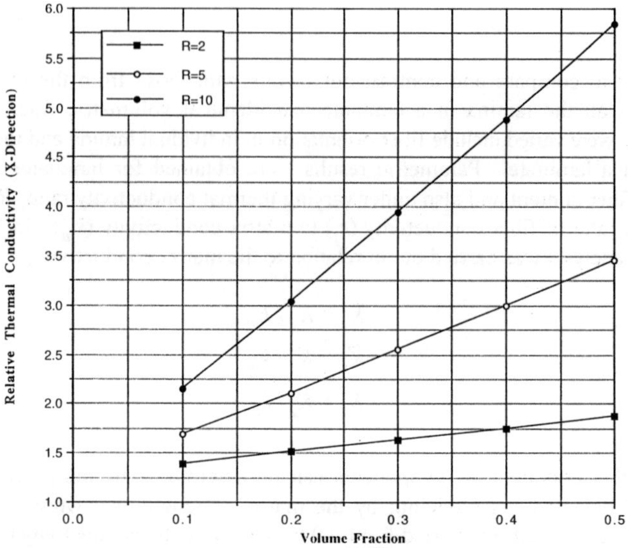

Fig. 1a. Relative thermal conductivity in the global x-direction as a function of fiber volume fraction for values of *R* from 2 to 10 from global conductivities model (random fiber distribution).

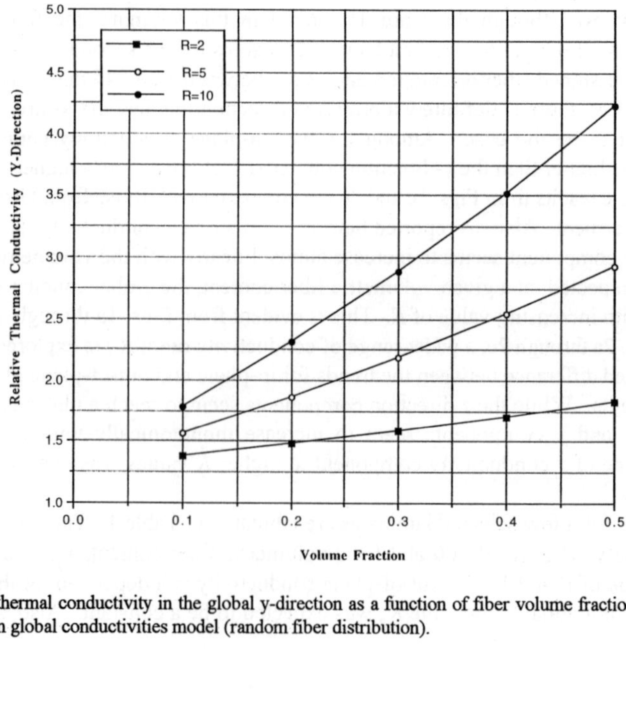

Fig. 1b. Relative thermal conductivity in the global y-direction as a function of fiber volume fraction for values of *R* from 2 to 10 from global conductivities model (random fiber distribution).

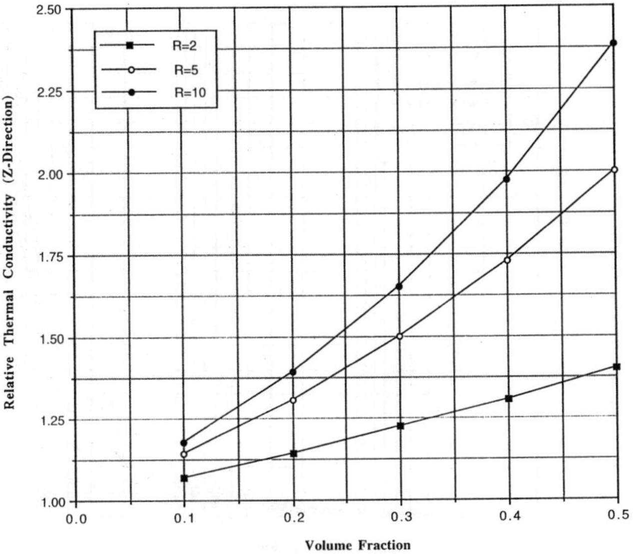

Fig. 1c. Relative thermal conductivity in the global z-direction as a function of fiber volume fraction for values of R from 2 to 10 from global conductivities model (random fiber distribution).

Table 1. Global Conductivity Results for Typical Laminates for $R = 100$ and $V_f = 0.5$

Laminate	$\overline{K}_x$	$\overline{K}_y$	$\overline{K}_z$
$[45/0/90]_{2s}$	30.399	30.399	2.922
$[45/0/90]_4$	30.399	30.399	2.922
$[45/0/70]_{2s}$	36.098	29.717	2.922
$[45/0/70]_4$	36.098	29.717	2.922
$[90/0_2/45]s$	35.424	23.530	2.922
$[45/0/45/90_2/30]$	29.515	34.542	2.922
$[0/45/90]_{2s}$	35.895	26.379	2.922

conductivity model is based on electrical circuits analogy and since the total resistance is independent of the sequence of the individual resistances, it was expected that the values of $\overline{K}_x$ and $\overline{K}_y$ would not vary between the laminates $[45/0/90]_{2s}$ and $[45/0/90]_4$. Additionally, the $\overline{K}_x$ and $\overline{K}_y$ are identical, because a lamina with an orientation angle of 0° (fibers oriented in x-direction) is exactly counter balanced by the lamina with an orientation angle of 90° (fibers oriented in y-direction). In order to illustrate the point about the *sequence* of individual resistances, consider,

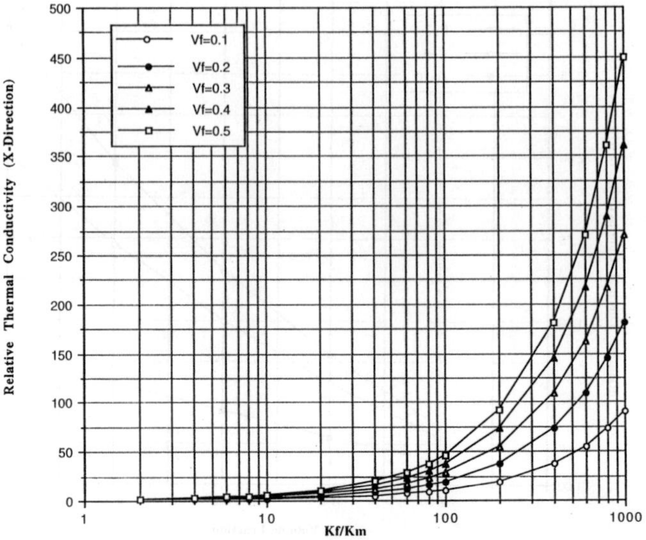

Fig. 2a. Relative thermal conductivity in the global x-direction as a function of K_f/K_m for fiber fraction values of 0.1 to 0.5 from global conductivities model (random fiber distribution).

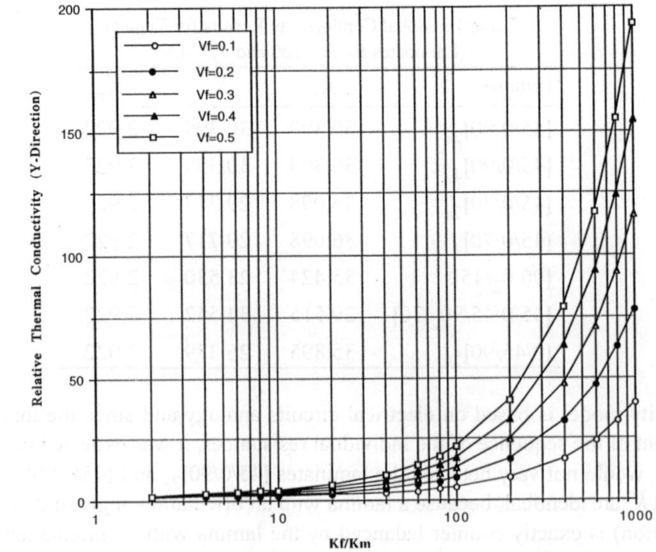

Fig. 2b. Relative thermal conductivity in the global y-direction as a function of K_f/K_m for fiber fraction values of 0.1 to 0.5 from global conductivities model (random fiber distribution).

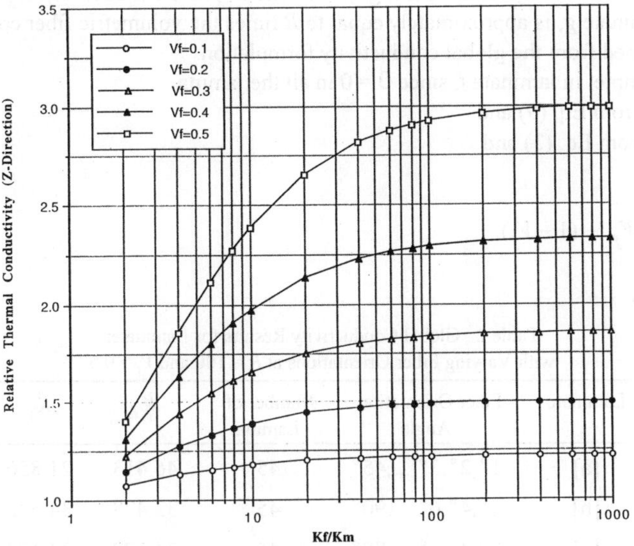

Fig. 2c. Relative thermal conductivity in the global z-direction as a function of K_f/K_m for fiber fraction values of 0.1 to 0.5 from global conductivities model (random fiber distribution).

$[45/0/90]_{2s}$ ≡ 45,0,90,45,0,90,90,0,45,90,0,45 ≡ lamina each with 0°, 45° and 90°,

and

$[45/0/90]_4$ ≡ 45,0,90,45,0,90,45,0,90,45,0,90 ≡ 4 lamina each with 0°, 45° and 90°.

This will make the $\overline{K}_x$ and $\overline{K}_y$ to be the same in both laminates. Similar invariance would have been obtained for $[45/0/90]_{2s}$ and $[45/0/90]_4$; however, now the $\overline{K}_x$ and $\overline{K}_y$ would not be identical. In fact, since 0 represents fibers oriented exactly in the x-direction, and while 70 does not correspond to fibers oriented exactly in the y-direction, the $\overline{K}_x$ value would be slightly higher than $\overline{K}_y$, which is the case.

The effect of the lamina fiber orientations on the global (laminate) thermal conductivity is reported in Table 2. It can be seen that for a 45-ply laminate, $\overline{K}_x$ is significantly higher than $\overline{K}_y$, if all the angles are between 1 and 45 degrees. But, if the angles are between 2 and 90 degrees instead, then only a slightly higher $\overline{K}_y$ is obtained, due to the fact that the last lamina (90°) is not counter balanced by a 0° lamina, which is the case in laminate *b*. In laminate *c*, there are 44 lamina and all of them are exactly balanced; hence, the values of $\overline{K}_x$ and $\overline{K}_y$ are identical, as they should be.

A comparison of laminates *d* through *g* will indicate that the results are not dependent on the sign of the fiber orientation angles, which was expected. Also, it was expected that when the orientation is exactly 0 degrees or 90 degrees, then the *other* ($\overline{K}_y$ or $\overline{K}_x$, respectively) conductivity will be the same as that of out-of-plane conductivity ($\overline{K}_z$). Further, since the assumed conductivity ratio, R, is high (100), the in-plane conductivity, $\overline{K}_x$ in laminate f and $\overline{K}_y$ in laminate g, is approximately equal to R times the volumetric fiber content, which can be explained from the global conductivity formulation.

For example, in laminate f, since $\theta = 0$ in all the lamina,

$K_x = k_x$ from Eq. (9) and

$k_x = k_1$ from Eq. (3) and

from Eq. (1),

$$\frac{k_1}{k_m} = \frac{k_f}{k_m} V_f + (1 - V_f)$$

so, $\overline{K}_x \doteq R^* V_f$

Table 2. Global Conductivity Results for Laminates with Varying Fiber Orientations at $R = 100$ and $V_f = 0.5$

Laminate	Fiber Orientation Angle	Number of Lamina	$\overline{K}_x$	$\overline{K}_y$
[a]	1°,2°,3°,...,45°	45	46.413	21.850
[b]	2°,4°,6°,...,90°	45	33.478	34.535
[c]	2°,4°,6°,...,88°	44	34.172	34.172
[d]	+45° in all	45	37.775	37.775
[e]	−45° in all	45	37.775	37.775
[f]	0° in all	45	50.500	2.922
[g]	90° in all	45	2.922	50.500

Note: $\overline{K}_z$ is 2.922 for all of the above laminates

5. Conclusions

The foregoing results indicate that it is possible to determine the global thermal conductivity components for a laminated composite. What is needed is the detailed thermal knowledge of individual lamina. When compared with the results reported in the literature, this methodology seems to yield similar results. The global conductivity model is now being experimentally validated. This model is also being explored for characterizing delaminations via pulsed video thermography (PVT) techniques.

6. Acknowledgments

The authors acknowledge the seed grant support provided by the Materials Technology Center at Southern Illinois University at Carbondale.

7. References

1. A. Demain and J. P. Issi, *Journal of Composite Materials* **27(7)** (1993) 668.
2. R. E. Taylor and B. H. Kelsic, *Journal of Heat Transfer* **108** (1986) 161.
3. D. T. Marx, *Materially Speaking* **10(2)** (1995) 6.
4. G. S. Springer and S. W. Tsai, *J. Composite Materials* **1** (1967) 166.
5. S. M. Grove, *Composites Science and Technology* **38** (1990) 199.
6. M. W. Pilling, B. Yates, M. A. Black and P. Tattersall, *Journal of Material Science* **14** (1979) 1326.
7. D. P. H. Hasselman, K. Y. Donaldson and J. R. Thomas, Jr., *Journal of Composite Materials* **27(6)** (1993) 637.
8. T. Pence, personal communication (1993).
9. J. A. Charles and D. W. Wilson, *Polymer Composites* **2(3)** (1981) 105.
10. J. P. Holman, *Heat Transfer* (McGraw-Hill, New York, 1990) 28.
11. F. P. Incropera and D. P. Dewitt, *Introduction to Heat Transfer* (John Wiley, New York, 1990) 84.
12. B. Gebhart, *Heat Conduction and Mass Diffusion* (McGraw-Hill, New York, 1993) 48.
13. D. E. Hayes, personal communication (1995).
14. B. D. Agarwal and L. J. Broutman, *Analysis and Performance of Fiber Composites* (John Wiley, New York, 1980) 341.

Physical Properties

of

Carbon Materials

ELECTRICAL AND THERMAL CONDUCTIVITIES OF CARBON FIBERS AND THEIR COMPOSITES

J-P. ISSI
Unité de Physico-Chimie et de Physique des Matériaux
Université Catholique de Louvain
1, Place Croix du Sud, B-1348 Louvain-la-Neuve, Belgique

ABSTRACT

After briefly depicting the particular situation of carbons and graphites, and describing the types of fiber candidates for electrical and thermal conductivity applications, we discuss the electrical resistivity data. Then we analyze the thermal conductivity data showing how with electrical resistivity it depends on structural perfection. On these lines the direct correlation between these two transport coefficients is emphasized in carbon fibers. Then the thermal conductivity of carbon fiber-polymer composites is presented and the various parameters influencing the thermal conductivity are briefly discussed.

1. Introduction

There is nowadays an increasing demand for high thermal conductivity materials to improve heat exchanges in practical devices. This is particularly true for space and airborne systems. The particular case of the carbon-carbon composites has been thoroughly discussed in that context at this conference[1]. Among the numerous advantages of using polymeric materials as heat exchangers are their low specific gravity, their chemical resistance, their ease of processing and energy saving. Also, of paramount importance for heat evacuation in electronic devices is the possibility to realize good heat conductors that are electrically insulating.

Recent work has shown that we are now in a position to tailor materials at the atomic, molecular and microstructural levels to obtain the desired properties. This is particularly true for the case of the thermal conductivity[2].

The materials we are interested in here have in common that they are either electrical insulators (polymers) or relatively poor electrical conductors (carbon fibers). Heat is transferred in polymers and in carbon materials at ordinary temperatures via lattice vibrations. In both classes of materials it is the strong c-c covalent bonding that is very efficient in transmitting the vibrations from atom to atom, provided the path is straight and not too short.

By adding electrically conductive entities, such as carbon blacks or fibers, to polymeric materials, a transition from insulating to electrically conductive behavior occurs for an amount of filler exceeding the percolation threshold. These composites find many practical applications such as for emi shielding.

In the following, after briefly overviewing the mechanisms of conduction, we will discuss some of the experimental data obtained by our group in Louvain-La-Neuve in view of understanding and improving the thermal conductivity of carbons and carbon fiber-polymer composites. We will first discuss the situation of carbon fibers relative to the electrical resistivity, thermal conductivity and then the relation between these two transport parameters. The effect of intercalation on the electrical resistivity will also be stressed. Then we shall turn to carbon fiber composites.

2. Carbon Fibers

2.1. Specific Situation of Carbons and Graphites

HOPG is the semimetallic while some carbons are narrow-gap semiconductors. This means that in either case these materials have very few charge carriers compared with ordinary metals and much more than in typical semiconductors. Also, HOPG and other carbons, according to structural order, have more than one type of charge carriers. This renders the analysis of electronic data rather complicated. By intercalation, one increases the carrier density and metallic behavior may result.

Concerning the thermal conductivity, the small density of charge carriers in carbons and graphites in combination with a relatively large lattice contribution leads to a negligible contribution of the charge carriers. Except at very low temperatures, heat is carried by the lattice in these materials. HOPG in-plane is, with diamond, the best known heat conductor.

2.2. Fibers Investigated

Among the fibers investigated with respect to electrical and thermal conductivities are the PAN-based, pitch-derived and vapor grown carbon fibers (VGCF). We may also consider carbon nanotubes, whether single or multiwalled, as a kind of carbon fiber. For basic studies, VGCF are the more interesting, since they are very similar to HOPG and have the highest structural perfection. They have thus, at least up to now, the highest electrical and thermal conductivities among carbon fibers. They can be intercalated for increased electrical conductivity. Their main drawback is that up to now they can only be obtained with limited lengths.

As will be shown below, the electrical and thermal conductivities of pitch-based carbon fibers depend dramatically on the in-plane coherence length, which in turn mainly depends on the heat treatment temperature. Some pitch-based carbon fibers have thermal conductivities exceeding $1000 Wm^{-1}K^{-1}$. These high thermal conductivities along with the fact that they may be obtained in a continuous form make them the ideal candidates for thermal management applications[1,3,4].

PAN-based fibers are also obtained as continuous fibers, but they are by far too disordered to attain the desired conductivity levels. One exception is the Celanese GY70, which has a room temperature thermal conductivity of almost $200 m^{-1}K^{-1}$, though it is heat treated at very high temperature[5].

Some nanotubes have electrical conductivities comparable to VGCF heat treated at high temperature[6,7] (Fig. 1). Their thermal conductivities have not been measured, and it is unlikely that such data will be available soon. However, by comparison with graphites, one may expect for such materials very high thermal conductivities.

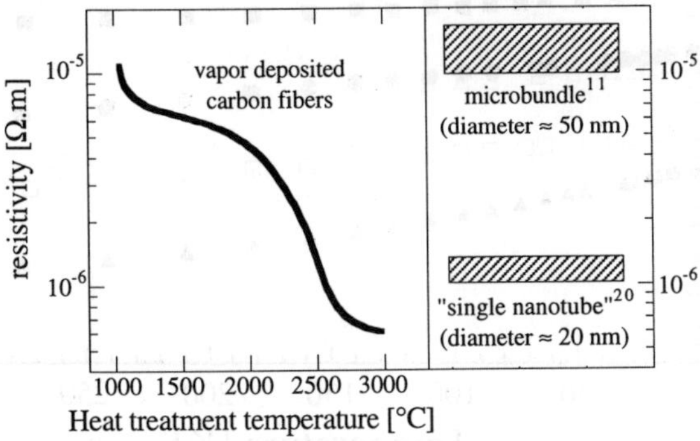

Fig. 1. Estimated room-temperature electrical resistivity of carbon nanotubes compared to those of VGCF heat treated at different temperatures.

We will show below how the electrical and thermal conductivities are directly related to the defect structure of the sample.

2.3. Electrical Resistivity

Bright and Singer[8] have measured the temperature dependence of the electrical resistivity of radial and random pitch-based carbon fibers heat treated at various temperatures from 1000°C to 3000°C. They have shown that the magnitude of the resistivity and its temperature variation depends on the HTT. We present in Fig. 2 the temperature dependence of the electrical resistivity of a series of pitch-based carbon fibers heat treated at various temperatures; the thermal conductivities are presented in Fig. 7b[9]. In Fig. 3 we present the room temperature resistivity of various experimental pitch-based carbon fibers versus in-plane coherence length[10]. We may see that there is a direct correlation between the two parameters, and, as expected, the electrical resistivity decreases with increasing structural perfection.

Energy band structure calculations have shown that carbon nanotubes can be either metallic or semiconducting, depending on the tubule diameter and chirality[11–13]. Using submicronic lithographic patterning of gold films with a scanning tunneling microscope, Langer et al.[6] measured the electrical resistivity of a single microbundle of 50 nm diameter down to 0.3K (Fig. 4). Also, later on, a single nanotube was measured by the same group down to 20 mK[(7)]. A room temperature electrical resistivity of $\approx 10^{-4}\Omega$cm was estimated for the single nanotube (Fig. 1). Above 2 K, the observed zero-field resistance was found to fit the simple two-band (STB) model derived by Klein[14–16] for semimetallic graphite. This fit yielded a carrier density an order of magnitude smaller, consistent with a larger resistivity,

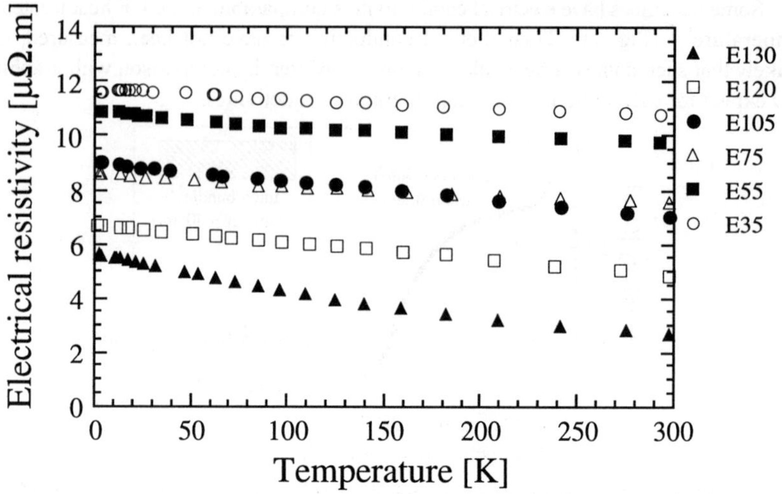

Fig. 2. Temperature dependence of the electrical resistivity of pitch-based carbon fibers heat treated at various temperatures.

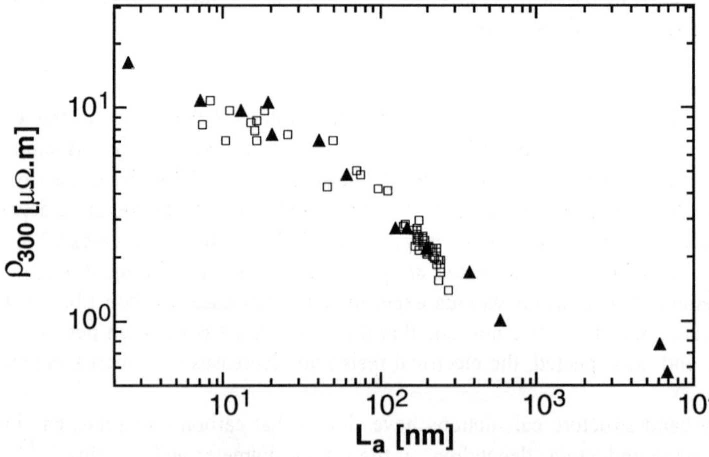

Fig. 3. Room temperature resistivity of various pitch-based carbon fibers versus in-plane coherence length.

than that observed in crystalline graphite. The difference was ascribed to the turbostratic stacking of the adjacent layers which is expected to reduce drastically the interlayer interactions, as is the case for disordered turbostratic graphite.

One can increase significantly the electrical conductivity of graphitic materials by intercalation of various species[17]. In Fig. 5 we present the effect of intercalation on the temperature variation of the electrical resistivity of a VGCF heat treated at 2900°C[18]. There is a dramatic decrease in resistivity due to intercalation for either a donor (Rb) or acceptor ($FeCl_3$) intercalate. The resulting intercalation compounds exhibit typical metallic behaviors with room temperature electrical resistivities 3 to 4 times that of pure copper.

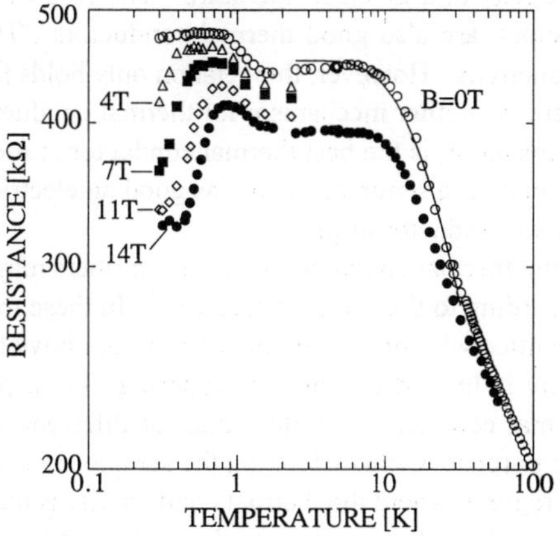

Fig. 4. Electrical resistance of a microbundle of nanotubes of approximately 50 nm diameter versus temperature at various magnetic fields. The continuous curve is calculated from a fit of the Klein 2D band model for graphite with an overlap Δ=3.7 meV and a Fermi level right in the middle of the overlap. (Reprinted from *Journal of Materials Research*, ref. 9).

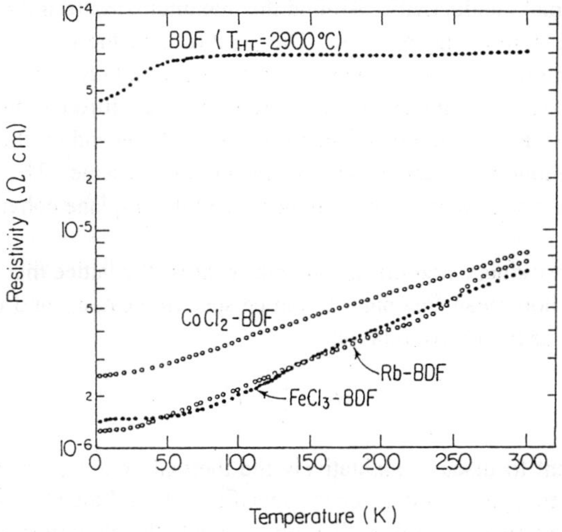

Fig. 5. Effect of intercalation by a donor (Rb) and acceptor ($FeCl_3$) intercalate on the temperature variation of the electrical resistivity of a VGCF heat treated at 2900°C.

2.4. Thermal Conductivity

Among the properties of solids, the thermal conductivity is generally not well known from the microscopic point of view. Very often the knowledge is limited to that of the Wiedemann-Franz relation. This relation states that the thermal conductivity is proportional to the electrical conductivity at a given temperature. The result is that metals, which are good electrical conductors, are also good thermal conductors. They have indeed a high electronic thermal conductivity. However, this relation only holds for pure metals.

For nonmetals there is another mechanism for thermal conduction. This explains why diamond, an electrical insulator, is the best thermal conductor at room temperature. This is also true for graphite, which as a semimetal is not as good an electrical conductor as metals, but is a very good thermal conductor in-plane.

It was shown that the thermal conductivity of carbon fibers may vary widely, about two orders of magnitude, according to their microstructure[5,9]. In these materials, the transport of heat is entirely via the lattice vibrations, the phonons. Well above the dielectric maximum, the thermal conductivity is limited by phonon scattering due to phonon-phonon resistive processes and the thermal conductivity is the same for different samples. We are in the intrinsic range. For the purpose of characterizing the sample investigated, we are interested in the low temperature regime, where the thermal conductivity is mainly limited by phonon-lattice defects and phonon-boundary scattering. In the lowest temperature range, phonon-boundary scattering dominates and the phonon mean free path is limited by the in-plane coherence length L_a. It was shown that for carbon fibers the in-plane coherence length coincides with the phonon mean free path. Thus, the larger the in-plane coherence length, the higher the thermal conductivity. Around the maximum, scattering of phonons by point defects will be the dominating process. The position and the magnitude of the thermal conductivity maximum will thus depend on the competition between the various scattering processes (boundary, point defect, phonon, etc.). So, for different samples of the same material the position and magnitude of the maximum will depend on the lattice defects and L_a, since phonon-phonon interactions are assumed to be the same. Thus, by measuring the thermal conductivity, one may gather information about the in-plane coherence length L_a and point defects.

From elementary kinetic arguments, one may express the lattice thermal conductivity as a product of a phonon mean free path, L, lattice specific heat, c_v, and phonon velocity, v. This is well known as the Debye relation:

$$\kappa = \frac{1}{3} c_v v L \qquad (1)$$

which is convenient to discuss qualitatively the thermal conductivity results of isotropic materials. For a given type of solid, since the specific heats and the phonon velocities are the same, the thermal conductivity is directly proportional to the phonon mean free path. In the lowest temperature range the in-plane phonon mean free path, L, is constant and is determined by phonon scattering on grain boundaries. For well-ordered graphites, the lattice thermal conductivity, κ, follows a T^n law with n = 2.3[(19)]:

$$\kappa = A\,L\,T^{n} \tag{2}$$

where A is a constant, which has roughly the same value for all graphitic samples, and L is directly related to the in-plane coherence length L_a. The fact that A should not vary significantly from one sample to another may be easily understood, since it depends on the vibrational modes of the crystal lattice, which do not vary from one sample to another provided their defect structure is not too much pronounced.

In carbon fibers the in-plane coherence depends mainly on the precursor and the heat treatment temperature (HTT), the higher the HTT the larger the in-plane coherence length, up to a certain limit.

In Fig. 6 we compare the room temperature thermal conductivity of various carbon fibers to that of other solids. It may be seen that VGCF and pitch-derived carbon fibers have thermal conductivities higher than that of copper when heat treated at sufficiently high temperatures. Also, there are orders of magnitude differences between the thermal conductivity of the highest conducting fiber, a benzene derived carbon fiber heat treated at 3000°C, and that of the lowest ones, PAN-based fibers. We do not present here the thermal conductivity of activated carbon fibers which have thermal conductivities much lower than the PAN fibers presented here[20].

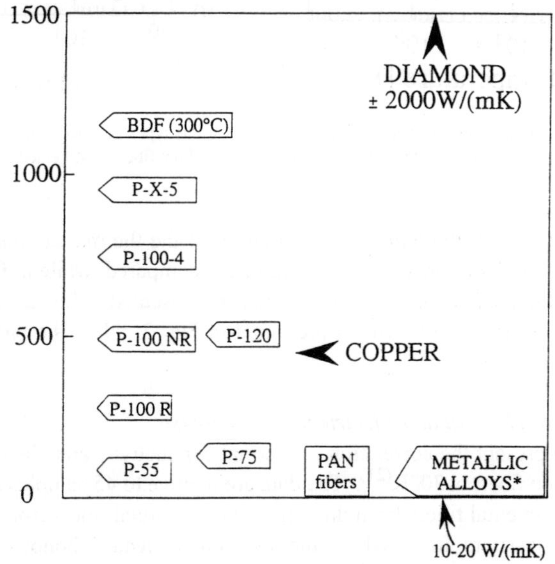

Fig. 6. Comparison of the room temperature thermal conductivity of various carbon fibers to that of other solids.

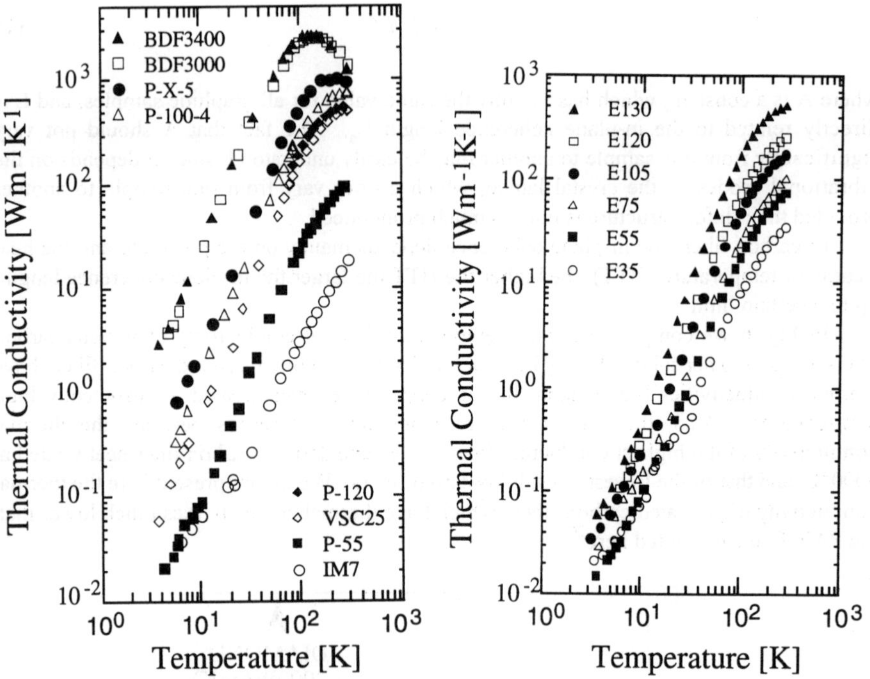

Fig. 7. Temperature variation of the thermal conductivity of various types of fibers. In Fig. 7a (left), fibers of various origins are compared, while in Fig. 7b (right), a series of du Pont fibers heat treated at various temperatures is presented.

In Fig. 7 we present the temperature variation of the thermal conductivity of various types of fibers. In Fig. 7a fibers of various origins are compared, while in Fig. 7b a series of du Pont fibers heat treated at various temperatures are presented. This last series is the same series for which electrical resistivity is presented in Fig. 2. Note the parallel behavior of electrical and thermal data.

2.5. Correlation of Thermal and Electrical Conductivities

In Fig. 8 we present the correlation of thermal conductivity and electrical resistivity in pitch-derived carbon fibers at 300K[(21)]. The data are relative to 45 samples among which the E-series and experimental fibers from du Pont and commercial fibers from Amoco.

Though the carriers involved in the two phenomena (phonons for the thermal conductivity and electrons and/or holes for the electrical resistivity) are different, the correlation is quite good.

The data may be fitted to the relation:

$$\kappa = \frac{44.10^4}{\rho + 258} - 295 \qquad \textbf{(3)}$$

where ρ is expressed in 10^{-6} Ωcm and κ in $Wm^{-1}K^{-1}$. So, for the same precursor, once the electrical resistivity is known, the thermal conductivity may be calculated. This may avoid delicate and time consuming thermal conductivity measurements.

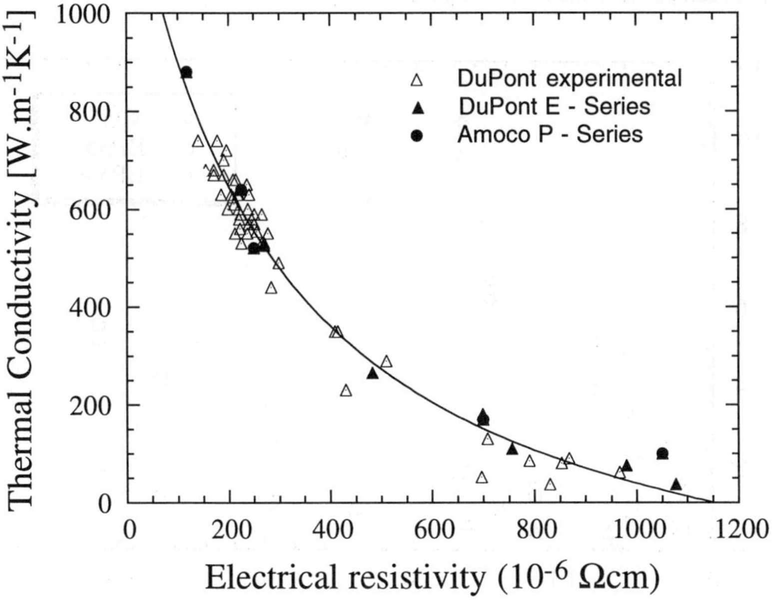

Fig. 8. Correlation of thermal conductivity and electrical resistivity in pitch-derived carbon fibers at 300K. (Reprinted from *Carbon*, vol. 31, Lavin et al, "The Correlation of Thermal Conductivity with Electrical Resistivity in Mesophase Pitch-based Carbon Fiber," p. 1002, with kind permission from Elsevier Science Ltd, The Boulevard, Langford Lane, Kidlington OX5 1GB, UK.)

3. Composites

3.1. Continuous Fibers

For unidirectional composites with highly conducting continuous carbon fibers the result is rather trivial, since it was checked experimentally that, as expected, the thermal conductivity of the composite along the fiber axis direction is equal to the product of the volume fraction and the thermal conductivity of a fiber[2,4]. However, in order to check that, thermal conductivity measurements had to be performed in a single fiber, or on a small bundle of fibers, and the results compared with those obtained for the composite as a function of fiber volume fraction. With these composites thermal conductivities superior to that of copper may be attained, since continuous fibers with thermal conductivities higher than 1000 $Wm^{-1}K^{-1}$ are now available.

3.2. Chopped Fibers

Fig. 9 shows the relation at 300K between the thermal conductivity of a chopped fiber-polycarbonate matrix composite versus fiber volume fraction. The heat flow is parallel to the plane of the composite in the form of a plate.

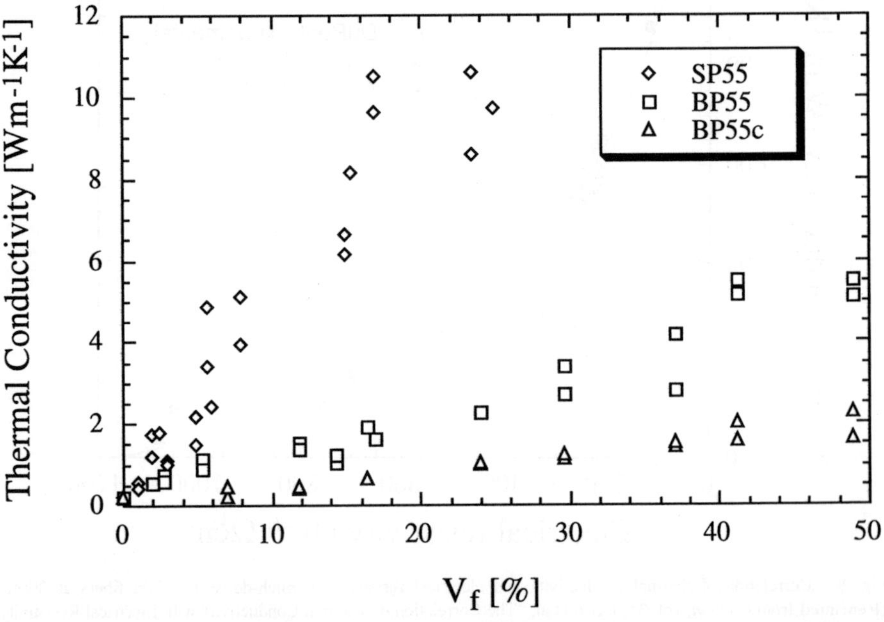

Fig. 9. Room temperature dependence of the thermal conductivity of a chopped fiber-polycarbonate matrix composite versus fiber volume fraction. The heat flow is parallel to the plane of the composite in the form of a plate.

For composites with short fibers, the situation is more complicated than for continuous fibers, since there are a large variety of parameters to control in order to improve the thermal conductivity. The dependence on one of these parameters is shown in Fig. 10 where we present the room temperature thermal conductivity of chopped fiber-polycarbonate matrix composites versus the thermal conductivity of the fibers[22]. We may see that for carbon fibers with a thermal conductivity of 550 $Wm^{-1}K^{-1}$, one may fabricate a composite with a thermal conductivity of ~10 $Wm^{-1}K^{-1}$.

Another important parameter is the fiber length for which results have been reported elsewhere[23,23]. Work is now in progress to determine the various factors, other than fiber length, conductivity and concentration, which control the thermal conductivities of the composite with short fibers.

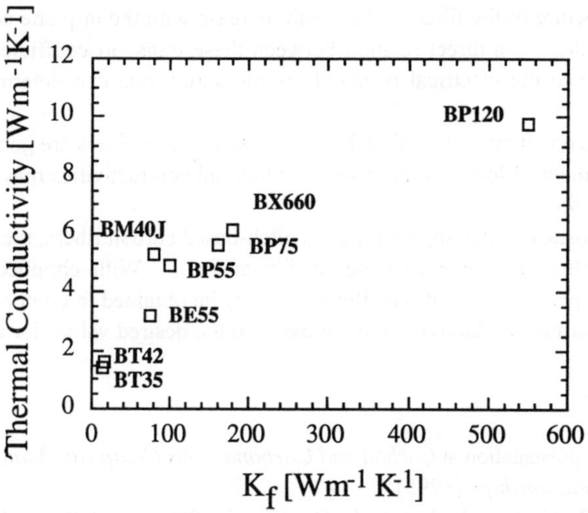

Fig. 10. Room temperature thermal conductivity of chopped fiber-polycarbonate matrix composites versus the thermal conductivity of the fibers.

In some cases, the presence of the fiber may have a direct effect on the properties of a relatively important portion of the matrix in its vicinity (nucleation, transcrystallinity). For composites which are relatively poor conductors, this may have an impact on their thermal conductivity. So, one should investigate this third phase, the interphase. While we started with a system of two constituents, we are left now with at least three different phases (the matrix itself may contain amorphous and crystalline phases): the matrix (excluding the interphase), the fibers (or carbon blacks) and the interphases (modified matrices).

3.3. *Composites as Heat Hyperconductors*

The temperature variations of the thermal conductivity below room temperature for different types of carbon fibers have been compared. By measuring single fibers, it was shown that by choosing adequately the precursor and by increasing the heat treatment temperature higher thermal conductivities are obtained. This is obtained by improving the microstructure and lattice perfection of the carbon fibers. It was also demonstrated that the thermal conductivity is directly correlated with the tensile modulus and electrical conductivity. In some cases, the thermal conductivities were found to be three times larger than that of copper around room temperature.

4. Conclusions

In conclusion, we may see that, though the electrical and thermal conductivities of carbon fibers are generated by different entities, charge carriers for the electrical resistivity and phonons for the thermal conductivity, these transport properties both depend dramatically on the structure of the fibers. They both increase with the in-plane coherence length. As a corollary, there is a direct relation between these transport coefficients for the same precursor and once the electrical resistivity is measured, one can determine the thermal conductivity.

Though less conductive than VGCF, pitch-based carbon fibers are good candidates for thermal management. Electrical resistivity and thermal conductivities may be used as tools for characterization.

With composites containing continuous pitch-based carbon fibers, thermal conductivities exceeding that of copper may be readily attained. With chopped fibers thermal conductivities superior to that of metallic alloys may be obtained in composites. Thus, one may tailor the thermal conductivity of composites to the desired values for applications.

5. References

1. A. Bertram, presentation at *Carbon and Carbonaceous Composite Materials: Structure Property Relationships* (1995).
2. J-P. Issi, B. Nysten, A. Jonas, A. Demain, L. Piraux and B. Poulaert in *Thermal Conductivity* **21**, eds. C. J. Cremers and H. A. Fine (Plenum Press, New York, 1990), p. 629.
3. G. Allen and J-P. Issi, *Proceedings of ECCM-1* (1985), p. 447.
4. B. Nysten and J-P. Issi, Composites **21** 1990, 339.
5. B. Nysten , L. Piraux and J.-P. Issi, *Thermal Conductivity* **19**, Proc. of 19th ITCCC, ed. D. W. Yarbrough (Plenum Press, New York, 1987), p. 341.
6. L. Langer, L. Stockman, J. P. Heremans, V. Bayot, C. H. Olk, C. Van Haesendonck, Y. Bruynseraede, and J.-P. Issi, *J. Mater. Res.* **9** (1994) 927.
7. L. Langer, V. Bayot, J.-P. Issi, L. Stockman, C. Van Haesendonck, Y. Bruynseraede, J. P. Heremans and C. H. Olk, in *Physics and Chemistry of Fullerenes and Derivatives*, eds. H. Kuzmany, J. Fink., M. Mehring and S. Roth (World Scientific, London, 1995), p. 565.
   L. Langer, V. Bayot, E. Grivei, J.-P. Issi, J. P. Heremans, C. H. Olk, L. Stockman, C. Van Haesendonck and Y. Bruynseraede, *Phys. Rev. Letters* (submitted).
8. A. A. Bright and L. S. Singer, *Carbon* **17** (1979) 59.
9. B. Nysten, J-P. Issi, R. Barton Jr., D. R. Boyington and J. G. Lavin, *J. Phy. D: Applied Physics* **24** (1991) 714 and *Phys. Rev.* **B44**, (1991) 2142.
10. B. Nysten, Ph.D thesis, Louvain-la-Neuve (1991).

11. J. C. Charlier and J. P. Michenaud, *Phys. Rev. Lett.* **70** (1993) 1858; J. C. Charlier. *Carbon Nanotubes and Fullerenes*, Ph.D thesis, Catholic University of Louvain (1994).
12. C. T. White, D. H. Roberston and J. W. Mintmire, *Phys. Rev.* **B 47** (1993) 494.
13. R. Saito, G. Dresselhaus, and M. S. Dresselhaus, *J. Appl. Phys.* **73** (1993) 494.
14. C. A. Klein, *J. Appl. Phys.* **33** (1962) 3388.
15. C. A. Klein, *J. Appl. Phys.* **35** (1964) 2947.
16. C. A. Klein, in *Chemistry and Physics of Carbon, vol. 2*, ed. P. L. Walker, Jr. (Marcel Dekker, New York, 1966), p. 217
17. See, e.g., *Graphite Intercalation Compounds*, Springer Series on Topics in Current Physics, Vol. II (1992).
18. T. C. Chieu, M. S. Dresselhaus, M. Endo and A. W. Moore, *Phys. Rev.* **B 27** (1983) 3686.
19. B. T. Kelly, *Physics of Graphite* (Applied Science, London, 1981).
20. S. L. di Vittorio, M. S. Dresselhaus, M. Endo, J-P. Issi, L. Piraux and V. Bayot, *J Mater Res.* **6** (1991) 778.
21. J. G. Lavin, D. R. Boyington, J. Lahijani, B. Nysten and J.-P. Issi, *Carbon* **31** (1993) 1001.
22. A. Demain and J-P. Issi, to be published.
23. B. Nysten, A. Jonas and J-P. Issi, in *Thermal Conductivity* **21**, eds. C. J. Cremers and H. A. Fine (Plenum Press, New York, 1990), p. 647.

THE STRUCTURE OF ELECTROCHEMICAL CARBON. A RAMAN STUDY

L. KAVAN
J. Heyrovský Institute of Physical Chemistry,
Dolejškova 3, CZ-182 23 Prague 8, Czech Republic

J. KASTNER and H. KUZMANY
Institute of Solid State Physics, University of Vienna,
Strudlhofgasse 4, A-1090 Vienna, Austria

ABSTRACT

Electrochemical carbon is formed by a cathodic carbonization of perhalogenated hydrocarbons. Electroreduction of C_xF_y with alkali metal (M = Li, Na, K) amalgams gives the best defined product: a solid composite of carbon and the equivalent amount of MF. Raman spectra of C-MF show an unusually strong band at 2000–2100 cm^{-1}, which was assigned to the C≡C stretching mode of polyyne. The position and relative intensity of this band depend on the type of fluorocarbon and reducing agent used. Optimum conditions were attained for electrochemical reduction of poly(tetrafluoroethylene) with K- or Na-amalgams. Poly(tetrafluoroethylene) oriented by mechanical stretching gives partly oriented polyyne chains with a conjugation length of sixteen carbon atoms. Electrochemical polyyne is stabilized against interchain crosslinking (graphitization) by interspersed nanocrystalline alkali metal fluoride (reaction byproduct). Polyyne decomposes spontaneously if the alkali metal fluoride is extracted from the composite, e.g., by leaching out with water. The final salt-free product is a highly disordered carbon material containing graphene links only.

1. Electrochemical Carbon

Electrochemical carbon (the term introduced in reference 1) can be prepared by anodic oxidation or cathodic reduction of suitable precursors. Electrochemical carbonization occurs under mild conditions (room temperature and pressure), hence it is a promising alternative to pyrolytic processes, common in the synthesis of carbons. The most carefully studied process was reductive carbonization of poly(tetrafluoroethylene) (PTFE) and other perhalogenated hydrocarbon.

A complete cathodic defluorination of PTFE was first demonstrated by Barker et al.[2] Analogous reactions of hexachlorobuta-1,3-diene[3], diiodoacetylene[4], polyvinylidene fluoride (PVDF)[5] and carbon tetrachloride[6] were also investigated. These reactions take place on platinum or carbon cathodes in cells with aprotic electrolyte solutions, e.g., tetraalkyl-ammonium salts in dimethylformamide or acetonitrile. The perhalogenated molecules (PTFE) seem to react directly on the electrode, whereas the partly halogenated ones (PVDF) are carbonized only indirectly, i.e., by the action of nucleophilic intermediates generated on the cathode[5,7]. The use of liquid electrolyte solutions introduces problems with impurities (humidity) and parasitic breakdown processes at the very negative potentials needed to complete the carbonization.

Jansta and Dousek[8] have solved the problem of impurities and the electrolyte/solvent instability by the preparation of electrochemical carbon *via* the reaction of PTFE with alkali metal (M = Li, Na, K) amalgams in vacuum. Their procedure is equivalent to a discharge of a short-circuited galvanic cell with amalgam anode, carbon cathode and fluoropolymer functioning as a cathodic active material:

$$\mathrm{M(Hg)|C\text{-}MF|C,\ PTFE} \quad (E_0 = 3.251\ \mathrm{V\ for\ 25°C\ and\ M = Li}).$$

C-MF denotes the reaction product, i.e., a conducting (e^-/M^+) mixture of carbon with a stoichiometric amount of alkali metal fluoride. In terms of the Jansta and Dousek's electrochemical model, the carbonization of PTFE takes place on the *in-situ* formed carbon cathode, which is continuously renewed as the reaction proceeds.

There have been speculations[1,8–10] that the electrochemical reduction of a linear perhalogenated hydrocarbon (e.g., PTFE) can lead primarily to a linear (sp-bonded) carbon chain, polyyne:

$$\left(-\mathrm{CF_2}-\mathrm{CF_2}-\right)_n \xrightarrow{4n\,e^-} \left(-\mathrm{C}\equiv\mathrm{C}-\right)_n + 4n\,\mathrm{F}^- \qquad \textbf{(1)}$$

This conclusion was supported by the X-ray diffraction analysis of C-LiF (from PTFE), which shows the average carbon-carbon distance to be 135 ± 4 pm[9]. It was interpreted as a superposition of graphitic (142 pm, 70–45%), diamond-like (154 pm, 0–15%) and also polyyne (≈ 120 pm, 30–40%) structures[9]. Other indirect arguments in favor of the polyyne structure are based on UV-Vis spectra, high chemical reactivity, electronic conductivity, etc.[1,10] Infrared spectra of C-LiF show weak absorption band at about 2000 cm^{-1}, which can be assigned to the (C≡C) stretching vibration of polyyne[10]. However, the IR spectrum changes rapidly if the sample is exposed to air. This makes the IR measurements, as well as their conclusions, somewhat problematic.

2. Raman Spectra of C-MF

The troublesome sample handling in infrared spectroscopy favors the use of Raman spectroscopy, where the measurement can easily be performed *in-situ*. We have designed an optical cell for sample preparation and Raman measurements, where any exposure of the sample to a reactive environment is safely excluded[11].

Raman spectra of the C-MF composite show strong bands of the (C≡C) stretch at about 2000 cm^{-1}. Spectral assignment to polyyne is unambiguous, since MF does not show any peaks in this region, and the same holds for the common carbon polymorphs graphite, diamond, and fullerene (Fig. 1). Unlike the reference carbon allotropes in Fig. 1, the C-MF exhibits broader Raman bands due to the poor crystalline ordering of C-MF[9].

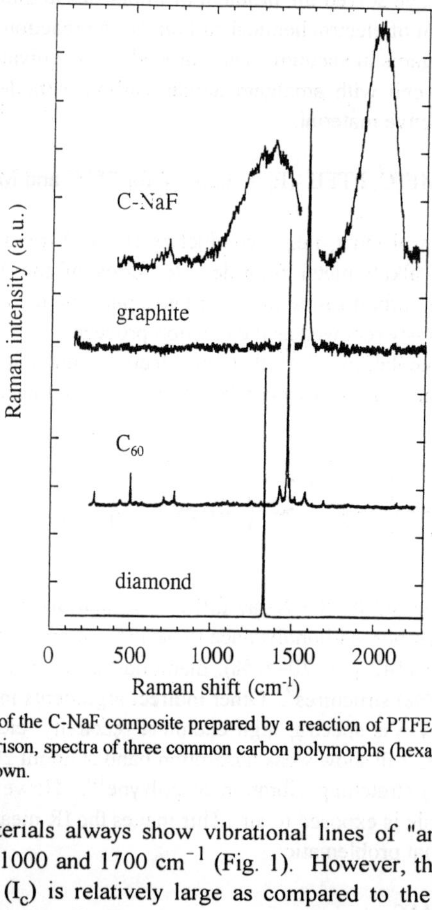

Fig. 1. Raman spectrum of the C-NaF composite prepared by a reaction of PTFE with Na amalgam at 25°C (λ_{exc} = 457.9 nm). For comparison, spectra of three common carbon polymorphs (hexagonal graphite, fullerene C_{60} and cubic diamond are also shown.

Our C-MF materials always show vibrational lines of "amorphous" (graphite-like) carbon between *ca.* 1000 and 1700 cm^{-1} (Fig. 1). However, the intensity of the polyyne band at 2000 cm^{-1} (I_c) is relatively large as compared to the intensity of the band of amorphous carbon (I_a). By dividing the areas of the two bands in Fig. 1 (curve C-NaF), we obtain $I_c/I_a = 1.6$. Our best sample showed $I_c/I_a = 2.2 \pm 0.2$.

3. Structural Assessments

The unusual structure of electrochemical carbon is documented not only by the high intensity of the Raman peak of C≡C vibrations, but also by its position. From there, we can calculate the number of triple bonds in an unperturbed polyyne chain (conjugation length, **n**)[11]. In C-NaF, the peak appears at 2006 ± 4 cm^{-1} (Fig. 1), which corresponds to **n** ≈ 6[11].

The best result was found for C-NaF from a PTFE foil pretreated by mechanical stretching. Here, the Raman peak appears at 1995 cm^{-1}, which translates into **n** ≈ 8, i. e., sixteen carbon atoms in a polyyne conjugation. Polarized Raman measurements indicate a preferential orientation of the polyyne chains in the stretching direction (1995 and 2007 cm^{-1} for parallel and perpendicular geometry, respectively).

Various other perhalogenated molecules react with liquid alkali metal amalgams to give carbon-halide mixtures similar to those from PTFE. Also Cs and eutectic Na-K alloy (both liquid at > 28.5°C) react under complete dehalogenation of the precursors. Fig. 2 shows several Raman spectra of the prepared C-MF composites. The products from perfluoro-n-alkanes, C_xF_{2x+2} (x = 2,3,6,9,20,24), poly(chlorotrifluoroethylene), and poly(tetrafluoroethylene-co-hexafluoropropene) exhibit weaker polyyne peaks than the corresponding products from PTFE. A limiting case is perfluorobenzene, C_6F_6, which gives no polyyne, but does produce graphite-like carbon (Fig. 2). The interpretation is self-evident: alkali metal amalgams are able to split the C-F (C-halogen) bonds, but not the C-C bonds[1]. The defluorination of C_6F_6 leads, therefore, to six-membered all-carbon rings, which subsequently condense into graphite-like clusters.

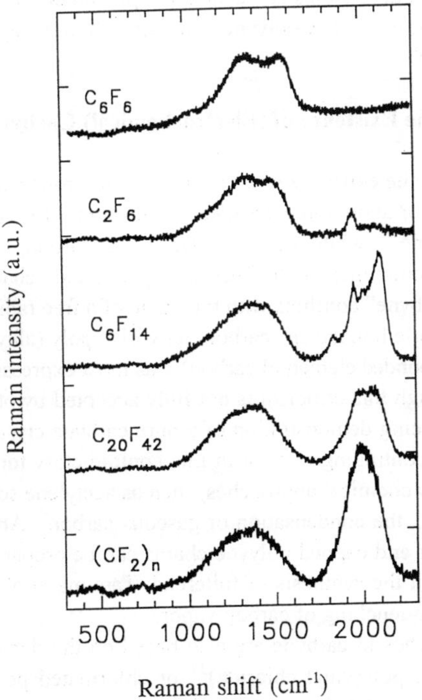

Fig. 2. Selected Raman spectra of C-MF prepared from different fluorocarbons and reducing agents at 25–30°C (λ_{exc} = 457.9 nm). From top to bottom: C_6F_6+Li(Hg), C_2F_6+Cs(1), C_6F_{14}+Cs(1), $C_{20}F_{42}$+Na(Hg), PTFE+Na(Hg).

The occurrence of a graphite-like carbon in the carbonization products from perfluoro-n-alkanes is caused by a subsequent crosslinking of the primarily formed polyyne chains. Their conversion to graphene can be schematically depicted as crosslinking of an array of parallel polyyne chains:

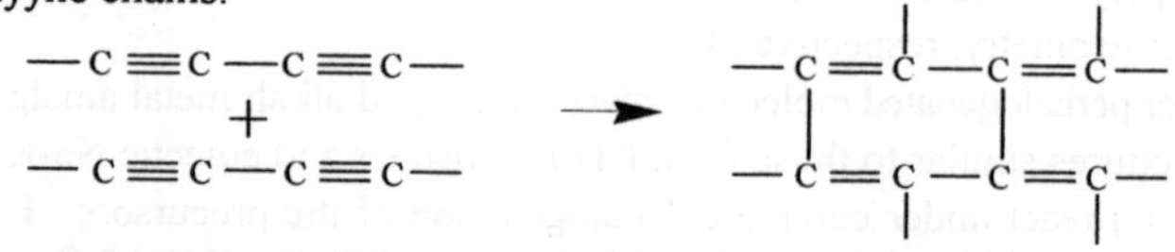

(2)

The crosslinking (eq. 2) is suppressed if there is a sufficiently large interchain distance. The conjugation lengths and polyyne contents in C-MF increase in the series Li < Na < K. Apparently, the bulky KF matrix suppresses mutual crosslinking more efficiently than the closely packed LiF matrix[1,11]. A slow degradation of polyyne (eq. 2) is, however, always detectable. It can be conveniently monitored through changes of electronic conductivity[12], electronic (UV-Vis) absorption[10], and the Raman spectrum[11]. The rate of these processes decreases, as expected, in the series Li > Na > K[1,10-12]. The mentioned properties change with time under a well-defined kinetics, which is valid in a time period of several years[1,10-12]. Crosslinking (eq. 2) accelerates at elevated temperatures and by the action of a reactive environment (air, water, etc.). The polyyne structure collapses rapidly after leaching out the MF with water (Fig. 3).

4. Some Aspects of the Existence of (Electrochemical) Carbyne

Discussions about the existence of a new carbon allotrope containing linear, all-carbon chains have continued for about three decades (reviews refs. 13–18). Such a polymorph has also been called “carbyne,” “white carbon,” “Ries crater carbon,” “polyacetylene” (this misleading term appears sometimes in old Russian papers), and “chaoite.” (Note that even the most popular term “carbyne” conflicts with the name of a free radical ⋮C–H; it appears also in connection with single bonded sp^3 carbon networks, poly (alkylcarbynes)[19].) The existence of carbyne (triple bonded elemental carbon) was most expressively questioned by Smith and Buseck[20]. Although this criticism is not fully accepted by others[14–18], ten years later, there is still no convincing demonstration of a pure carbyne crystal. Recent studies in this field[21–28] evidence a continuing interest in this contradictory topic. The preparations are focussed both on purely chemical approaches, such as acetylene scaffolding[21], as well as on physical methods, e. g., the condensation of gaseous carbon. Among the latter, Lagow et al.[22] have reported that end capped polyyne chains can be prepared by gas-phase reactions similar to those used in the synthesis of fullerene; Tanuma et al.[27] claimed a synthesis of carbyne crystal by fast quenching of carbon vapor.

Chemical approaches to carbyne are also based on the dehydrohalogenation of polyvinylidene halides[15,30], poly(vinyl chloride)[31] or chlorinated polyacetylene[29], and on the reductive dehalogenation of halogenated hydrocarbons[1,4,5]. Fig. 4 compares the Raman spectra of C-NaF (from PTFE) with “carbyne” made by electroreduction of diiodoacetylene[4].

Also shown are the spectra of electrochemically made carbons from C_4Cl_6 and CCl_4 where no carbyne formation was reported[3,6].

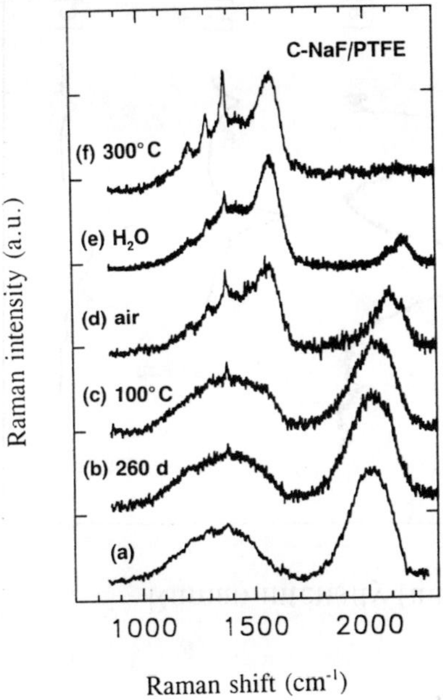

Fig. 3. Raman spectra (λ_{exc} = 457.9 nm) of C-NaF (from PTFE) after various treatments. The original sample was prepared by a PTFE/Na(Hg) reaction as a thin layer (0.54 μm) on top of PTFE foil. (a) C-NaF, measured immediately after the preparation; (b) sample a after 260 days of storing in vacuum at room temperature; (c) sample b treated for 12 hours at 100°C; (d) sample c exposed to air; (e) sample d washed with water for 20 min.; (f) sample e dried at 300°C in vacuum.

Other carbynes made by dehydrohalogenation of PVDF[30] or chlorinated polyacetylene[29] are demonstrated on Fig. 5. These products show the Raman lines of the C≡C stretch with lower intensity and higher frequency (the polyyne contents and conjugation lengths are relatively smaller). The same holds for carbynes made by other methods[4,15,31,32]. Although the experimental conditions are not identical for all the reported spectra of carbynes [4,15,29–32] the unique properties of our material are evident.

5. Acknowledgment

This work was supported by the Grant Agency of the Czech Republic (contract No. 203/93/0251) and by the COST action D4/0004/94.

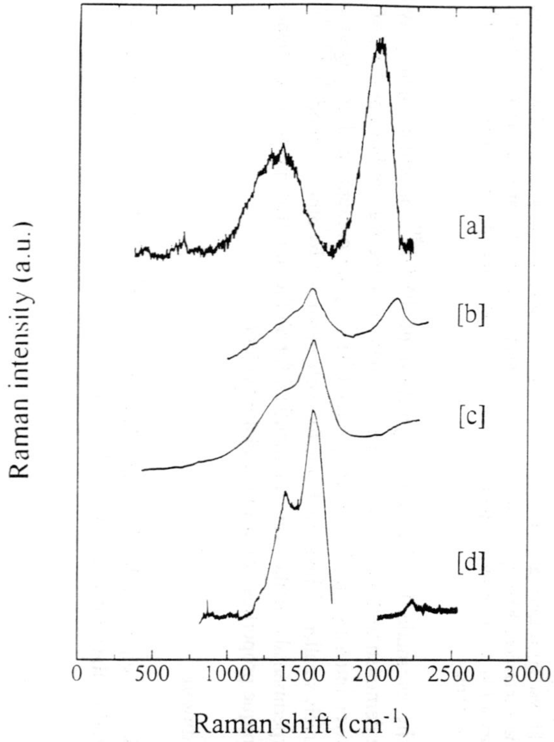

Fig. 4. Raman spectra of several electrochemical carbons. (a) C-NaF composite prepared by a reaction of PTFE with Na amalgam at 25°C (λ_{exc} = 457.9 nm); (b) carbon made from diiodoacetylene (λ_{exc} = unknown)[4]; (c) carbon made from hexachlorobuta -1,3-diene (λ_{exc} = 640 nm)[3]; (d) carbon made from carbon tetrachloride (λ_{exc} = 488 nm)[6]. The reference spectra are adapted from the literature[3,4,6].

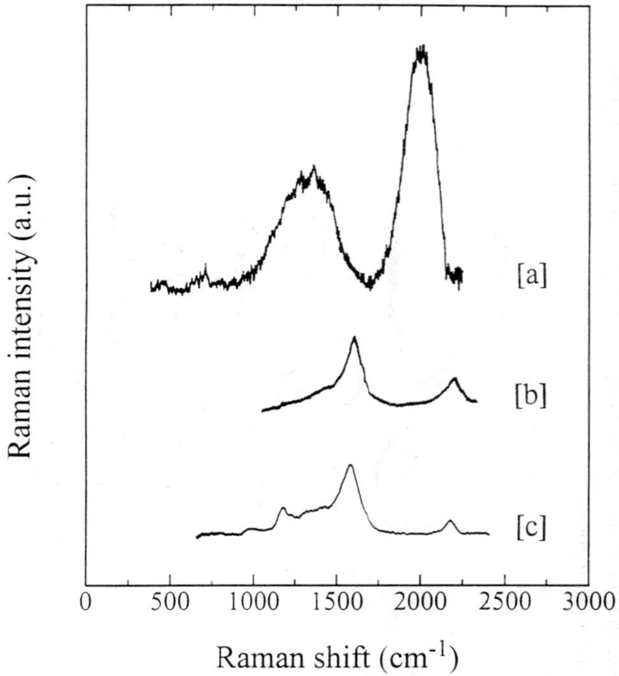

Fig. 5. Raman spectra of several materials declared as "carbynes" at λ_{exc} = 457.9 nm. (a) C-NaF (as in Fig. 4); (b) carbyne made by dehydrofluorination of PVDF[30]; (c) carbyne made by dehydrochlorination of chlorinated polyacetylene[29]. The reference spectra are adapted from the literature[29,30]. (For carbyne made by reduction of C_2I_2 see Fig. 4, curve (b).)

6. References

1. L. Kavan, in *Chemistry and Physics of Carbon, vol. 23,* ed. P. A. Thrower (M. Dekker, New York, 1991).
2. D. J. Barker, D. M. Brewis, R. H. Dahm and L. R. J. Hoy, *Electrochim. Acta* **23** (1978) 1107; *Polymer* **19** (1978) 856; *J. Mat. Sci.* **14** (1979) 749; *Adhesion* **4** (1980) 215.
3. H. Nishihara, H. Harada, S. Kaneko, M. Tateishi, K. Ararmaki and R. W. Murray, *J. Chem. Soc. Faraday Trans.* **87** (1991) 319; 1187.
4. M. Kijima, Y. Sakai and H. Shirakawa, *Chem. Lett.* **1994,** 2011; *Synth. Metals* **71** (1995) 1837.
5. M. Kijima, T. Toyabe and H. Shirakawa, *Chem. Lett.* **1995,** 553.
6. R. D. Herrick, A. S. Kaplan, B. K. Chinh, M. J. Shane, M. J. Sailor, K. L. Kavanagh, R. L. McCreery and J. Zhao, *Adv. Mater.* **7** (1995) 398.
7. A. A. Pud, D. E. Mikulina and G. S. Shapoval, *Macrom. Reports* **A32** (1995) 621.
8. J. Jansta and F. P. Dousek, *Electrochim. Acta* **18** (1973) 673; *Carbon* **18** (1980) 433.
9. L. Červinka, F. P. Dousek and J. Jansta, *Phil. Mag.* B **51** (1985) 603.
10. L. Kavan and F. P. Dousek, *Synth. Metals* **58** (1993) 63.
11. J. Kastner, H. Kuzmany, L. Kavan, F. P. Dousek and J. Kürti, *Macromolecules* **28** (1995) 344.
12. L. Kavan, F. P. Dousek and K. Micka, *J Phys. Chem.* **94** (1990) 5127.
13. A. M. Sladkov, *Soviet Scientific Review Sect. B* **3** (1981) 75.
14. P. B. Heimann, J. Kleiman and N. M. Salansky, *Carbon* **22** (1984) 147; *Nature* (London) **306** (1983) 164.
15. Y. P. Kudryavtsev, S. E. Evsyukov, M. B. Guseva, V. G. Babaev and V. V. Khvostov, *Russian Chemical Bulletin* **42** (1993) 399.
16. C. T. Pillinger, *Phil. Trans. R. Soc. Lond.* **A 343** (1993) 73.
17. F. J. M. Rietmeijer, *Meteoritics* **28** (1993) 242.
18. R. B. Heimann, *Diamond Rel. Mat.* **3** (1994) 1151.
19. G. T. Visscher, D. C. Nesting, J. V. Badding and P. A. Bianconi, *Science* **260** (1993) 1496; G. T. Visscher and P. A. Bianconi, J *Am. Chem. Soc.* **116** (1994) 1805; R. D. Rieke and T. A. Chen, *Chem. Mater.* **6** (1994) 576.
20. P. P. K. Smith and P. R. Buseck, *Science* **216** (1982) 986; **229** (1985) 486.
21. F. Diederich and Y. Rubin, *Angew. Chem. Int. Edn. Eng.* **31** (1992) 1101; F. Diederich, *Nature* **369** (1994) 199.
22. R. J. Lagow, J. J. Kampa, H. C. Wei, S. L. Battle, J. W. Genge, D. A. Laude, C. J. Harper, R. Bau, R. C. Stevens, J. F. Haw and E. Munson, *Science* **267** (1995) 362.
23. Y. Ando, *Carbon* **33** (1995) 171.
24. G. Burkhard, K. Dan, Y. Tanabe, A. B. Sawoka and K. Yamada, *J Appl. Phys.* **33** (1994) 5875.
25. R. B. Heimann, S. Fujivara, Y. Kakudate, Y. Koga, T. Komatsu and Nomura, *Carbon* **33** (1995) 859.

26. T. Komatsu, M. Nomura, Y. Kakudate, S. Fujiwara and R. B. Heimann, *Macrom. Chem. Phys.* **196** (1995) 3031.
27. S. Tanuma, V. Palnichenko and N. Satoh, *Synth. Metals* **71** (1995) 1841; *J. Mater. Res.* **10** (1995) 1120.
28. K. Yamada, G. Burkhard, K. Dan, Y. Tanabe and A. B. Sawoka, *Carbon* **32** (1994) 1197.
29. K. Akagi, M. Nishiguchi and H. Shirakawa, *Synth. Metals* **17** (1987) 557.
30. V. V. Bedryugin, Yu. P. Kudryavtsev, S. E. Evsyukov, Yu. V. Korshak, P. P. Shorygin, and V. V. Korshak, *Dokl. Akad. Nauk SSSR* **305** (1989) 362.
31. M. Shimoyama, H. Niino and A. Yabe, *Makrom. Chem.* **193** (1992) 569.
32. M. Nakamizo, R. Kammereck and P. L. Walker, *Carbon* **12** (1974) 259.

SAMPLE PREPARATION FOR EVALUATING CARBONACEOUS FIBERS

P. DVOŘÁK and M. KALOČ*
Department of Mineral Raw Materials and Geoinformatics
**Department of Ironmaking and Coking*
VŠB - Technical University of Ostrava, Czech Republic

ABSTRACT

Any material under microscopical evaluation requires a different method of sample processing. Sample processing is especially influenced by the method of evaluation—whether the coal matter is evaluated under transmitted light or under reflected light, regardless of whether the coal matter or coke (blast furnace, resinous, or foundry cokes), or other material which was produced from the raw materials mentioned above, is to be evaluated.

Preparations that are usually used for such studies require, in principle, using similar methods with only small differences. They ask for the methods that make it possible to understand material properties of an appropriate product. They require the utilization of procedures fulfilling, at the end, the requirements of the evaluation methods.

Basically, the methods used to prepare samples of coal, coke electrodes and so-called carbonaceous materials or residual products after combustion must differ to a certain extent. The different evaluation methods require special treatment of samples meeting all requirements of the evaluation methods, and are based upon the properties of the materials being evaluated. For evaluating electrodes, resin, and carbonaceous fibers, polished sections and grain sections are used.

1. Introduction

Each material which is being microscopically evaluated requires a different manner of specimen processing. This is especially influenced by the methodology of evaluation, namely, whether the evaluation is carried out by the optical or electron microscope. It is necessary to know if the investigation is to be done in transmitted or reflected light. This will influence the choice of methods of specimen preparation, as well as whether coal matter, coke, or a material with high carbon content is to be evaluated.

To fulfill different requirements of microscopical techniques, various improvements of specimens used for microscopical evaluation are carried out. For optimal microscopy, the following forms of treatments can be used: thin sections, lump polished sections, grain polished sections, polished thin sections, oriented polished sections, series sections and spatial polished sections. For comparison, their characteristics are stated briefly here:

- Polished section: a specimen whose one plane is ground and polished.
- Lump polished section: a specimen formed from a solid piece of a sample; a polished plane could be selected to cover the greatest number of components.
- Grain polished section: a milled grained or non-cohesive material which is bound together with artificial or native resins into the shape further treatable. It can be used for reflected light, both polarized and non-polarized.
- Thin section: a specimen whose preparation includes, first, the preparation of the polished section up to the stage of pre-polishing. A plane thus prepared is adhered (e.g., through Canadian balm) to the underlying glass. The remaining part of the specimen is ground into a thin slice (as thin as possible) and after the pre-polishing phase, it is covered with a cover glass which is also stuck by Canadian balm or another adhesive. It is used for transmitted light, polarized or non-polarized.
- Polished thin section: a combination of a thin section and a polished section. First, a thin section is prepared without covering with a glass slip. Its upper plane is treated in a mode similar to that in the polished section. It is used for transmitted and reflected light, polarized as well as non-polarized.
- Spatial polished section: usually prepared in the cubic form on which three planes, normal to one another, are ground and polished. No solidification of the specimen is carried out.
- Series section: First, a polished section is prepared. After evaluating and documenting the plane, the specimen is ground and polished once more. A newly originated polished plane is evaluated and documented again. This process is repeated on parallel planes throughout the specimen.

A carbonaceous material, i.e., a material of a rather high carbon content, is a material that is non-transparent and opaque under transmitted light. For this reason, it can be microscopically evaluated by means of the technique of polished sections that enables observation under reflected light. From the above descriptions, it fits into the treatment categories of lump and grain polished sections.

Before preparing the polished section, the basic operation is the choice of a specimen. It must be representative of the matter under examination. After selection, the specimen can be further treated by cutting and subsequent grinding and polishing. Solid and compact pieces may be cut, ground, and polished without any further treatment. Crumbly specimens need to be impregnated with some medium before cutting to enable their further treatment, similar to porous specimens. The latter must be perfectly prepared, their planes must have a high-mirror finish without scratches and grooves, i.e, without relief.

Solidification of specimens can be implemented with cold or heated epoxy resin, Carnauba wax, or Canadian balm.

Cutting is possible using a diamond disk saw.

For grinding, 200 silicon carbide is used (60–80 μm), 400 (22–32 μm), 600 (10–15 μm), and 1000 (7–10 μm).

For polishing, 2 μm or 1 μm alumina is used in water suspension or in powder on a cloth pad (velvet, billiard cloth). The last polishing phase can be, if required, enhanced by grains less than 1 μm.

Loose material is treated to so-called grain polished sections by mixing together with a relevant amount of binder, pouring it into a form and setting. It can also be placed, together with a suitable amount of binder, into the specimen press, where it is set. At the preparation by the form of a mixture alone, when gravity separation of represented grains can occur, it is more suitable to treat by grinding and polishing the side that is normal to the specimen base. With preparation in the press, gravity separation of grains cannot be assumed.

In the case of fibrous material, the required specimen preparation is determined by the goal. For common evaluation of fiber properties, the fibers should be placed in a manner that allows the evaluator to see them in the longitudinal and transverse directions. Therefore, fibers are to be adhered with a small amount of binder at first, partly recumbent and partly erect, so that it may be possible to meet the given requirement. After setting, the form used is filled with a cementing agent.

The specimen prepared in this manner can be treated in the same fashion as a normal lump polished section. When grinding, attention must be paid to avoiding excessive grinding of laying fibers which could result in the loss of the specimen.

2. Conclusion

Finally, it is necessary to state that only a specimen prepared well is a guarantee of good results.

PHOTOCHEMICAL AND THERMAL TRANSFORMATION OF LIGNITE

B. MIKULÁŠKOVÁ, L. LAPČÍK, JR.*, L. LAPČÍK**,
I. MAŠEK, and O. SALYK
Faculty of Chemistry, Technical University of Brno, Veslařská 230,
CZ-637 00 Brno, Czech Republic

M. KALOČ
Faculty of Metallurgy and Engineering, Technical University of Ostrava,
CZ-708 33 Ostrava - Poruba, Czech Republic

ABSTRACT

Photochemical and thermal transformations of South-Moravian lignite were studied. As the photochemical form of activation inductively coupled low temperature physical plasma was applied. In both cases of our study the total content of humic and fulvic acid, FT-IR and EPR spectra, free volume and sorption capacities for cationic and anionic sorbates were analyzed.

1. Introduction

A shortage of fossil raw material resources for organic chemicals is becoming a potential barrier to the economic and social development in many countries. The present systems of production of organic chemicals are based mostly on the conversion of fossil raw materials by non-biological methods, but these materials, particularly crude oil and natural gas, are gradually being exhausted. Their prices are expected to increase and pressure for cheap, renewable alternative resources will gradually rise[1].

The latter mentioned fact is a reason for the more detailed research organized in many laboratories at universities and industrial research centers. Expected results should be used in the next century also in the complex utilization of phytomass, which seems to be the most promising potential raw material for the future. In this research we concentrated our effort on searching dry way of South-Moravian lignites (SML) transformations and their direct unconventional applications.

Coals of low degrees of coalification are economically extremely important raw materials. All of them have a long-time perspective. Lignite belongs to this group of low-rank caustobiolites.

---

* Also with the Faculty of Technology, Technical University, CZ-762 72 Zlín, Czech Republic

** Author to whom all correspondence should be addressed

There are the following fundamental characteristics of *lignite from the point of view of chemistry.* Lignite is defined as an organic substrate with mineral inclusions situated on the transformation route from phytomass to a dehydrated, dehydrogenated and deoxidized carbon-type complex and water. It seems to be a very complicated mixture of low and high molecular weight organic compounds–complexes of cyclic-aromatic-aliphatic hydrocarbons and their derivatives. Water is the dominating inorganic compound of lignite. Its content varies from 40 to 60 wt.%. The higher amount of water is localized in free spaces (micropores, pores, microcracks and original intercellular spaces of the network system of material). Microscopic mineral bodies of inorganic compounds of Si, Fe, Al, S and others represent a minority part of the system, as well as macroscopic inclusions of the same type. The last one should be separated out of the organic part of lignite by simple physical procedure.

From the *point of view of physical chemistry* the lignite represents a solid colloidal system with a different surface morphology depending on the degree of coalification, the properties of which are determined by compositions of parental phytomass (content of saccharides, hemisaccharides and lignin) and the deepness of the pressure and thermal conditions in the time of transformation. The original lignin is characteristic with its ability to swell in certain liquids and vapors and transition through a plastic state. It has a strongly developed pore structure. In a very rough scheme the porous structure of lignite is characterized by a complicated branched system of cavities and voids which used to be divided into four main groups:

1. microvoids of diameter ranging from 0.45 to 1.2 nm,
2. mesovoids or transitory voids of diameter ranging from 1.2 to 30 nm,
3. macrovoids of diameter ranging from 30 to 7500 nm,
4. rough voids with a diameter over 7500 nm.

From the viewpoint of the total surface of matter the target constitution is given by microvoids, the lowest by macro- and rough ones. Until present a universal method of exact expression of the total surface of carbonaceous materials had not yet been developed. This is resulting from very complicated composition of a void-type system of the matter created by spatial elements of a different size and shape.

There were also determined content of other chemical elements (Fe, Mg, Si, P, K, Ca, Ti, V, Cr, Mn, Co, Ni, Cu, Ga, Ge, Se, As, Sb, Sr, Y, Zr, Te, Ce, W, Ta, Hg, Pb, Bi, Th and U) present in a very low percentage of their value. For this reason they could not influence the practical applicability of assumed products.

From the petrologic point of view, SML represents so-called brown coal mesophase, i.e., a matter situated in the transition state of development, characterized by a very low degree of coalification. The xylolitic phase represents the main part of the matter and is characterized by a preserved vegetable structure. It can be easily split and the resulting planes are characterized by a dull and corned appearance. Some of this part contains gel-like spots of dark color without the original vegetable structure. There are presented also lenses and fragments of textinite, frequently corpuscle of resin like, but rarely corpohuminite and fusivite.

Lignite are those low-rank coals which were subjected to the least amount of metamorphic change during the coal-forming process. They contain a greater fraction of a volatile matter from the original peat material and contain less fixed carbon than high-rank coals (e.g., bituminous and anthracite). Lignite should be supposed as a raw material for the following utilizations:

1. production of some individual chemical compounds,
2. production of mixture-type modificants for direct application without separation of individual components,
3. microbiologic and fermentation ways of transformations, such as feeding materials for animals,
4. synergetic component for mixtures with characteristic non-additive effects,
5. sorption material for ecological applications,
6. component of purpose-made composites (a specific-type system).

South-Moravian lignite as an energetic raw material seems to be improper for the following reasons:

1. low combustibility (combustion of an organic matter out of the specific reason seems to be negation of the fundamental principle of the nature),
2. very high content of water.

1.1. Hypothesis of the Research Work

A working hypothesis of this work could be defined in the following way:

1. to study plasmachemical transformation of SML in continuously works appearance, where the matter is subjected to extremely high temperature discharge during very short time period and continuously separated in a cyclones system.
2. the study of thermally governed carbonization in a static regime and a comparison of properties of the product with the properties of that received in plasmachemical transformation.

2. Experimental

2.1. Starting Material

South-Moravian lignites are outputted at present only from a single locality in Mikulčice mine. Typical composition is given in Table 1, where the content of the other 62 determined minority elements is not included.

Table 1. Typical Composition of the South-Moravian Lignite

Compound	Content (wt.%)
H_2O	48–50
ash	11.6–13.2
combustible part	39.1–40.2
H	~ 2.15
C	25–27
O	10–11
S total	0.8–0.9
S organic	0.4–0.5
S pyrite	0.20–0.35
N	~ 0.3
halogens	< 0.01

There was used as a subject of the study a sample of SML containing as a main part of xylitic and detritic components, xylitic components were created by hemiphaseous xylitic coal fragments of the range about 40–120 x 30 x 10 mm. The detritic part should be characterized as a hemiphaseous xylodetric coal. In this matter are spoiled lightly laminar fragments of xylitic character about 1 mm thick. These inclusions are a little bit more gel-like and are characterized by an intensive resin type glass. The total content of the first one does not exceed 40 wt. % and the second one is less than 30 wt. %. The mineral inclusions are characteristic sand-type particles. In general, the studied material can be defined as a brown-coal hemiphase with a very low degree of coalification and a well preserved parental plant texture.

The starting material was after crushing in the mill selected in the vibration separator according to diameter magnitude in two classes: below and over 1 mm. Smaller fractions were used for the next study in the plasmachemical reactor.

2.2. *Thermal Activation of SML in the Statistic System*

The statistic activation process is comprised of four periods: drying, pyrolytic carbonization, activation by the action of mixture of gaseous products, and steam. The thermodynamic temperature of activation was 800°C and time lag of material in the hot zone was 120 minutes.

2.3. *Photochemical Activation in the Fluid Plasmachemical Reactor*

Plasmachemical treatment of SML was carried out in the continuous reactor (Fig. 1) consisting of feeding part, gases mixture streamer, discharge chamber, post-discharge product cooler and cyclone system separator.

The speed of both the solid and gaseous reactants were kept at a constant value. As a source of plasmachemical activation, torch-high frequency discharge (1 MHz, 500 W) was applied.

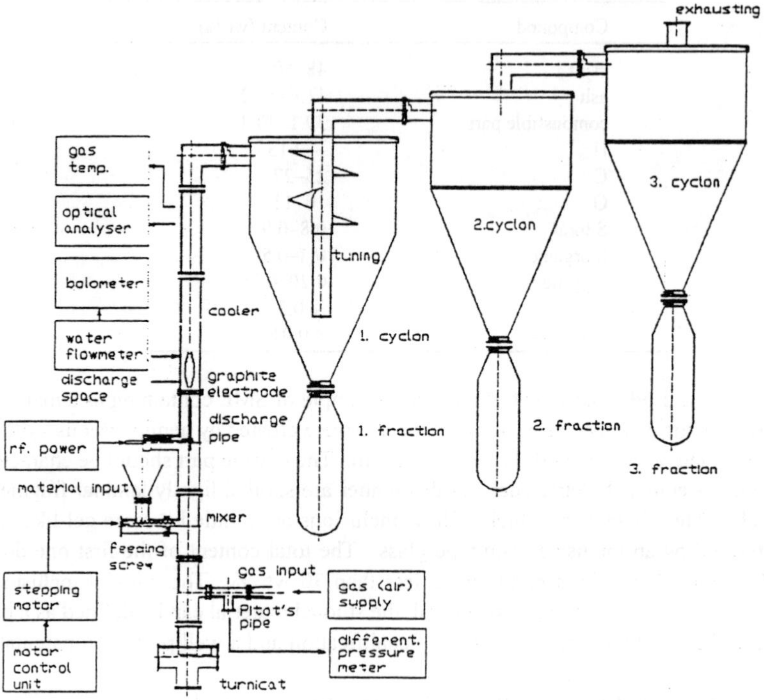

Fig. 1. Scheme of the low temperature physical plasma reactor.

2.4. Electron Paramagnetic Resonance Spectroscopy

A Bruker 200D X-band EPR spectrometer equipped with an Aspect 2000 computer and the corresponding Bruker Standard Program were used to obtain the spectra. All measurements were performed at the room temperature of 293K in quartz measuring tubes. All samples for the measurement were in the form of dust particles about 0.1 mm in size. All measurements were performed at the central field of 336 mT, sweep width of 500 mT, gain of 6300 and the time constant of 100 ms. As an EPR standard, the weak pitch of Varian was used, having the g-factor of 2.0028.

2.5. Fourier-Transform Infrared Spectroscopy (FT-IR)

A Nicolet Impact 400D FT-IR spectrometer equipped with an IBM compatible personal computer was used for measuring infrared spectra. The standard KBr disk matrix method was used for the measurements. KBr of IR purity was purchased from Fluka.

2.6. *Determination of the Inner Free Volume and Total Surface Area*

For determination of the inner free volume and of the total surface area of studied lignites the one point method based on BET (Brunauer-Emmet-Teller) isotherm was used. The measurements were realized by a standard procedure on Flow Sorb II 2300 apparatus from Micromeritics and were checked by standard calibration procedure. Each sample was evacuated and dried at 100°C to the constant weight prior to the measurement.

2.7. *Equilibrium Adsorption of Phosphates and Sulfates*

For plasmachemically treated samples collected in cyclone 1 the equilibrium adsorption isotherms of phosphate and sulfate adsorption were determined. For comparison, the same isotherms were also determined for the commercially available activated char coal samples. All samples were dried at 100°C for one hour prior to the measurements. For determination of phosphate concentration in studied solutions a spectrophotometric method based on reaction of phosphate anions with ammonium salt of molybdenate at acidic conditions was used. Then the absorbance of the created yellow complex was measured at the wavelength of 410 nm on a Hitachi U3300 spectrophotometer. The concentrations of sulfates in the studied samples were determined by titration of the measured solutions with $Pb(NO_3)_2$ on dithizone.

2.8. *Determination of Humic and Fulvic Acids*

The extraction procedure used in this study was taken from F. J. Stevenson's monography[2].

3. Results and Discussion

It was found that the thermal process of SML transformation and the plasmachemical treatment lead to different products. The polarity of the system decreases with increased temperature of activation and time of reaction. In general, if the mean thermodynamic temperature of the processes exceeds 400°C, a characteristic decrease of humic and fulvic acid content can be observed (Tables 2a and 2b).

Table 2a. Thermal Transformation of SML (time of reaction 120 min, gas 232)

Value of mean temperature in reactor (°C)	Humic acid content (wt.%)	Fulvic acid content (wt.%)
300	19.2	1.9
500	9.0	0.0
700	0.0	0.0
900	0.0	0.0
1100	0.0	0.0

Table 2b. Plasmachemical Transformation of SML
(gas 232, time of reaction 0.1–2 s, HDF 2.5 MHz, 4kW)

Mean temperature in discharge zone (°C)	Humic acid content* (wt.%)	Fulvic acid content* (wt.%)
300	19.2	1.9
500	9.1	1.1
700	3.2	0.8
900	0.5	0.3
1100	0.0	0.0

*mean values from 10 experiments

Also, the characteristic changes in FT-IR spectra (Fig. 2) were observed. The dominating adsorption caused by humic acid is decreasing in the same way as a decomposition of saccharide, hemisaccharide and lignite characteristic units is proceeding.

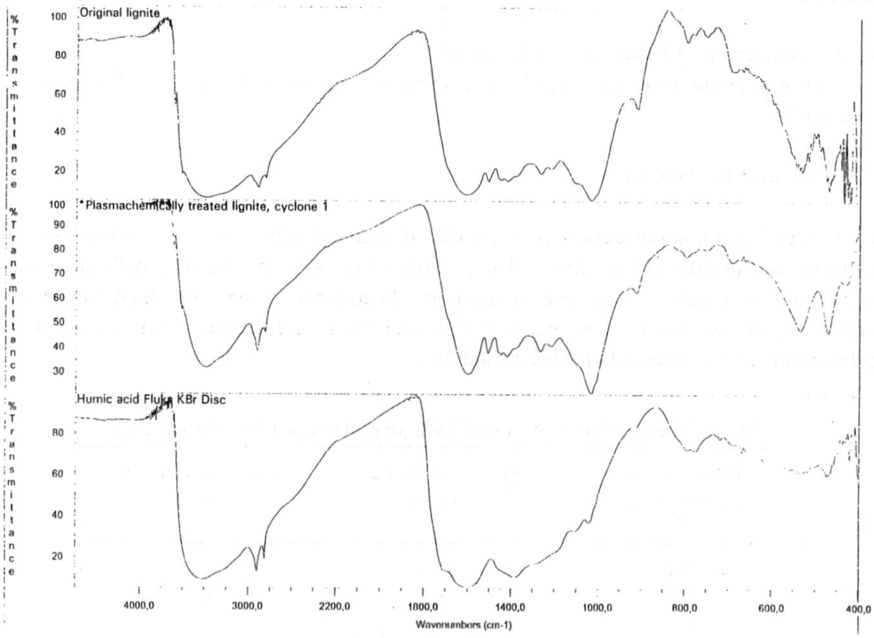

Fig. 2. FT-IR spectra of the original lignite, plasmachemically treated lignite from cyclone 1 and humic acid.

The following methods of product characterization were applied:

- chemical elements content (determined by X-ray fluorescence spectrometry-RFS),
- free surface and total volume of voids,
- sorption capacity to NH^{4+}, Cr^{3+}, Cd^{2+}, Cu^{2+}, Fe^{2+}, Pb^{2+} and Zn^{2+} (concentration coexisting solution range from 10^{-5} to $5.\ 10^{-3}$ mol/L),
- EPR spectroscopy used for study of free radicals (Figs. 4 through 6).

From the chemical point of view, the most reactive system in lignite matter seems to be chemicelluloses and polyuronides. Hemicelluloses[3] are heteropolysaccharides with degree of polymerization lower than that of cellulose (usually 100–200). The majority are amorphous and possess branched structures with neutral or acidic side chains. The solubility (as well as other properties) of hemicelluloses depends strongly on the degree of branching and on the chain length. Their crystallinity relates directly to degree of crystallinity, but native hemicelluloses are mostly amorphous[4,5]. The isolation of hemicelluloses from lignite substrate by simple extraction with water is possible only in a few cases. During continuous coalification the content of hemicelluloses rapidly decreases. After plasmatic or thermal treatment of SML the content of these was less than 1 wt. % (after their isolation by 0.01N KOH and chromatographic separation from humic and fulvic acids). This is quite understandable, because of the fact that hemicelluloses containing predominantly pentose and hexose units are easily hydrolysable and transformable to 2-furaldehyde and other very reactive furan derivatives[6,7] thus able to proceed condensation reactions.

Oxidative conditions in plasma reactor at a middle thermodynamic temperature (below 400K) leads to an intensive creation of adsorption centers in lignite. This fact results in a greater activity of adsorption of ammonium ions on such a substrate. These results are documented in Fig. 3.

Even the sorption capacity is not extremely high, a fact that resulting material after application can be used in agriculture seems to be economically interesting. The sorption capacity of thermally produced products is one order lower. During the past two decades the significant developments of both the theory of adsorption and the technology were observed. As an example, adsorption of metal ions by hydrous oxides and adsorbents such as activated char coal were the center of attention to a number of investigators interested in environmental problems. The maximum extent of adsorption, achievable under a given set of conditions, is determined by the equilibrium established between the solution and adsorbed phases. In the cases where all the relevant variables except the metal ion concentration are essentially constant, correlation of metal ion adsorption data can be made by fitting the data with Langmuir and Freudlich equations. For the beginning of our work we used the Langmuir isotherm in the form:

$$(c / a) = [1 / (a_m + b)] + c / a_m = A + Bc, \tag{1}$$

where the individual symbols have their usual meaning[8].

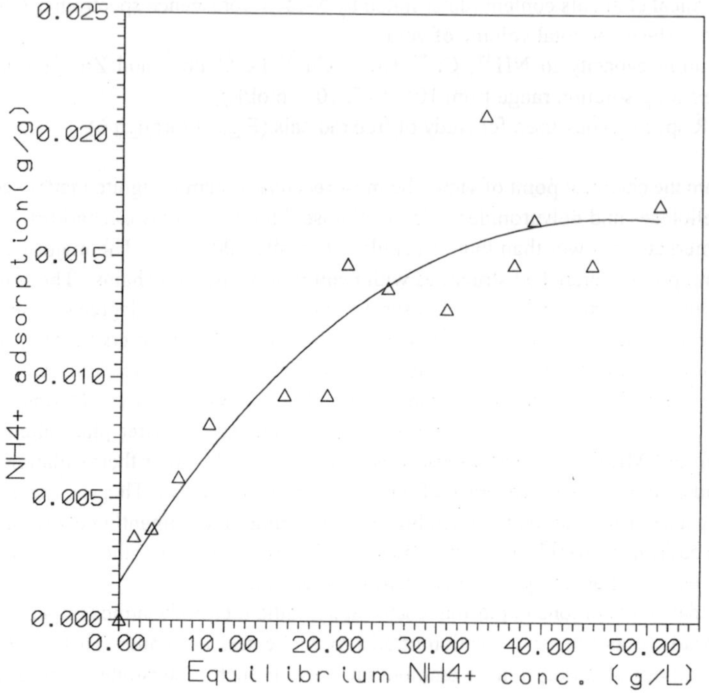

Fig. 3. Adsorption isotherm of NH_4^+ ions on plasmachemically treated lignite (T = 298K, p_{air} = 101,325 kPa, aqueous solution of NH_4Cl).

Values for maximum adsorbed amount a_m calculated from Langmuir isotherm (1) summarized for 6 samples are given in Table 3. The specific area and total free voids volume of the studies samples are given in Table 4. The samples are:

- Sample 1: plasmachemically treated lignite worked out by the gliding discharge method,
- Sample 2: plasmachemically treated lignite from the high frequency discharge unit (shown in Fig. 1) separated in cyclone 1,
- Sample 3 : the same lignite as sample 2, separated in cyclone 2,
- Sample 4: original SML,
- Sample 5: original SML from another date of mining,
- Sample 6: original SML from another date of mining.

Table 3. Best-fit Parameters of Maximum Adsorbed Amount a_m Calculated from Langmuir Adsorption Isotherm (1) of some Studied Samples

	a_m [μ.mol.g^{-1}]						
Sample	Al^{3+}	Cd^{2+}	Cr^{3+}	Cu^{2+}	Fe^{2+}	Pb^{2+}	Zn^{2+}
1	248	246	232	1463	617	199	336
2	248	251	225	1418	682	193	387
3	345	244	233	1550	483	195	377
4	257	218	126	450	379	164	144
5	250	227	139	231	377	168	110
6	200	212	82	83	200	135	303

Table 4. Specific Area and Total Free Voids Volume of the Studied Samples

Sample	Specific area [m^2 . g^{-1}]	Total free voids volume [cm^3 . g^{-1}]
1	2.28	0.00 12
2	2.93	0.0016
3	2.6	0.0014
4	3.33	0.0018
5	3.25	0.0018

It is quite clear from Table 3 that plasmachemically treated lignite has better adsorption properties to the studied metal cations. The mined form of SML seems to be for sorption properties less valuable. The first transformation of it seems to be crushing and selection according to diameter. The selected fractions could be separately applied for specific applications. The inner free volume and total surface was determined by one point method based on BET isotherm (N_2):

$$\frac{p}{a(p_o - p)} = \frac{1}{a_{\max}.c} + \frac{c-1}{a_{\max}.c} \cdot \frac{p}{p_o} \qquad (2)$$

where: a is the amount of sorbed N_2 in a known with amount of lignite (m^3 kg^{-1}) or (kg kg^{-1}) or (mol kg^{-1})

a_{max} is the amount of adsorbate bounded in the monomolecular layer of the absorbent

p is the pressure of vapor of adsorbate

p_o is the saturated pressure of adsorbate at given temperature.

Equation (2) may be rewritten in the linearized form:

$$y = \left| \frac{p}{a(p_o - p)} = A + B\frac{p}{p_o} = \right| A + Bx \qquad (3)$$

It was found that experimental data fit equation (3) with good accuracy. It was observed that specific area of plasmachemically treated samples was lower than the original ones. For this phenomenon we do not have a reasonable explanation yet.

From the practical point of view for the potential application of our adsorption materials, prepared from South-Moravian lignite in municipal waste water purification stations, the adsorption isotherms of several phosphates and sulfates were determined. Results of these experiments are summarized in Table 5.

Table 5. Determined Maximum Adsorbed Amounts a_m [mol g^{-1}] of Some Phosphates and Sulfates (commonly present in municipal waste waters) on Plasmachemically Treated Lignite Collected from Cyclone 1 and on Activated Char Coal

Sorbate	Lignite from cyclone 1	Activated charcoal
H_3PO_4	32.5	56.8
Na_2HPO_4	54.8	61.5
NaH_2PO_4	24.3	44.4
K_3PO_4	12.3	76.3
K_2HPO_4	47.2	47.0
KH_2PO_4	68.3	66.8
$(NH_4)_3PO_4$	13.8	18.8
$(NH_4)_2HPO_4$	87.0	98.2
$NH_4H_2PO_4$	23.5	27.5
Na_2SO_4	521.0	403.0
K_2SO_4	827.0	827.0
$MgSO_4$	133.0	98.7
$Al_2(SO_4)_3$	7.9	7.6
$CuSO_4$	13.4	13.7
$MnSO_4$	1.0	1.6
$CdSO_4$	0	0
$NiSO_4$	0	0
$(NH_4)_2SO_4$	0	0

It is known that the plasmachemical treatment of the material generates a high concentration of radicals both in their bulk as well as at their surfaces. For this reason we studied these phenomena in more detail using electron paramagnetic resonance (EPR) spectroscopy.

In Fig. 4 are shown EPR spectra measured on a solid lignite, A) original sample of lignite (prior to the plasmachemical treatment), B) plasmachemically treated lignite, fraction from cyclone 1, C) plasmachemically treated lignite, fraction from cyclone 2. It is evident that the original lignite sample is diamagnetic, thus there is neither the concentration of free radicals nor of paramagnetic transition metals observable by EPR spectroscopy. After the plasmachemical treatment in our apparatus, the vigorous change appeared in measured EPR spectra (see Fig. 4, B, C). Measured spectra were characteristic with two regions, i.e., the region around the magnetic field of 336 mT, characteristic by g-factor of 2.0023193, what is the value close to the g-factor of a free electron; by the region of the magnetic field of 282.85 mT with the g-factor of 2.3791 and finally by the regions of the magnetic field of 161.24 mT with the g-factor of 4.1735 and the magnetic field of 97.286 mT with the g-factor of 6.9172.

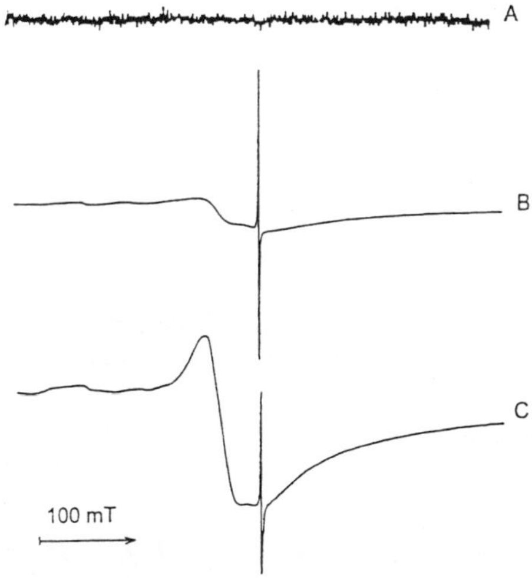

Fig. 4. EPR spectra of: A) original lignite, B) plasmachemically treated lignite, fraction from cyclone 1, C) plasmachemically treated lignite, fraction from cyclone 2.

Region with the g-factor of 2.0023 is characteristic for free radicals, in our case localized most probably on carbon atoms. It is possible to propose intensive delocalization of these radicals in the whole volume of the sample. Interesting is also the finding that these radicals have a very long half-time of decay, they are extremely stable and their concentration is still high also after the one month after plasmachemical treatment. Other three regions recognized in measured EPR spectra at the lower intensities of the magnetic fields are the regions characteristic for the EPR resonance of paramagnetic elements[9,10]. Very specific is the signal at the g-factor of 4.1735 attributed to the resonances of Fe^{3+} ions present in tetragonal spatial configuration and the g-factor of 6.9172 attributed also to the iron ions present in octahedral configuration.

Observed EPR signal in the region of g-factor of 2.3791 with the implication of the doublet hyperfine structure of the parallel component of g-tensor is characteristic for the Cu^{2+} ions, which are present in the samples after the plasmachemical treatment most probably from our plasmachemical reactor. Interesting is also the comparison of the EPR line intensities for the various fractions of the treated materials. As it is clear from Fig. 4, B, C, the intensity of the organic free radical is approximately ten times higher in the fraction from cyclone 1 (Fig. 4, B), then from cyclone 2 (Fig. 4, C). This fact suggests the importance of controlling the particle size distribution of the treated samples. More details of the measured spectra are shown in Figs. 5 and 6. As inserts, the regions characteristic for the paramagnetic iron ions are shown.

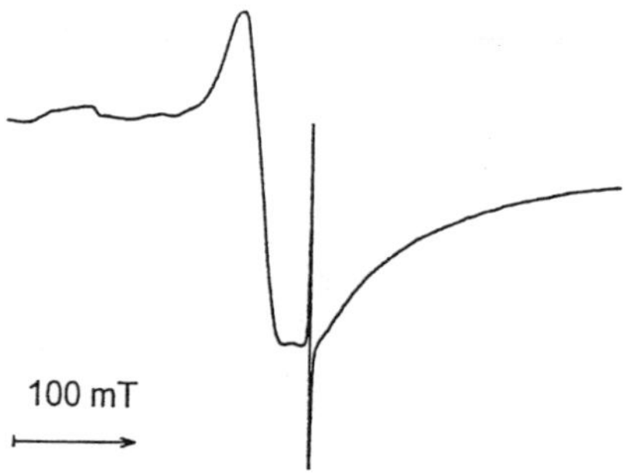

Fig. 5. EPR spectrum of plasmachemically treated lignite, fraction from cyclone 1.

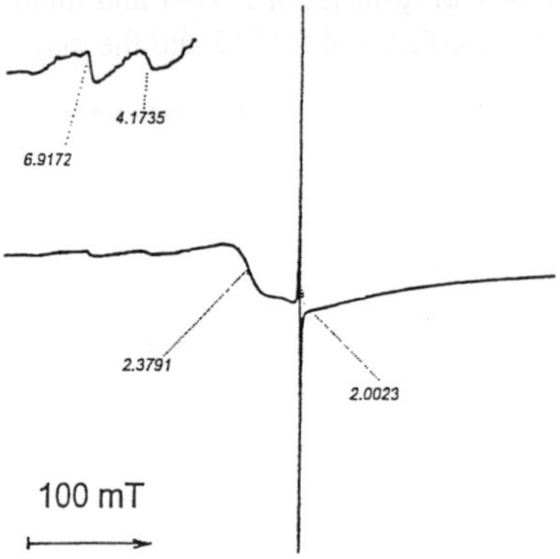

Fig. 6. EPR spectrum of plasmachemically treated lignite, fraction from cyclone 2.

4. Conclusions

We studied thermal and plasmachemical transformations of South-Moravian lignite. It was found that these processes give products applicable for adsorption of some metal ions. These processes are associated with the decreasing content of humic and fulvic acids upon the appropriate increase of the temperature in the reactor. The plasmatic process leads to the production of free radical moieties in the bulk as well as at the surface of the treated lignite dust particles. Interesting is the fact that the intensity of the plasma treatment reflected in corresponding EPR spectra is very sensitive on the size and the shape of the particles. This is because of the influence of the size and shape of the particles on the resulting density and homogeneity of the plasmatic field and the total aerodynamic conditions in the reactor.

5. Acknowledgments

The authors would like to express their gratitude to Dipl. Ing. J. Šesták and Southern Moravian Lignite Mines, Ltd. (Jihomoravské doly, s.r.o.), Hodonín for the financial support of this research.

6. References

1. A. Blažej and M. Košik (1993). *Phytomass a Raw Material for Chemistry and Biotechnology* (Ellis Horwood, New York).
2. F. J. Stevenson (1982). *Humus Chemistry. Genesis, Composition, Reactions* (Wiley - Interscience, New York), p. 26.
3. A. Blažej, L. Šutý, M. Košík, P. Krkoška, and E. Golis (1975). *Chemistry of Wood* (Alfa, Bratislava) in Slovak.
4. T. E. Timell (1964). *Adv. Carbohydrate Chem.* **19** 247, *ibid.* **20** , 409.
5. M. S. Dudkin (1964). *Khim. Dreves.* **7**, 18.
6. A. Dunlop & F. Peters (1953). *The Furans* (Van Nostrand Reinhold, New York).
7. E. F. Morozov (1979). *Production of Furfurol* (Lesnaja Preimyselnosť, Moscow), in Russian.
8. W. J. Moore (1972). *Physical Chemistry* (4th edition, Prentice-Hall, Englewood Cliffs, New Jersey).
9. A. Abragam and B. Bleaney (1970). *Electron Paramagnetic Resonance of Transition Ions* (Clarendon Press, Oxford).
10. L. Lapčík, Jr. and Z. Šimek (1996). *Cement and Concrete Research* **26**, 237.

ANALYSIS OF HOPG SURFACES USING SCANNING TUNNELING MICROSCOPY: AN ALTERNATE INTERPRETATION[1]

S. R. PALMER, D. TANDON, E. J. HIPPO, and J. SANDERS
Department of Mechanical Engineering and Energy Processes
Southern Illinois University at Carbondale
Carbondale, Illinois, 62901-6603, USA

ABSTRACT

This paper reports scanning tunneling microscopy (STM) images of HOPG basal planes that may provide an alternate interpretation to many that have been previously reported. Most STM images of HOPG in the literature show a trigonally symmetric series of bumps protruding from the surface. These protrusions most frequently have been interpreted as carbon atoms. The images reported in this paper also show a trigonally symmetric array of protrusions with exactly the same spacing as those reported in the literature. However, in addition to the protrusions, a ring system of six-carbon atoms can be seen that surrounds each protrusion. This observation leads us to believe that the protrusions may not be carbon atoms at all, but rather a tunneling artifact at the centers of the hexagonal rings. The ring centers have also been imaged as depressions. Possible explanations of these observations are presented in the paper.

1. Introduction

The study of carbon materials has been of great interest for many decades. Many microscopic techniques exist that help in the surface studies of these materials. Scanning tunneling microscopy (STM) excels over other microscopy techniques because of its very high (atomic level) resolution. It can provide information about both surface topography and electronic states at the atomic level. Many forms of carbon, including diamonds[1–3], fullerenes[4,5], isotropic porous carbons[6–8] and glassy carbons[9], have been studied extensively by STM. However, it is the study of highly ordered pyrolytic graphite (HOPG) that demonstrates the scanning tunneling microscope's ability to resolve atomic images of carbon materials.

The (0001) basal plane of HOPG has a very regular structure that is conductive and chemically inert. These factors allow individual atoms at the surface to be imaged under atmospheric conditions and account for the use of HOPG as an STM calibration material[9–11]. Typically, the 142 pm carbon-carbon bond length and the 250 pm hexagonal ring center-to-center distance[9–11] observed for the HOPG basal plane, along with 335 pm spacing between basal planes[12], can be used to calibrate the instrument. Because HOPG is used for calibration, it is essential that the interpretation of its STM images receive detailed attention. It is

---

[1]The authors acknowledge that the interpretations set forth in this paper are not accepted by all researchers in the field. It is our intention, via the publication of this manuscript, to contribute to the debate regarding the interpretation of STM images of the HOPG basal plane.

believed that the images presented in this paper lead to an interpretation that is somewhat different from those previously reported[12,13] and widely accepted.

Graphene layers in HOPG are staggered in such a way that every other carbon atom (α) is directly over another carbon atom in the graphene layer below[10]. It is at these points of direct carbon atom overlap where basal planes are held together by weakly attractive van der Waals forces. According to the model of Soler et al., at the points where there is an atom directly beneath another atom, there is an increased overlap of the atoms' electron orbitals. This overlap causes the electron orbitals of these surface atoms to be smaller than the electron orbitals of the surface carbon atoms (β) that do not have another carbon atom directly beneath[12]. Thus, every other carbon atom on the surface of HOPG should be drawn slightly lower relative to the other atoms at the surface[10]. Hence, the surface of HOPG is expected to show a slight corrugation in the vertical direction (z-axis) due to the alternating attracted and unattracted carbon atoms at the surface. This argument has led to the claims that only every other carbon atom is visible in the STM image of HOPG basal plane[12–14]. However, according to Selloni et al.[15], there is a very small difference in the local density of states (LDOS) of the adjacent inequivalent atoms of a graphite ring. This suggests that all six carbon atoms of the hexagonal rings should be visible.

Although only slight vertical corrugations are expected, the corrugations observed are most often much larger than expected and giant corrugations (up to 1000 pm) have been imaged by STM for HOPG. Initially, these giant corrugations were attributed to the anomalies in surface electron state and deformation of the graphite surface itself. However, Tiedje et al.[16] reported that the giant corrugations result due to tip contamination as a result of adsorption of soft carbonaceous material at the end of the tip. Several other researchers also proposed various reasons for corrugations. Yao and Jiao[17] observed varying images of graphite with Pt/Ir tip, on the sample area, without any variations in the bias voltage, tunneling current or the scan rate. They concluded that the differences in the images were due to tip effects rather than the sample itself.

In a recent review article attention is drawn to the possibility that in layered structures such as graphite, the Fermi surface structures at surface Brillouin zone boundaries exhibit standing wave solutions. The result is that the corresponding STM image does not reflect the true atomic positions, and the corrugation does not reflect the atomic corrugation[18]. Indeed, the variability in the vertical size of these corrugations strongly implies that they are artifacts and not representative of the true surface structures. Although many explanations have been presented for the apparent surface corrugation of HOPG as viewed by STM, it should be noted that this issue has not been fully resolved and a completely satisfactory explanation has yet to be found.

According to the Soler et al. model[12], the alternate depressed clouds are not imaged by the microscope's tip, and this explains why a trigonally symmetric structure is observed. This argument leads to an interpretation of HOPG basal plane images in which only every other carbon atom protrudes from the surface. The Park and Quate[13] STM image of HOPG with trigonal symmetry and alternate carbon atom image is the one most frequently regarded as the correct surface topography of HOPG. In this paper, we have referred to this interpretation as the "three-carbon atom" model. Initially, this interpretation appeared to fit images obtained in our laboratory; but as more images were collected, especially those plotted in 3-dimension, doubts regarding this interpretation were generated. Images were obtained that led us to an interpretation in which all six-carbon atoms of the hexagonal rings are visible in every image. This is referred to as the "six-carbon atom" model in this paper. It should be noted that low z axis settings often exaggerate the protrusions to the extent that the hexagonal ring structure is obscured. This effect will be discussed.

The "three-carbon atom" model leaves many questions unanswered. For example, if only alternate carbon atoms (β) are imaged by the STM tip, because the tip cannot tunnel to the slightly depressed electron clouds (less than 0.1 nm) of the carbon atoms (α) in the same plane, then how can the tip tunnel to the atoms at the bottom of a pore/pit that are more than 1 nm deep? The regions assigned to alternate carbon atoms in the "three-carbon atom" model are too big to be carbon atoms. The "three-carbon atom" model assigns almost the same tunneling efficiency to the depressed atoms and the center of the ring.

The asymmetry of the widely accepted three-carbon atom model is not explained by the single imaged wave function theory of Tersoff[19]. Graphite slab calculations explain this asymmetry to a small extent[14,15]. First principle calculations reveal even a larger asymmetry[20,21]. Asymmetry is also observed as a result of electronic effects of the tip on sample surface, which creates new surface states that affect tunneling[10]. Thus, there is considerable confusion about the STM image interpretation; and it has been accepted that even for highly regular structures like graphite, the STM image understanding is not final. In this paper, we have made an attempt to provide an alternate interpretation of STM images of HOPG.

2. Experimental

All STM nanographs were obtained using a Nanoscope II scanning tunneling microscope manufactured by Digital Instruments of Santa Barbara, CA. Pristine samples of HOPG were obtained from Union Carbide. For all images, Pt/Ir (80/20) tips manufactured by Digital Instruments were used. Instrument resolution was within ±1 pm. Bias voltages of 13.1–20.14 mV and a current of 1.99 nA were used. All images were obtained in constant current mode.

In a separate set of experiments, a series of images was obtained by varying the bias voltage from 5 to 100 mV in steps of 5 mV (set point current 1.99 nA). The set point current was varied from 0.5 to 3.5 nA in steps of 0.5 nA (bias voltage 20.14 mV). These experiments were performed to study the effect of the bias voltage and set point current on the topography of the electron states.

3. Results and Discussion

Fig. 1a is a top view raw image of HOPG basal plane and shows the centers of the rings as depressions (black areas) surrounded by six-carbon atoms (white areas). Fig. 1b is the low

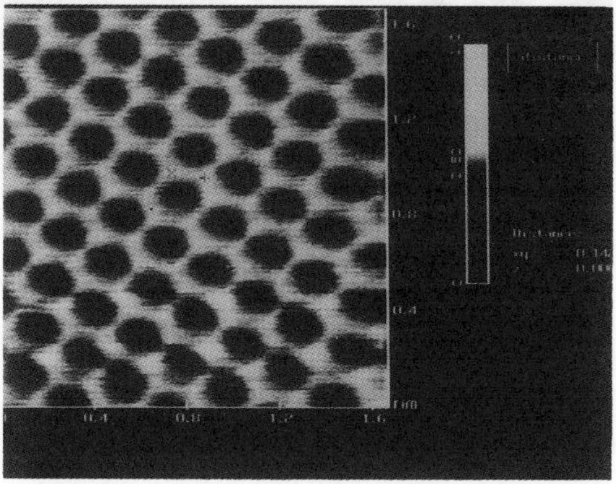

Fig. 1a. Topview of HOPG basal plane (raw image).

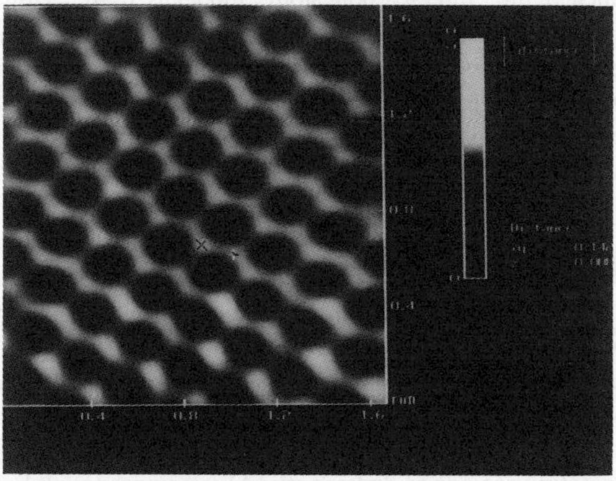

Fig. 1b. Low pass filtered image of Fig. 1a.

pass filtered image of Fig. 1a and is similar to the images reported by Magonov et al. and Hoffman et al.[22,23]. From the two figures (1a and 1b), it can be clearly seen that filtering does not change the features of the raw image, but just sharpens the image by reducing the noise. For this reason, the images presented in this paper were subjected to low pass filtering.

Fig. 1c is a 2 nm x 2 nm top-view image of an HOPG basal plane. This image is similar to the images most frequently reported in the literature in which the white spots are attributed

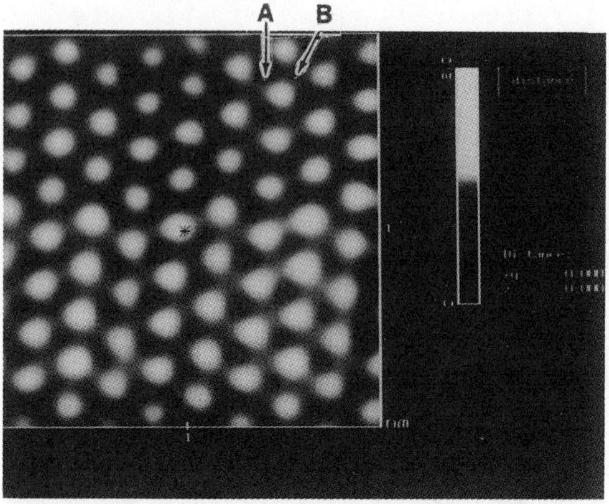

Fig. 1c. Topview of HOPG basal plane (sample 1) 1.7 x 1.7 nm.

to the alternating raised carbon atoms. The distance between adjacent white spots is 250 pm, the expected distance if the "three-carbon atom" model is used. However, due to the geometry of a hexagonal system such as that predicted for graphene sheets composed of hexagonal rings, the 250 pm measurement is also that expected for a ring center-to-ring center measurement. Thus, if the white spots are ring centers, this image and those in the literature also fit the "six-carbon atom" model. To obtain clearer nanographs and to help identify which model is being observed in these images, they were replotted in 3-D using two pitch angles (30° and 60°).

Fig. 1d shows the three-dimensional view of Fig. 1c at a pitch angle of 60°. This image clearly shows a hexagonal ring structure in which all six-carbon atoms are visible. The hexagonal periodicity follows the pattern expected for the C-C sp^2 sigma bonds in a graphene sheet. The white spots in Fig. 1c that were previously thought to be carbon atoms now align with the centers of the hexagonal rings in Fig. 1d. Using this information, it now appears that the gray border around the white spots in Fig. 1c represents the carbon atoms.

It can be seen in Fig. 1c that the gray border surrounding the white spots has variable tone. There are alternating regions of darker gray (A) interspaced with lighter gray regions (B). This variation in tone indicates that there is vertical resolution between alternate carbon atoms in the hexagonal rings. The darker gray regions (A) are thought to represent the carbon atoms that have another atom beneath them in the graphene layer below, while the lighter gray regions (B) are thought to represent the carbon atoms that do not have another atom beneath them in the graphene layer below. Thus, this interpretation of the STM image is entirely consistent with the prediction of interlayer van der Waals attractive forces at alternating carbon atom positions.

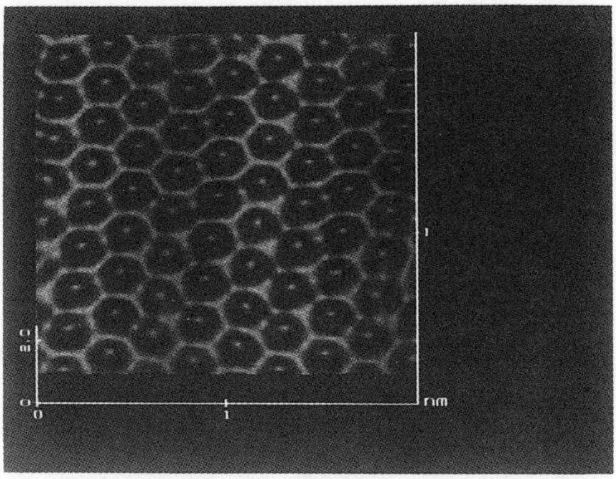

Fig. 1d. 3D surface view of Fig. 1a at 60° pitch angle.

The lateral distance between the darker and the lighter gray areas (A-B) is 141 ± 1 pm, which corresponds extremely well with the theoretical C-C bond length for a hexagonal graphite ring system (142 pm). The vertical separation of adjacent carbon atoms in each ring was measured at 30 to 40 pm. The distance between the centers of two white areas is 250 ± 1 pm, and again this is the distance predicted from theory for the hexagonal ring center-to-center measurement. If Fig. 1d is examined more closely, the centers of the rings have a raised dome like appearance.

Fig. 1e is another top view image of the HOPG basal plane using a different sample and tip from Figs. 1a and 1c. This image is similar to the image in Fig. 1c, except that the apparent topography has been inverted, i.e., what was white in Fig. 1c is now black in Fig. 1e. Fig. 1f is the 3-D image of Fig. 1e at a 60° pitch angle. The image shows the hexagonal pattern of the graphene planes with the lengths of sides of the hexagons corresponding to the theoretical carbon to carbon bond lengths (142 pm). Once again, the centers of the rings are separated by a distance of 250 pm. The black areas of Fig. 1e appear as hexagonal depressions in Fig. 1f. A comparison of Figs. 1d and 1f demonstrates that the centers of the hexagonal rings can be imaged both as depressions and as raised domes. At this time, it is uncertain why both depression and dome topographies are observed for the same material. However, it is believed that this phenomenon may be due to variability in tip/surface interactions or tunneling to different orbitals.

One possible explanation for the apparent protrusion in the middle of the ring could be of simple geometry. When the tip is directly above the center of the ring, it is equidistant from all six-carbon atoms. When the tip moves away from the center, it becomes closer to some of the carbon atoms but further away from the others. If we sum all of the distances between the tip and the six-carbon atoms, we find that the shortest summed distance is when

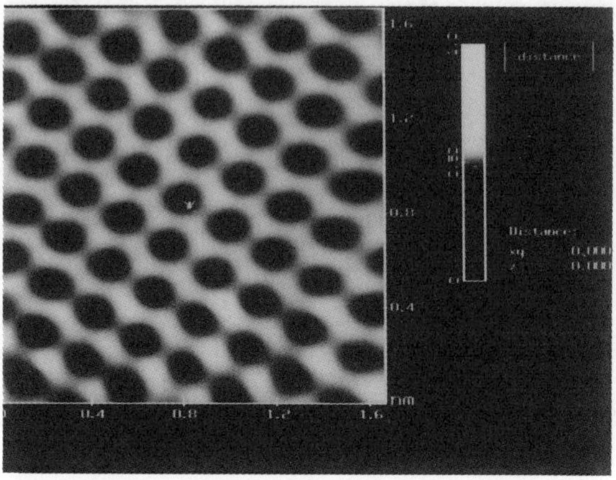

Fig. 1e. Topview of HOPG basal plane (sample 2) 1.7 x 1.7 nm.

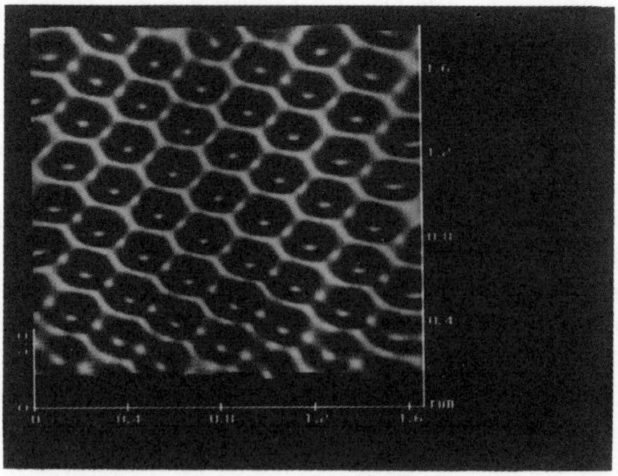

Fig. 1f. 3D surface view of Fig. 1c at 60° pitch angle.

the tip is above the ring center. Since tunneling efficiency reduces exponentially with distance, it seems logical that the ring centers should be imaged as protrusions. Thus, the protrusions seem to result from tunneling artifacts, rather than from surface features (alternate carbon atoms). This, however, does not explain the ring centers being imaged as depressions. A possible explanation for the ring centers being imaged as depressions could be selective tunneling to different orbitals.

Figs. 2a and 2b are 3-dimensional images of an enlarged portion of the HOPG basal plane observed in Fig. 1c. Fig. 2a is imaged at a pitch angle of 30°. The hexagonal patterns are clearly observed around the centers of the rings, which have a torus-like appearance. Fig. 2b is imaged at a pitch angle of 60°. The hexagonal patterns of carbon atoms are once again evident, and an even clearer torus-like appearance at the center of the ring is observed.

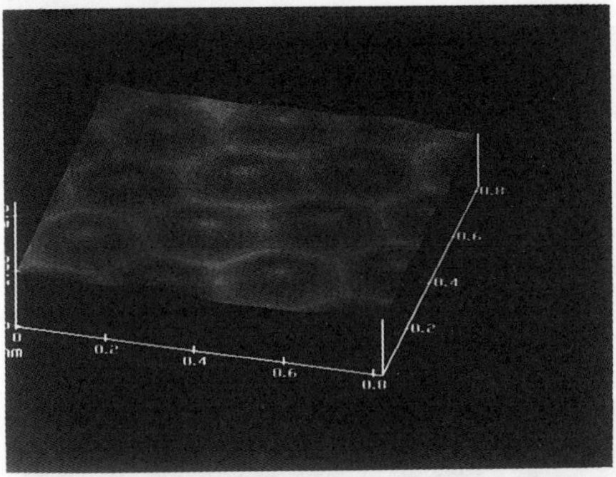

Fig. 2a. 3D surface view of enlarged portion of Fig. 1a (pitch angle 30°).

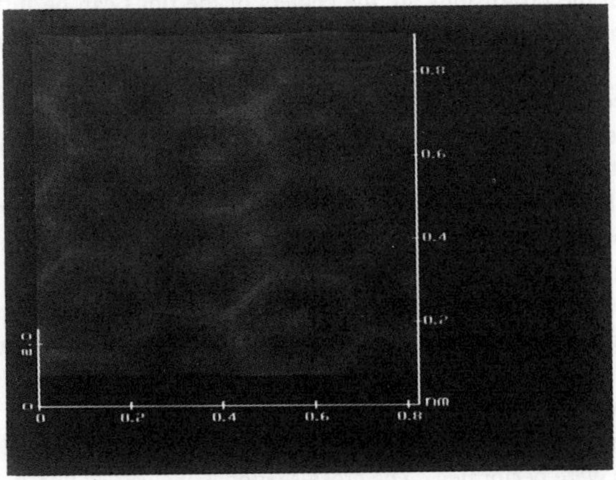

Fig. 2b. 3D surface view of enlarged portion of Fig. 1a (pitch angle 60°).

Fig. 2c is the 3-dimensional view of an enlarged portion of Fig. 2b showing a single hexagonal ring. There is very good agreement between the actual image and that predicted for a graphite ring system. The bond lengths vary from 140 to 142 pm, and the bond angles vary from 118 to 122°. Once again, the measurements taken from Fig. 2c are consistent with the "six-carbon atom" interpretation of the STM images of HOPG basal planes.

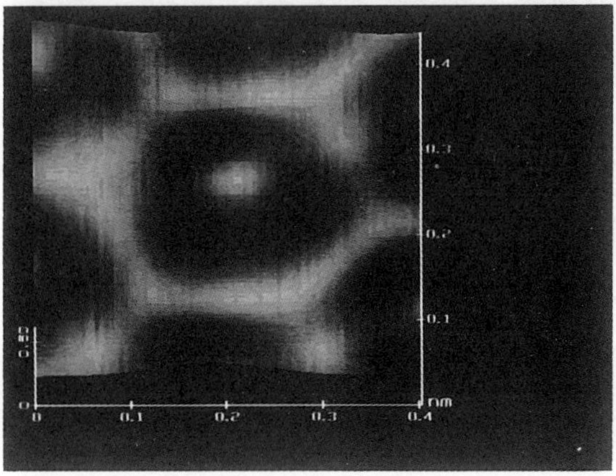

Fig. 2c. 3D surface view of enlarged portion of Fig. 2b, showing a single hexagonal ring.

The fact that the protrusions are the centers of the rings can be further illustrated by examining the series of Figs. 3a through 3d. Figs. 3a through 3d were obtained from the same data array of an HOPG image by resetting the z axis during image processing, i.e., the only difference between Figs. 3a through 3d is the way they have been plotted. The raw data is identical. It can be seen that when a low z axis setting is used, the protrusions are exaggerated to the extent that the hexagonal ring structure is obscured. This condition appears most frequently in the literature[24]. As the z axis setting is increased, the protrusions become less prominent, and the hexagonal ring (showing six-carbon atoms) becomes much clearer. In the transition from Fig. 3a to Fig. 3d, it can be seen that the protrusions in Fig. 3a are located at the ring centers in Fig. 3d. The images presented here clearly demonstrate that all six-carbon atoms of the hexagonal ring in the HOPG are visible with the STM, and that it is the different z-setting that obscures the hexagonal ring. It is suggested that the "three-carbon atom" model, as interpreted by other researchers, should undergo revaluation.

Fig. 4 is the top view of the image in Fig. 2a and shows the lines along which cross-sectional analyses were performed. The cross-sectional analysis of line A shows the center-to-center distance of hexagonal rings (point 1 to point 2) as 250 pm. The diameter of the white areas (210 ± 20 pm) suggests that they are too big to be carbon atoms. The vertical

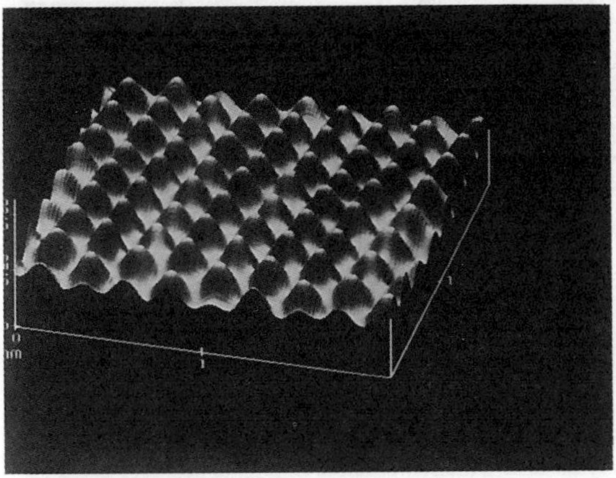

Fig. 3a. 3D image of HOPG with z-axis set at 0.5 nm.

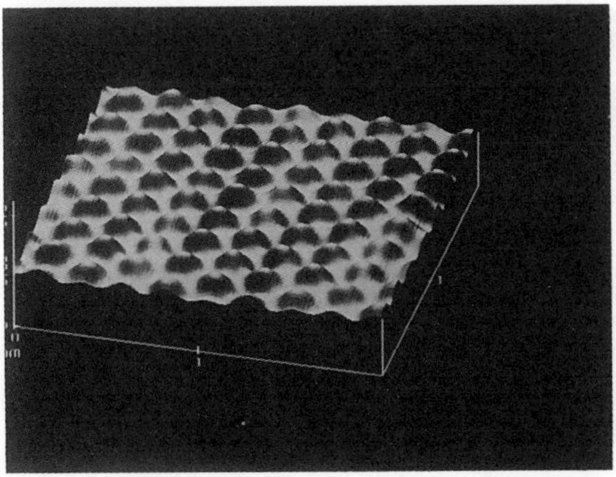

Fig. 3b. 3D image of HOPG with z-axis set at 1.0 nm.

distance between white and lighter and darker areas is 90 to 140 pm. As mentioned earlier, the vertical distance between lighter and darker gray areas measures 30 to 40 pm. The analysis of line B shows the length of the arm-chair configuration (point 3 to point 6) to be 430 pm and the carbon-to-carbon distance (point 4 to point 5) to be 141 pm. This agrees very well with the theoretical measurements of 426 and 142 pm, respectively.

A series of experiments was performed in which the bias voltage was varied from 5 to 100 mV at 5 mV intervals (keeping the set point current at 1.99 nA). The images obtained

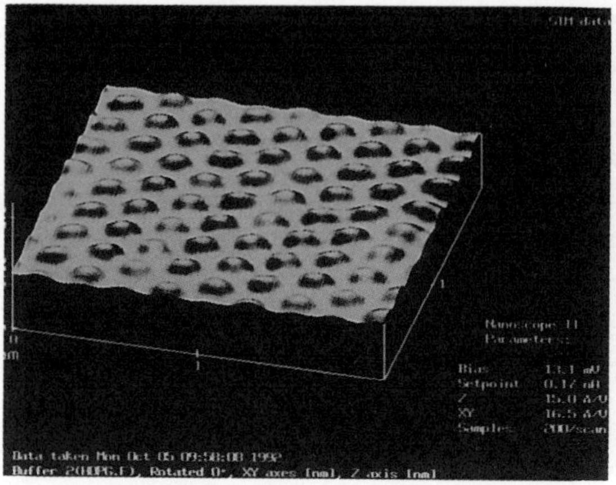

Fig. 3c. 3D image of HOPG with z-axis set at 2.0 nm.

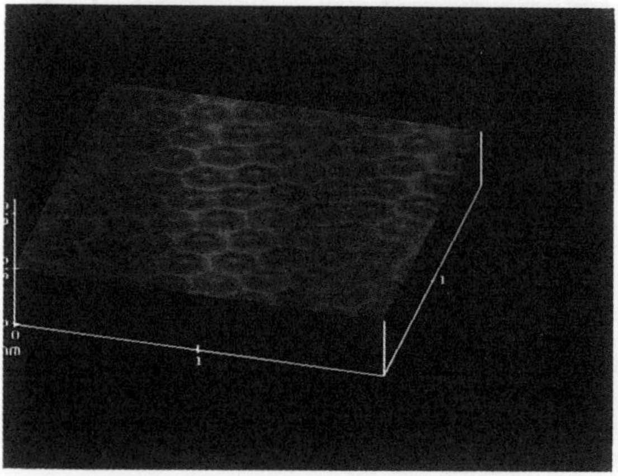

Fig. 3d. 3D image of HOPG with z-axis set at 6.0 nm.

showed the center of the rings both as the raised domes and depressions. There was no correlation between the bias voltage and the topography of the center of the ring. Different tips were used to image the surface (series 1, 2, 4), and the same tip was used to study different areas of HOPG (series 2, 3). Fig. 5 shows the vertical height between the center of the ring and the mean height of the graphene sheet (carbon atom being the reference point, i.e., z = 0). It is clear from Fig. 5 that although the dome topography occurs more often than the depression topography, both occur with reasonable frequency regardless of the bias voltage.

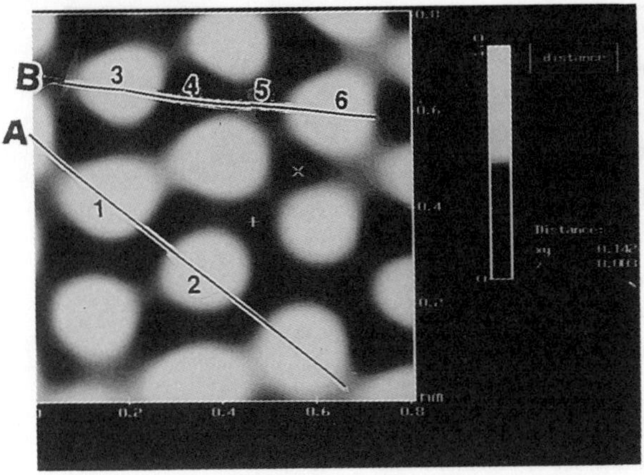

Fig. 4. Topview of the image in Fig. 2a, showing lines along which cross-sectional analyses were performed.

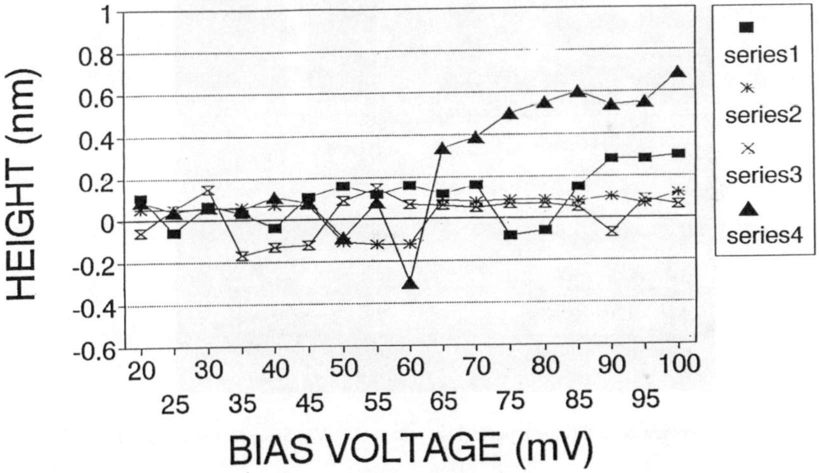

Fig. 5. Effect of bias voltage on the topography of the center of the hexagonal ring. Average carbon atom height being the reference point, i.e. z = 0. (Set point current 1.99 nA).

We have observed images with the domes changing to depressions with a change in the bias voltage. If the bias is changed back, the depressions also change back to domes. Fig. 6 shows such a phenomenon where the image in Fig. 6a has centers as raised domes (bias voltage = 60 mV) and when the bias voltage is increased to 80 mV the centers of the rings

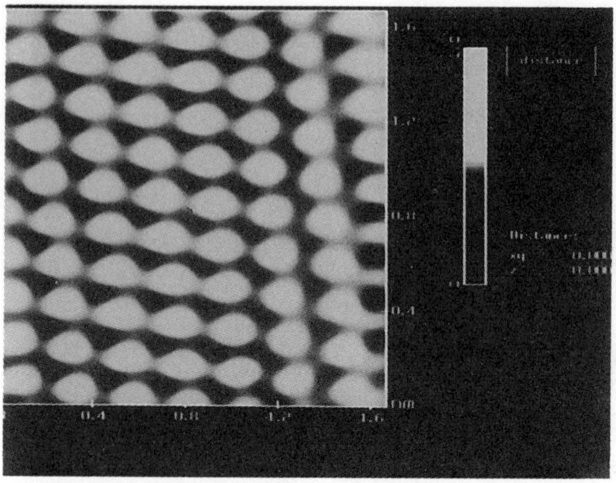

Fig. 6a. Topview of HOPG basal plane (bias voltage 60 mV).

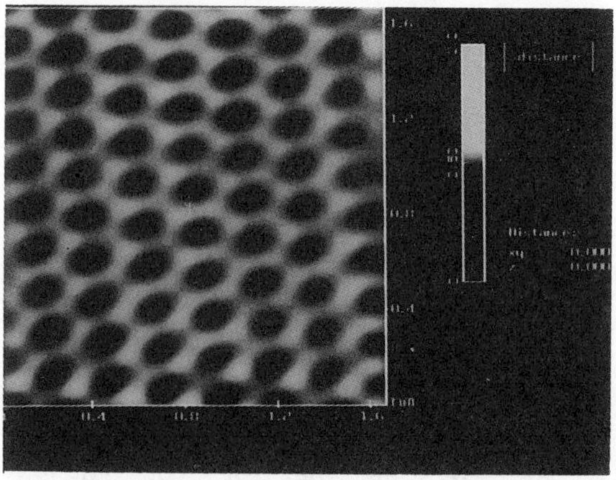

Fig. 6b. Topview of HOPG basal plane (bias voltage 80 mV). Same area as in Fig. 6a.

become depressions (Fig. 6b). Again, when the bias is changed back to 60 mV, the ring centers appear as apparent domes. Figs. 6c and 6d show the images in Figs. 6a and 6b in 3-dimensional mode (60° pitch angle), and it can be seen from these figures that the domes in Fig. 6c overlap the depressions in Fig. 6d. No correlation exists between the bias voltage and the topography of the center of the ring. Similar images were also obtained by varying the set point current from 0.5 to 3.5 nA at intervals of 0.5 nA (bias voltage was kept at 20.14

mV). Once again, no correlation between the topography of the center of the rings and current was observed.

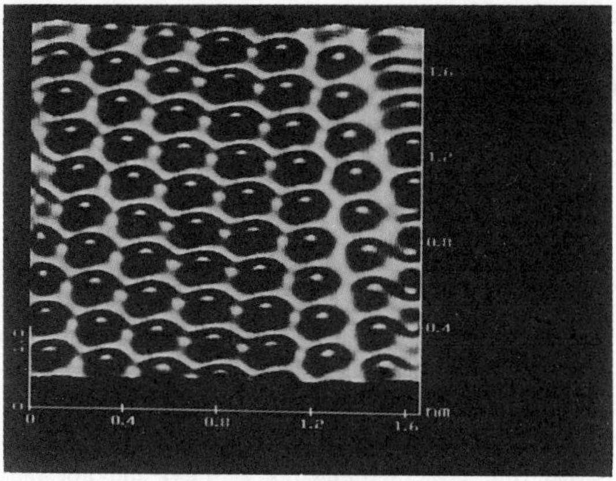

Fig. 6c. 3D image of Figure 6a.

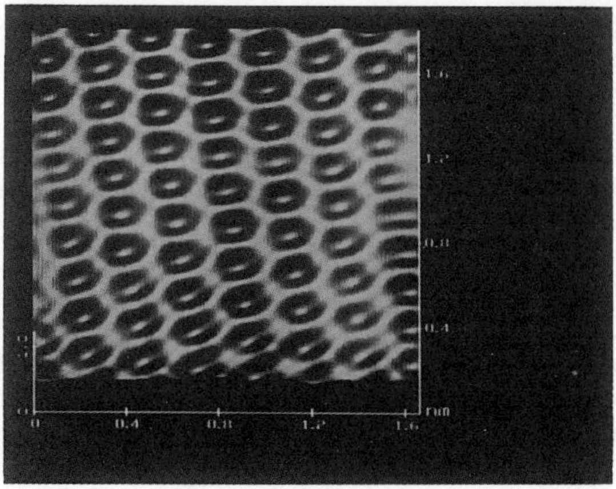

Fig. 6d. 3D image of Figure 6b.

By comparing the data for series 1 through 4 in Fig. 5, obtained under identical set point current and bias voltage conditions, it can be seen that z-height measurement is quite variable. Since the relative position of the tip to the surface and the tip type were the only differences between series 1 through 4, it seems likely that the nature of the tip-surface interaction is quite variable and hence is responsible for the variable shape of the ring centers. Perhaps the tip surface interaction is influenced by transient migrations of atoms at the tip surface. This in turn could create transient tunneling conditions which result in the variable and often inverted topographies that have been observed. The occurrence of the ring centers as depressions and domes could also be due to tunneling to different orbitals (σ and π) on the surface, which could be a function of operating conditions and tip. More detailed investigation is required to determine the exact cause of the topographical variabilities that have been observed.

4. Conclusions

In conclusion, it appears that a final understanding of STM images, even for well defined surfaces like HOPG, is still to be achieved. A great deal of confusion surrounds the issue of the apparent corrugation of HOPG surfaces when viewed with a scanning tunneling microscope. Indeed, many STM images of the HOPG surface could have been misinterpreted and fitted to a "three-carbon atom" model instead of a "six-carbon atom" model. It is hoped that the images provided here have clearly indicated that the corrugation is due to apparent raised and/or depressed ring centers (tunneling artifact) and not raised carbon atoms (structural features), as has been reported in the literature. The observation of all six-carbon atoms in the hexagonal rings of HOPG is significant, for it changes the way the STM images of a calibration material are interpreted. No relationship has been found between the shape of the center of the ring and bias voltage or set point current.

5. Acknowledgments

The authors would like to thank AMOCO corporation and New Energy Development Organization (NEDO), Japan, for financial support. Also, the input of Professor Harry Marsh is sincerely appreciated.

6. References

1. H. G. Busmann, H. Sprang, I. V. Hertel, W. Zimmermann-Edling and H. J. Guntherodt, *Appl. Phys. Letters* **59** (1991) 295.
2. T. Tsuno, T. Imai, Y. Nishibayashi, K. Hamada and N. Fujimori, *Jpn. Journal of Appl. Phy.* **30** (1991) 1063.
3. W. Zimmermann-Edling, H. G. Busmann, H. Sprang and I. V. Hertel, *Ultramicroscopy* **42-44** (1992) 1366.
4. R. J. Wilson, G. Meijer, D. S. Bethune, R. D. Johnson, D. D. Chamblis, M. S. de Vries, H. E. Hunziker and H. R. Wendt, *Nature* **348** (1990) 621.
5. J. L. Wragg, J. E. Chamberlin, H. W. White, W. Kratschmer and D. R. Huffman, *Nature* **348** (1990) 623.

6. N. H. Cho, D. K. Veirs, J. W. Ager III, M. D. Rubin, C. B. Hopper and D. B. Bogy, *Journal of Appl. Phys.* **71**(5) (1992) 2243.
7. W. P. Hoffman, V. B. Elings and J. A. Gurley, *Carbon* **26** (1988) 754.
8. H. N. Lei, A. Metrot and M. Troyon, *Carbon* **32** (1994) 79.
9. S. Gauthier et al., *Journal of Vacuum Science and Technology*, **A6**(2) (1988) 360.
10. I. P. Batra and S. Ciraci, *Journal of Vacuum Science and Technology* **A6**(2) (1988) 313.
11. L. L. Soethout, *Journal of Microscopy* **152-1** (1988) 251.
12. J. M. Soler, A. M. Baro, N. Garcia and H. Rohrer, *Phys. Rev. Letters* **57** (1986) 444.
13. S. Park and C. F. Quate, *Appl. Phys. Letters* **48** (1986) 112.
14. A. Selloni, P. Cornevali, E. Tosatti, and C. D. Chen, *Phys. Rev.* **B31** (1985) 2602.
15. I. P. Batra, N. Garcia, H. Rohrer, H. Salemik, E. Stroll, and S. Ciraci, *Surf. Sci.* **181** (1987) 126.
16. T. Tiedje, J. Varon, H. Deckman and J. Stokes, *Journal of Vacuum Science and Technology* **A6**(2) (1988) 372.
17. J. E. Yao and Y. K. Jiao, *Journal of Vacuum Science and Technology* **A8**(1) (1990) 508.
18. N. J. DiNardo, *Nanoscale Characterization of Surfaces and Interfaces* (VCH, New York, 1994).
19. J. Tersoff, *Phys. Rev. Lett.* **57** (1986) 440.
20. D. Tomanek, S. G. Louie, H. J. Mamin, D. W. Abraham, R. E. Thomson, E. Ganz, and J. Clarke, *Phys. Rev.* **B35** (1987) 7790.
21. D. Tomanek and S. G. Louie, *Phys. Rev.* **B37** (1988) 8327.
22. S. N. Magonov, H. J. Cantow and J. B. Donnet, *Polymer Bulletin* **23** (1990) 555.
23. W. P. Hoffman, W. C. Hurley, P. M. Liu and T. W. Owens, *Journal of Mater. Res.* **6**(8) (1991) 1685.
24. J. B. Donnett and R. Y. Qin, *Carbon* **30**(5) (1992) 787.

EFFECT OF STACKING DISORDER ON THE X-RAY DIFFRACTION PATTERNS OF GRAPHITE POLYTYPES

Z. WEISS
Central Analytical Laboratory, Technical University of Ostrava, 708 33 Ostrava-Poruba, Czech Republic

M. RIEDER
Department of Geochemistry, Mineralogy and Mineral Resources, Charles University, 128 43 Praha, Czech Republic

S. ĎUROVIČ
Institute of Inorganic Chemistry, Slovak Academy of Sciences, 842 36 Bratislava, Slovak Republic

1. Introduction

Among the different types of defects in layer structures, translation stacking faults are particularly frequent in natural and synthetic materials alike. This disorder is due to an irregular translation of the layers, parallel to their own planes. The occurrence of translation stacking faults is not limited to graphite or simple structures like ZnS, CDS and SiC, which are based on close-packed layer structures; more complicated layer structures, such as phyllosilicates or high Tc-superconductors, also exhibit translation stacking faults. Stacking faults play an important role in the solid-state transformations and in the formation of polytype structures.

2. Translation Stacking Faults

In order to explain the terminology of translation stacking faults, let us examine the situation in close-packed layer structures. If atoms in a plane in a close-packed layer structure occupy position *A*, the succeeding one can assume either position *B* or position *C*. Suppose it takes position *B*. The third plane can assume either position *C* or *A*. The combination *AB* can be followed equally well by *A* or *C*, and it would not be surprising if an occasional plane "forgets" to consult the atoms with which it is not in contact, and settles down in a "wrong or aperiodic" position. Therefore, an ordered layer sequence, e.g., *ABABAB*... in polytype structure *2H* can change into *ABABCABAB*... sequence if the combination *AB* is followed by *C* instead of *A* layer and if one translation stacking fault occurs (Fig. 1a).

The possible stacking fault configurations in close-packed structures have been classified in several different ways, and there is considerable confusion over the nomenclature (e.g., Pandey and Krishna[1]). Based on the mechanism of formation, stacking faults can

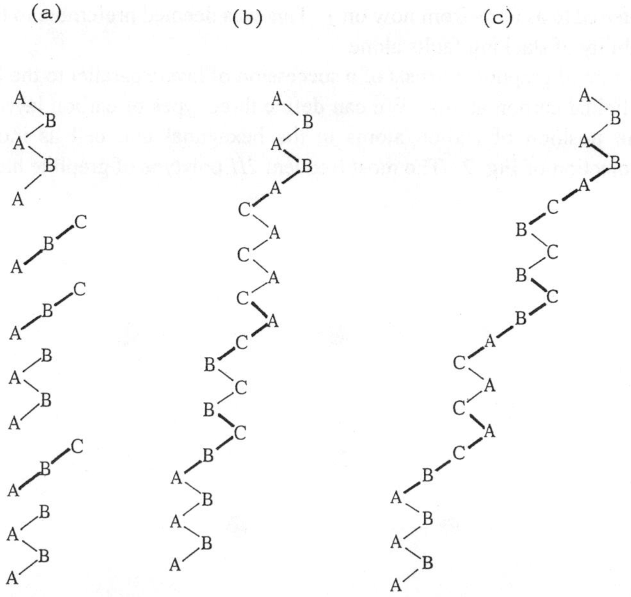

Fig. 1. Representation of the basic types of faulted *2H* structures with the following heterogeneous sequences of layers (different from the surrounding structure): (a) *ABC* (structures of the first type), limiting cases are *2H* and *3R* polytype; (b) *ABCB, BCAC* and *CABA* (structures of the second type), limiting cases are *2H* and a polytype with the sequence *ABCBCACAB;* (c) *ABCAC, CABCB* and *BCABA* (structures of the third type), limiting cases being *2H* and a polytype with the sequence *ABCACABCBCAB*.

be divided into growth and deformation types. During the layer-by-layer growth of a crystal with a close-packed structure, it is possible for a new layer to start growing "incorrectly" as mentioned above (e.g., the combination *AB* is followed by *C* instead of *A* layer in the *2H* structure). The growth of such a crystal by addition of successive close-packed layers is governed by the rule that each layer is different from the preceding two except when a fault occurs; then the new layer is identical with the second last layer. As an example, the sequence ...*ABAB(C)BCB*... represents a *2H* structure with a growth fault at the fifth layer (shown in Fig. 1b).

A deformation fault results when one part of an initially unfaulted crystal slips past the other across the close-packed plane through the partial slip vector 1/3<100>. Thus the sequence ...ABAB/CACA... represents a 2H structure with a deformation fault after the fourth layer (Fig. 1c). The oblique line indicates the slip plane across which the two crystal halves have slipped. In the present paper, we focus on the structures with the basic types of faults and their simple combinations as seen in Fig. 1. In order to quantify the effect of the presence of heterogeneous sequences of layers (different from the surrounding structure) in a periodic polytype on its powder diffraction pattern we use the concentration of such sequences (referred to as CHS from now on). This was deemed preferable to the concentration or probability of stacking faults alone.

The structure of graphite consists of a succession of layers parallel to the basal plane of hexagonally linked carbon atoms. We can define three types of carbon layer A, B, and C with different position of carbon atoms in the hexagonal unit cell as indicated in the horizontal projection of Fig. 2. The most frequent *2H* polytype of graphite has a hexagonal

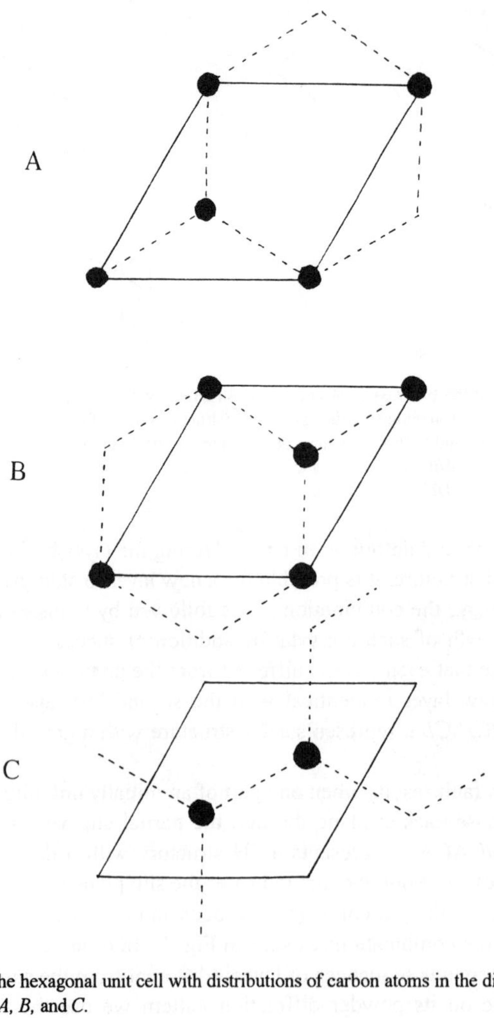

Fig. 2. XY projection of the hexagonal unit cell with distributions of carbon atoms in the different types of graphite layers occupying positions *A*, *B*, and *C*.

symmetry and its structure is represented by the symbol *ABABAB*.... (Fig. 3a). It is found that other graphite polytype *3R* has the rhombohedral stacking that can be represented by the stacking symbol *ABCABC*.... (Fig. 3b). The binding between successive layers is relatively weak and, therefore, nearly perfect graphite structures with the following types of disorder can be observed:

- translation stacking disorder
- turbostratic disorder (in-plane rotations and translation)
- tilt disorder (deviation from parallelism between basal planes)
- interstitial strength along the hexagonal axis (local deformation of the graphite layers)

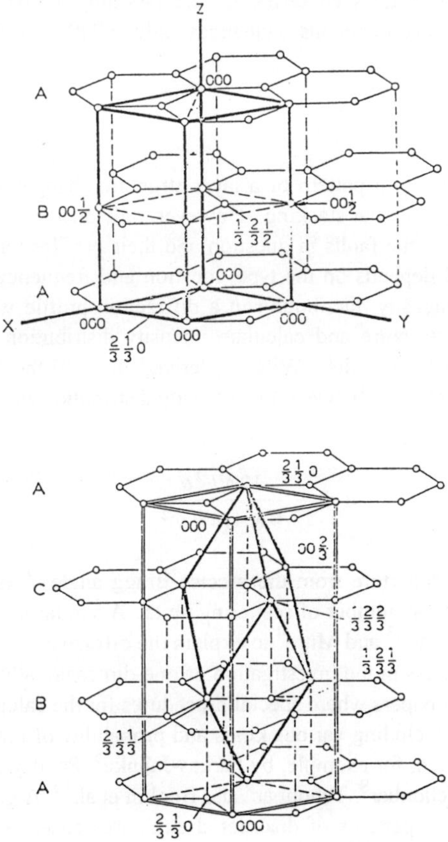

Fig. 3. Structure of two graphite polytypes *2H* (upper) and *3R* (down) with the layer sequences *ABABAB*... and *ABCABC*... respectively.

We have focused only on the first one and X-ray powder diffraction patterns were calculated for the *2H* polytype of graphite with the following various concentrations of several selected CHS, giving rise to an infinite series of periodic or non-periodic structures:

- *ABC* (structures of the first type, Fig. 1a), limiting cases are *2H* and *3R*
- *ABCB, BCAC* and *CABA* (structures of the second type, Fig. 1b), limiting cases are *2H* and a polytype with the sequence *ABCBCACAB*
- *ABCAC, CABCB* and *BCABA* (structures of the third type, Fig. 1c), limiting cases being *2H* and a polytype with the sequence *ABCACABCBCAB*

The limiting cases in all such series are *2H* (starting polytype, CHS = 0%) and a polytype built by the heterogeneous sequences only (CHS = 100%), both necessarily periodic.

3. Diffraction Patterns

Generally, the diffraction pattern of a layer structure changes if translation stacking faults are introduced into a regular stacking. However, only diffractions for $h - k \neq 3n$ (hex. indexing) are affected by the faults in question and their profiles tend to be diffuse. The amount of their spread depends on the type, location and frequency of the faults. Many authors correlate the intensity distribution in a diffraction profile with the probability of stacking faults in the structure and calculate intensity distribution as a function of the probability of layer stacking faults. Wilson[2] derived in 1949 the following "smearing" function to express the relation between the intensity distribution $I(u)$ and the probability α of faults:

$$I(u) = \frac{12NF^2\alpha}{9\alpha^2+(4\pi u)^2}, \qquad (1)$$

where u measures the departure from the precise Bragg angle, F is a resultant scattered amplitude and N is the total number of scattering units. A similar approach was applied by many authors, such as Ruland[3] and Mitra[4] to explain the diffusivity of observed diffractions.

An important progress in the investigation of one-dimensionally disordered structures is represented by a set of papers where special procedures for the calculation of the intensity distribution were used, including various kinds and probability of layer stacking disorders. This approach was applied, for example, by Farkas-Jahnke[5], Pandey et al.[6], Babkevich and Nikolin[7], Plançon and Tchoubar[8], Tchoubar[9] and Bookin et al.[10] A general approach to the calculation of diffraction patterns of disordered structures based on the OD theory was proposed by Dornberger-Schiff[11]. These ideas were utilized in this research to develop the *technique of a super-large unit cell* for modeling of diffraction patterns of periodic structures (e.g., *2H* polytype) with a variable concentration of stacking faults expressed by CHS. The calculations were performed under the following assumptions:

- the scattering power is the same for all close-packed layers and the crystal is free from distortions
- there is no change in the lattice spacing at the faults
- a finite number of layers (cca one hundred) were included, corresponding to a finite size of crystallites (cca 300 Å).

As the first step of the calculations, the binary symbol (*Hägg* symbol rewritten as binary form) of the whole structure was derived, including the random distribution of heterogeneous sequences. For example, the binary symbol corresponding to 96-layers sequence

ABABABCABABABABCABABABCABABCABABABABCABABABCABAB
ABABABABCABABCABABABABABABABABCABABABABABABABCAB

is

101011110101011110101111011110101011110101111010
101010111101111010101010101011110101010101011110

where 10 *ABC* heterogeneous sequences are included. Regular 96-layers sequence *ABCABC*... (polytype *3C*) contains 32 *ABC* heterogeneous sequences and, therefore, CHS = 10/32 = 31.25%. Using the binary symbol, atomic coordinates were generated for all atoms in the 96-layers large hexagonal unit cell ($a = 2.461$Å, $c = 321.98$Å).

As the second step, X-ray powder diffraction patterns were calculated using a modified DIFK program[12] where after generation of indexes $hk\varsigma$, the intensity for all points $hk\varsigma$ were calculated according to the equation:

$$I(hk\varsigma) = æ V^{-2} m Lp |F^2| \ , \tag{2}$$

where æ includes physical constants, V is the volume of the unit-cell, m is the multiplicity, Lp is the Lorentz-polarization factor and F is the structure factor (temperature factor was neglected). The following modified Lorentz intensity function was used to approximate diffraction profiles of the whole pattern:

$$I(\theta) = \sum \frac{a_1 I_i(hk\varsigma)}{\pi H_i(\theta)} \left[1 + a_2 \left(\frac{\theta - \theta_i}{H_i(\theta)} \right)^2 \right]^{-a} \tag{3}$$

where $I(hk\varsigma)$ is calculated intensity according to (2) at position θ_i, a, a_1, and a_2 are optional coefficients and the summation is over all contributing $I(hk\varsigma)$ the pattern. The $H_i(\theta)$ is assumed to be angle dependent:

$$H_i^2(\theta) = b_1 \tan^2\theta + b_2 \tan\theta + b_3 , \tag{4}$$

where b_1, b_2, and b_3 are halfwidth coefficients.

4. Results of Calculations

To describe changes in the XRD powder patterns of the first, second and third type structures with various concentrations of heterogenous sequences (defined earlier) in the periodic *2H* structure, we have focussed on the files corresponding to 010, 002, 011 and 012 diffractions of the original *2H* structure (CHS is equal to zero). Calculated diffraction pattern of periodic *2H* polytype containing all diffractions in range of 2θ between 22.0 and 65.0° is given in Fig. 4.

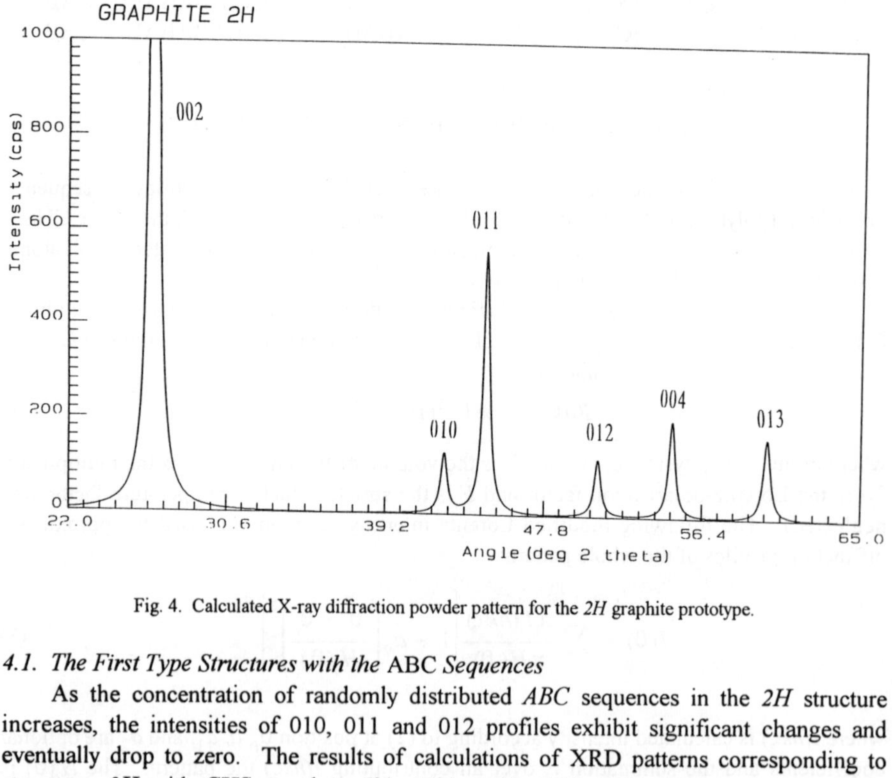

Fig. 4. Calculated X-ray diffraction powder pattern for the *2H* graphite prototype.

4.1. *The First Type Structures with the* ABC *Sequences*

As the concentration of randomly distributed *ABC* sequences in the *2H* structure increases, the intensities of 010, 011 and 012 profiles exhibit significant changes and eventually drop to zero. The results of calculations of XRD patterns corresponding to structures *2H* with CHS equal to 6%, 19%, 31%, 56%, 69%, 81% and 100% (3R), respectively, enable us to draw the following conclusions:

1. The intensity profile functions of the original 010, 011 and 012 diffractions each respond differently when the CHS of *ABC* sequences is increased (Fig. 5). The calculated profiles for selected CHS are given in Figs. 6 and 7. Relatively high sensitivity to CHS is exhibited by the 011 and 012 profiles; already 19% *ABC* sequences in the *2H* structure causes a significant decrease of the peak intensity and a splitting of the original 011 profile into more diffuse profiles so that a "011-band" with different modulation can be observed. The peak intensities of 010, 011 and 012 profiles exhibit systematic decreasing as the CHS increased and dropped to zero for CHS equal to 81%, 81% and 69%, respectively (Fig. 5).

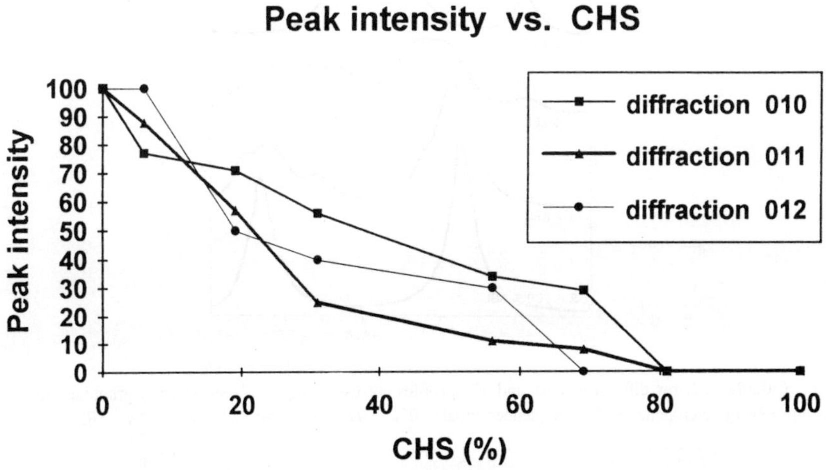

Fig. 5. Intensity at the maximum of 010, 002 and 011 diffraction profiles (peak intensity) as a function of concentrations of heterogeneous *ABC* sequences (CHS) in the *2H* structure of graphite.

2. The resulting shape of 010, 011 and 012 diffraction profiles depends not only on the concentration of *ABC* sequences but also on their position in the *2H* structure. Two different random or semi-random distributions of sequences with the same CHS produce profiles of the "011-band" that are only slightly different from each other, but the same percentage of *ABC* sequences grouped into blocks cause a significantly different profile (Fig. 8).

4.2. The Second Type Structures with the ABCB, BCAC *and* CABA *Sequences*

Turning our attention to the XRD powder patterns corresponding to the second type of faulted structures, we see a picture different from that involving the structures of the first type. The calculations involved the structure *2H* with concentrations of *ABCB, BCAC* and

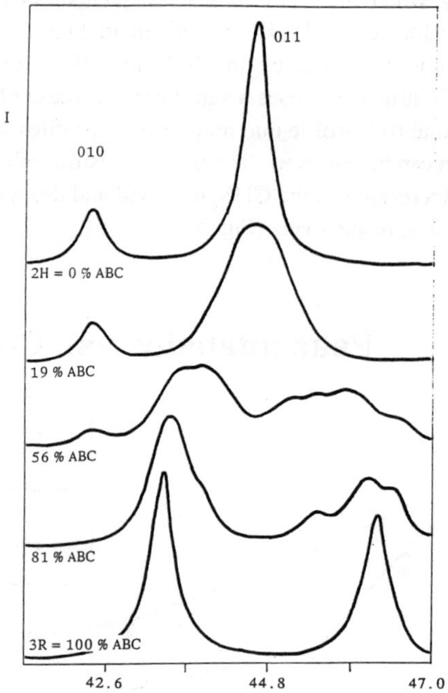

Fig. 6. Calculated X-ray diffraction 010, and 011 profiles corresponding to *2H* structure of graphite with various concentrations of heterogeneous *ABC* sequences equal to 0% = *2H*, 19%, 56%, 81%, and 100% = *3R*.

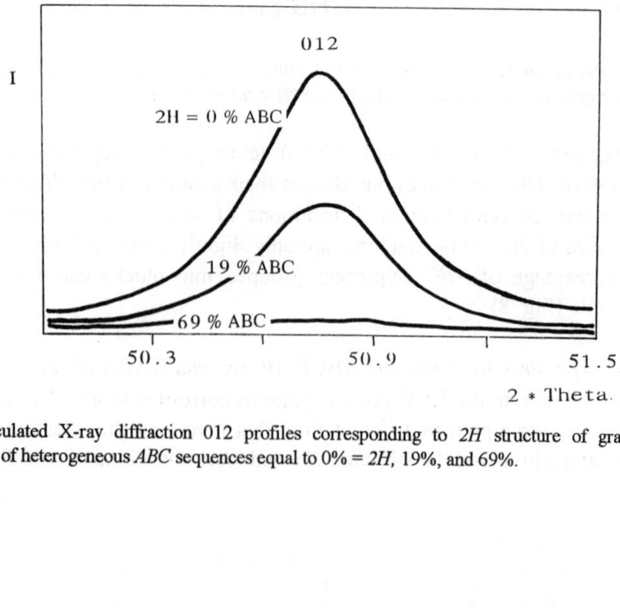

Fig. 7. Calculated X-ray diffraction 012 profiles corresponding to *2H* structure of graphite with various concentrations of heterogeneous *ABC* sequences equal to 0% = *2H*, 19%, and 69%.

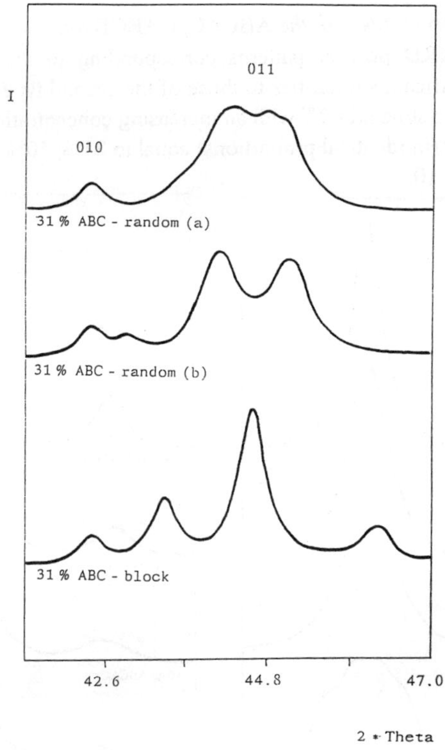

Fig. 8. Calculated X-ray diffraction 010, and 011 profiles corresponding to *2H* structure of graphite with the concentration of heterogeneous *ABC* sequences equal to 31 random or semi-random distributions of sequences (a and b) and for *ABC* sequences grouped into block.

CABA sequences (in identical proportions) equal to, respectively, 20%, 40%, 60%, 80%, and 100% of the *ABCBCACAB* polytype. As the concentration of randomly distributed *ABCB, BCAC* and *CABA* sequences in the *2H* structure increases, the following changes are observed:

1. The peak intensity of the 012 profile gradually drops to zero.
2. The intensity of the 011 profile decreases and the original peak broadens into a "011-band" with a different modulation (Fig. 9). The weakest modulation occurs for CHS equal to approximately 60%, but becomes more expressive for CHS increasing further. However, there appear new diffraction lines corresponding to periodic *ABCBCACAB* polytype (CHS = 100%). Of course, these diffractions are sharp for CHS = 100%.
3. If the concentration of randomly distributed *ABCB, BCAC* and *CABA* periodic sequences in the *2H* structure increases, the intensity of 010 profile slightly increases.

4.3. *The Third Type Structures with the* ABCAC, CABCB *and* BCABA *Sequences*

The calculated XRD powder patterns corresponding to the third type of faulted structures exhibit significant similarities to those of the second type. The calculated XRD patterns corresponding to structures *2H* with an increasing concentration of *ABCAC, CABCB* and *BCABA* sequences (in identical proportions) equal to 25%, 50% and 75%, respectively, are summarized in Fig. 10.

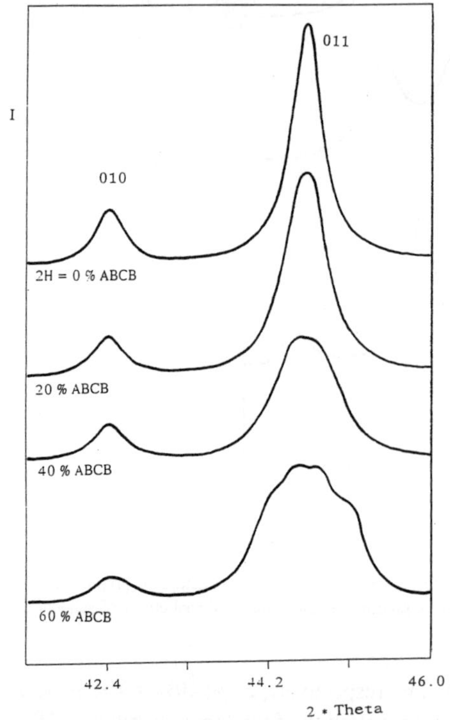

Fig. 9. Calculated X-ray diffraction 010, and 011 profiles corresponding to *2H* structure of graphite with various concentrations of *ABCB, BCAC* and *CABA* sequences (in identical proportions) equal to 0% = *2H*, 20%, 40%, and 60%.

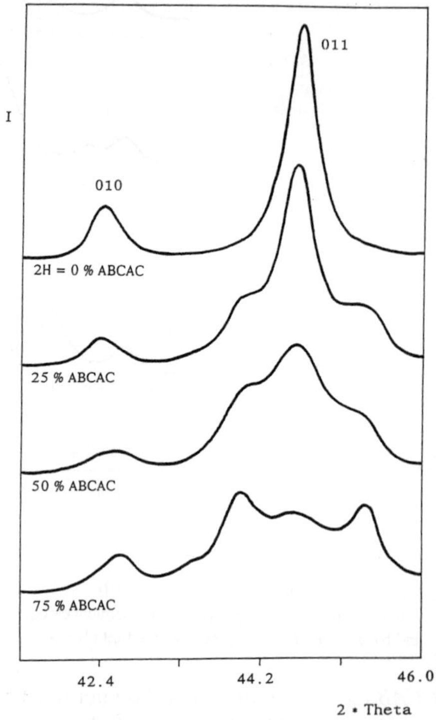

Fig. 10. Calculated X-ray diffraction 010, and 011 profiles corresponding to *2H* structure of graphite with various concentrations of *ABCAC, BCABA* and *CABCB* sequences (in identical proportions) equal to 0% = *2H*, 25%, 50%, and 75%.

5. References

1. D. Pandey and P. Krishna (1982). Polytypism in close-packed structures. *Current Topics in Material Science,* ed. E. Kaldis, **vol. 9**, 416-491.
2. A. J. C. Wilson (1949). X-ray optics. *Methuen,* London.

3. W. Ruland (1967). Fourier transform methods for random-layer line files. *Acta Cryst.* **22**, 615-623.
4. G. B. Mitra (1963). Structure defects in kaolinite. *Z. Kristallogr. and Kristallgeom.* **119**, 161-175.
5. M. Farkas-Jahnke (1983). Structure determination of polytype. *Crystal Growth and Characterization of Polytype Structures*, ed. P. Krishna, **vol. 7**, 163-211.
6. D. Pandey, V. K. Kabra and S. Lele (1986). Structure determination of one-dimensionally disordered polytype. *Bull. Mineral.* **109**, 49-67.
7. A. Yu. Babkevich and B. I. Nikolin (1992). Application of the Monte Carlo techniques to the investigation of one-dimensionally disordered structures. *Computer Physics Communications* **73**, 139-144.
8. A. Plançon and C. Tchoubar (1977). Determination of structural defects in phyllosilicates by X-ray diffraction. I. Principle of calculation of the diffraction phenomenon. *Clays and Clay Minerals* **25**, 430-435.
9. D. Tchoubar (1990). Structural characteristic of carbon. *X-ray Diffraction by Disordered Lamellar Structures*, eds. V. A. Drits and C. Tchoubar. Springer-Verlag, Berlin, Heidelberg, New York.
10. A. S. Bookin, V. A. Drits, A. Plançon and C. Tchoubar (1989). Stacking faults in kaolin-group minerals in the light of real structural features. *Clays and Clay Minerals* **37**, 297-307.
11. K. Dornberger-Schiff (1966). Lehrgang uber OD-strukturen. *Akademie-Verlag*, Berlin.
12. L. Smrčok and Z. Weiss (1993). DIFK91: A program for the modeling of powder diffraction patterns on a PC. *J. Appl. Cryst.* **26**, 140-141.

THE MODEL OF PREDICTION OF THE DISTRIBUTION CURVES OF SEGREGATION OF ELEMENTS AND IMPURITIES IN STEEL

J. DOBROVSKÁ and V. DOBROVSKÁ,
VŠB - Technical University Ostrava,
17. listopadu, 708 33 Ostrava, Czech Republic

K. STRÁNSKÝ and A. REK
Military Technical Institute for Protection,
Rybkova 2, Brno, Czech Republic

ABSTRACT

This paper presents the original method for estimation of the distribution curves of the dendritic segregation of alloying elements and of impurities in metals as well as in metal alloys and its application. The method deals with the relations among the following parameters:

- the solid/liquid partition coefficient of an element including its statistical distribution,
- the segregation equation according to Scheil[1],
- the degree of dendritic segregation of an element in solid state after solidification.

Application of this method enables us to estimate the distribution curve of elements in real metal or a metal alloy. Application of this method was qualitatively verified by comparing the calculated and measured distribution curves of dendritic segregation of manganese and silicon in steel.

1. Introduction

The heterogeneity and defects in steel and in alloys, affecting markedly their useful properties, are intimately associated with the solidification process. In this respect the segregation phenomena play the dominant role for origination of defects in the primary casting structure. The extent of dendritic segregation varies with the chemical composition (especially with C-content) and with the kinetics of crystallization. The dendritic segregation encountered in the course of solidification brings about heterogeneity in concentration with possible origination of non-equilibrium phases or inclusions.

These facts associated with crystallization of steels and alloys are most recently the reason for creating a model that would enable us to predict the extent of the dendritic segregation of elements in relation with both composition of the system considered and the kinetics of crystallization.

The problems investigated in this article are split into two fields:

1. Evaluation of the mathematical models of dendritic segregation of elements going out from solution of the diffusion equation,
2. Proposal of a new mathematical and statistic description of the cumulative curves of dendritic segregation of elements.

2. Mathematical Models of Dendritic Segregation

The diffusion of elements inside a solid phase is the limiting phenomenon of micro-segregation because the diffusion coefficient in solid phase is 3 to 5 times smaller than that in the melt.

In relation with the extent of application of the diffusion processes in a melt and even in a growing crystal there can occur examples given as follows:

a)	b)	c)	d)	e)	f)
D_L complete	D_L complete	D_L complete	D_L partial	D_L suppressed	D_L partial
D_S complete	D_S suppressed	D_S partial	D_S suppressed	D_S suppressed	D_S partial

Here, D_L refers to the diffusion in liquids and D_S is the diffusion in solids. In case of e) it is a theoretical example of a diffusionless crystallization and in the case of f) is completely a general example.

In the individual examples cited above the concentration of added element can be expressed as a dependence on the proportion of solid phase, namely as a function of the following variables:

$$C_S = f(k, C_0, D_S, t_f, \lambda_1, \lambda_2)$$

where k is the partition coefficient of element between the solid and liquid phases [-]

C_0 is the average concentration of an admixture [wt -%]

D_S is the diffusion coefficient in solid phase [$m^2\ s^{-1}$]

t_f is the local time of solidification [s]

λ_1 (λ_2) is the distance among the primary (secondary) branches of dendrites [m].

At first the model of equilibrium crystallization is given here, this model assumes complete equalization of concentration of the coexisting phases and this phenomena can be described by means of the diagram of state[1]:

$$C_s = \frac{k \cdot C_0}{g_s \cdot (k - 1) + 1} \tag{1}$$

However, under actual conditions the crystallization is always a non-equilibrium phenomenon, as the diffusion processes, taking place in the solid and liquid phase, are not able to equalize the concentration in the solid phase to a level shown in the equilibrium diagram. For that reason the other models refer already to non-equilibrium crystallization.

The models most frequently used are those of Scheil[2] and Pfann[1,3], who assumed a condition without diffusion in solid state. In addition, the directional solidification, heat abstraction and growth of crystals was assumed to be unidirectional. Distribution of admixtures can be then analytically expressed as follows:

$$C_s = k \cdot C_0 \cdot (1 - g_s)^{k-1} \tag{2}$$

Brody and Flemings[4] have analyzed for the first time a model in which complete diffusion in liquids and the incomplete revert diffusion were assumed:

$$C_s = k \cdot C_0 \cdot \left[1 - (1 - 2 \cdot \alpha \cdot k) \cdot g_s\right]^{\frac{k-1}{1-2 \cdot \alpha \cdot k}} \tag{3}$$

The dimensionless constant α combines the coefficient of diffusion of the segregating element D_S in the solid phase with local time of solidification t_f and with the mean distance of the dendritic axles λ.

$$\alpha = \frac{4 \cdot D_S \cdot t_f}{\lambda^2}$$

Clyne a Kurz[5] have derived a more suitable mathematical formulation that asymptotically approaches the Scheil's equation and the equilibrium equation for infinitely small and infinitely high coefficients of diffusion:

$$C_s = k \cdot C_0 \cdot \left[1 - (1 - 2 \cdot \Omega \cdot k) \cdot g_s\right]^{\frac{k-1}{1-2 \cdot \Omega \cdot k}} \tag{4}$$

where Ω is defined:

$$\Omega(\alpha) = \alpha \cdot \left[1 - \exp\left(-\frac{1}{\alpha}\right)\right] - \frac{1}{2} \cdot \exp\left(-\frac{1}{2 \cdot \alpha}\right)$$

Moreover, some modifications of these models are still applied[6–8].

All the mathematical models were processed into a computer program providing both tabular and graphic output. The models were tested with the help of cumulative curves of elements (Mn, Si, S, P) obtained experimentally for two steel grades with different C-contents (0.06 and 0.4 wt %) and with different cooling rates (3 °C/min and 30 °C/min)[9]. An example of a cumulative curve of manganese passing through by some models is illustrated in Fig. 1.

Testing has verified two features:

- Carbon is the limiting element from the viewpoint of the effect on the crystallization mechanism of steel. The degree of interaction of carbon with the other elements is rather high and surpasses (especially with the low- and medium alloyed steels) the interaction among other elements. In the case of steel with lower C-content, the interaction among other additional elements was much more expressive.

- The mathematical models of dendritic segregation resulting from solution of a diffusion equation enables us to predict with reasonable accuracy the maximum concentration of additional elements in steel. However, these models do not allow qualitative description of the measured distribution curves of the concentration of elements.

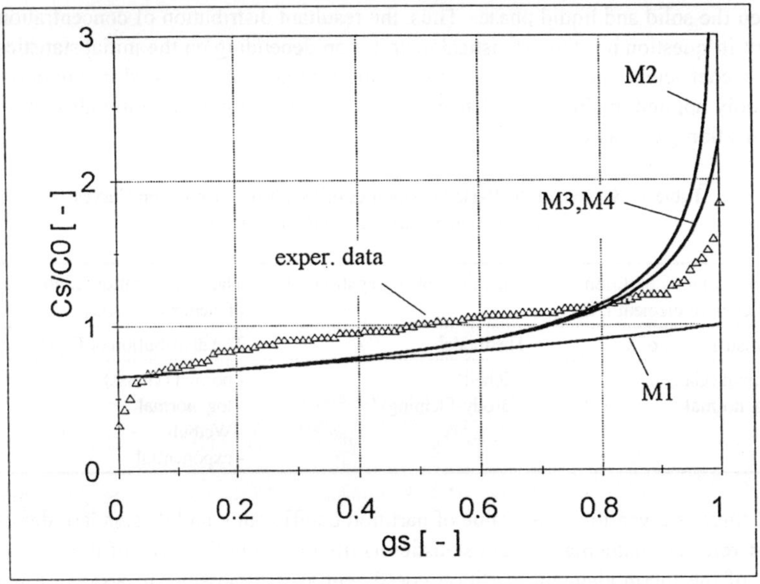

Fig. 1. Example of a cumulative curve of manganese passing through by some mathematical models. Model of equilibrium crystallization: M1. Models of non-equilibrium crystallization: M2, suppressed diffusion in the solid phase; M3, M4, partial and extensive diffusion.

3. Mathematical and Statistical Model of Description of the Cumulative Curves of Dendritic Segregation of Elements

The principal reasons why the above-cited models fail in qualitative description of the measured distribution curves of concentration of elements are as follows:

- The models dealing with segregation of elements are developed for monocrystals and/or for directed solidification, zone melting, single-dimensional temperature field and other special examples of crystallization.
- The partition coefficients encountered in such models are considered to be constant, whereby simultaneously the occurrence of dendrites of higher orders is neglected.

For these reasons a new mathematical and statistical model has been proposed for calculation of the distribution curves of dendritic-segregating element in actual alloys[10]. This model is going out from precondition saying the partition coefficient in the volume in question is not constant, but it should have a certain statistical distribution. The probability of such a distribution is transferred in the course of crystallization, for example by means of the existing models of dendritic segregation, to distribution of the percentage of element between the solid and liquid phase. Thus, the resultant distribution of concentration of the element in question has then statistical distribution depending on the initial statistics of the partition coefficient, the mechanism of dendritic segregation expressed by the derived and commonly applied methods and depending also on the marginal and initials conditions of crystallization (see Table 1).

Table 1. Scheme of the Basic Possibilities of Dendritic Segregation Curves Determination of Elements in Real Systems

Thermodynamic partition coefficient of element I: k	The model of segregation: M	The concentration contribution of element I: C_S/C_0
The distribution of k	The model	The distribution of C_S/C_0
- normal (Gauss)	- Scheil[1]	- normal (Gauss)
- log. normal	- Brody-Flemings[4]	- log. normal
		- Weibull
		- exponential

If there is given the mean value of partition coefficient k and its standard deviation σ_k incl. of relevant mathematical and statistic distribution. On the basis of the known distribution of the thermodynamic partition coefficient k, it is possible to arrange a matrix of discrete values k and the corresponding probability of occurrence of these values f_K in a given distribution. This matrix has two columns (k, f_k) and in their lines the pair values of the magnitude in question are compiled. The number of lines of the matrix is optional and it varies with the desired accuracy of the numerical calculation.

For the sake of further procedure let us assume the application of the Scheil's model expressed by Eq. (2). Now, the number of columns in matrix will be enlarged so as to cover the interval of the growing proportion of solid phase g_S within 0 and 1. The relevant dendritic segregation can be calculated by integration of Eq. (2) at the intervals g_S with following application of the theorem of mean value of integral number. For the definite integral of Eq. (2) in relation with magnitude g_S there can be written:

$$C_s/C_0 = [(1-g_s)^k]_{gs^{(i)}}^{gs^{(j)}} \; / \; \left(g_s^{(i)} - g_s^{(j)}\right) \tag{5}$$

The matrix compiled in this way gives information about the probability of occurrence of the individual grades of dendritic segregation, i.e., concentration C_S/C_0 of the element in question in the system.

Now, it is relatively easy to find the distribution of the calculated values of concentrations C_S/C_0 as the probability of their occurrence is known in the system of question. According to the character of distribution of the partition coefficient k, its standard deviation σ_k and the chosen model of the dendritic segregation (in this case of Scheil type) we can get for C_S/C_0 the statistical distribution called normal, logarithmic normal, the Weibull's one and with $\sigma_k = 0$ even the exponential distribution. Information on the most probable form of obtained distribution can be given by application of test of goodness of fit.

The given procedure is demonstrated by matrix I (Table 2) that was calculated with the help of $k = 0.8$ and $\sigma_k = 0.08$ with normal distribution and the calculated values of dendritic segregation were approximated also with normal distribution with the mean value of $C_S/C_0 = 1.000$ for $g_s = 0.5$ and standard deviation $\sigma_s = 0.258$, with error in determination of C_S/C_0 - value and of $\delta = 4.73 \times 10^{-10}$.

The course of calculated values of dendritic segregation plotted into graph, i.e., the distribution curve of dendritic segregation is shown in Fig. 2 (curve b). Comparison of this theoretical curve (for $k = 0.8$ and $\sigma_k = 0.08$) with the curve of concentration experimentally obtained for manganese (whose partition coefficient in iron is approaching the value chosen for calculation, curve c in Fig. 2) a qualitative coincidence of the two curves was rather good.

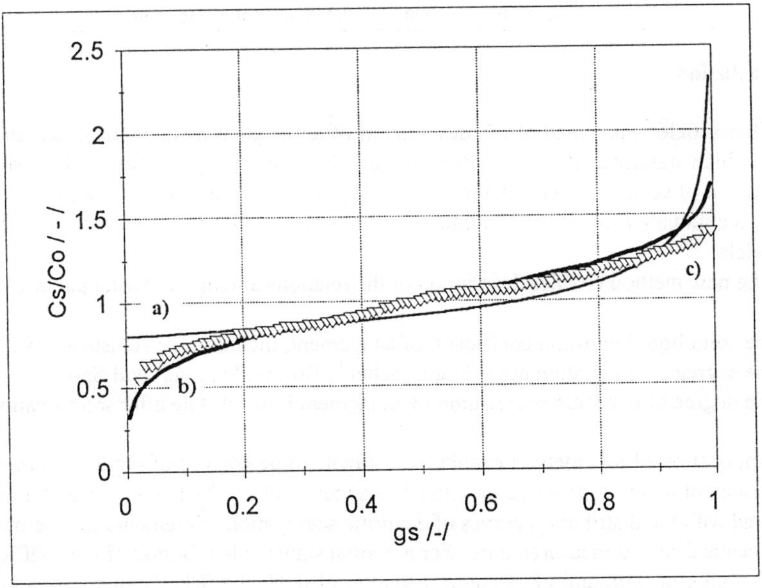

Fig. 2. Prediction of the distribution curves of dendritic segregation of elements. a) calculated distribution curve of dendritic segregation C_S/C_0 (g_s) according to the Scheil's model for $k = 0.80$ and $\sigma_k = 0.00$; b) calculated distribution curve of dendritic segregation C_S/C_0 (g_S) according to the prediction model for $k = 0.80$ and $\sigma_k = 0.08$, c) distribution curve of dendritic segregation for manganese determined experimentally.

Table 2. Matrix I of the Inlet and Calculated Parameters of Dendritic Segregation of Elements

Set - on parameters		Calculated parameters C_S/C_0 interval $g_S^{(I)}$ - $g_S^{(j)}$									
k	f_k	0.0 - 0.1	0.1 - 0.2	0.2 - 0.3	0.3 - 0.4	0.4 - 0.5	0.5 - 0.6	0.6 - 0.7	0.7 - 0.8	0.8 - 0.9	0.9 - 1.0
0.55	0.006	0.563	0.592	0.626	0.668	0.720	0.789	0.884	1.031	1.308	2.818
0.65	0.100	0.662	0.688	0.719	0.756	0.802	0.860	0.940	1.059	1.274	2.239
0.75	0.394	0.760	0.781	0.806	0.836	0.871	0.916	0.976	1.063	1.212	1.778
0.85	0.394	0.857	0.871	0.888	0.907	0.930	0.958	0.996	1.048	1.134	1.413
0.95	0.100	0.952	0.958	0.964	0.971	0.979	0.989	1.001	1.019	1.046	1.122
1.05	0.006	1.047	1.041	1.035	1.028	1.019	1.009	0.996	0.979	0.954	0.891

Notes: Inlet data - $k = 0.80$, $\sigma_k = 0.08$, normal distribution
- probability f_k
- model of dendritic segregation by Scheil, Eq. (2)

Calculated data - dendritic segregation $C_S/C0$ and its probability f_k in the system distribution curve of concentration of element C_S/C_0 (g_s) with parameters C_S/C_0 $(g_s{=}0.5) = 1.000$, $\sigma_c = 0.258$, normal distribution

4. Conclusion

This article was aimed at the designation of an original mathematical and statistical model which describes the cumulative curves of dendritic segregation. The reason for designation of such a model was the lack of a qualitative description of the experimentally measured cumulative curves of admixtures in steels and in alloys by the existing mathematical models.

The new method enables description of the relations among following parameters:

- the solid/liquid partition coefficient of an element, including its statistical distribution,
- the segregation equation according to Scheil[6], Brody-Flemings[1] and Šmrha[7],
- the degree of dendritic segregation of an element in solid state after solidification.

Application of this method enables an estimate to be derived of the distribution curve of the element in a real structure of a metal or a metal alloy. These theoretical curves were compared with the distribution curves of dendritic segregation of elements as determined by experimental data and measurements. For a normal statistic distribution of the coefficient k, a mean value of 0.08 and of standard deviation of 0.08, the Scheil's model produces the normal statistic distribution of the concentration C_S/C_0 (see heading of Table 2). There can also be achieved distribution curves of dendritic segregation that describe real behavior of the constitutive elements such as manganese and silicon, whose partition coefficient in iron is approaching the value assumed for calculation.

5. References

1. W.G. Pfann, *Zone Melting* (J. Wiley & Sons, Inc. New York 1966)
2. E. Scheil, *Z. Metallkd.*, **34**, 1942, p.70
3. W.G. Pfann, *J. Metals*, **4**, 1952, p.747
4. H. D. Brody and M.C. Flemings, *Trans. AIME*, **236**, 1966, s.615
5. T. W. Clyne and W. Kurz, *Metall. Trans*. A, **12A**, 1981, p.965
6. T. Matsumiya, H. Kajioka, S. Mizoguchi, Y. Ueshima and H. Esaka, *Trans. ISIJ*, **24**, 1984, p.837
7. I. Ohnaka, *Trans. ISIJ*, **26**, 1986, p.1045
8. S. Kobayashi, T. Nagamichi and K. Gunji, *Trans ISIJ*, **28**, 1988, p.543
9. J. Dobrovská, *CSc. (PhD) thesis*, VŠB-Technical University, Ostrava, 1992
10. V. Dobrovská, J. Dobrovská, A. Rek and K. Stránský, *Metal Materials* **33**, 1995.

Carbon Fibers

and

Carbon Particulates

DETERMINATION OF STATISTICAL SIZES OF ANISOTROPIC OPTICAL TEXTURES IN PITCH COKES

L. BROŽ and I. SÝKOROVÁ
Institute of Rock Structure and Mechanics,
Academy of Sciences of the Czech Republic,
V Holešovičkách 41, 182 09 Praha 8, Czech Republic

ABSTRACT

Parameterization of anisotropic optical textures of pitch cokes by the geometric mean size (GMS) and geometric size dispersion (GSD) is suggested The effective sizes of anisometric textures were estimated by extrapolation of the log-normal distribution of isometric texture sizes.

Mechanical and physical properties of a coke are closely related to its optical properties. To assess such relationships, a parameterization of the experimental optical texture distribution is desirable. This is equivalent to finding an adequate distribution function and its argument values.

The commonly used texture classes were defined by Marsh et al.[1,2] For the quantitative treatment, the classification was simplified as shown in Fig. 1. The textures in coke samples were assessed by a manually operated point counter (Swift, Great Britain) on the microscope MeF (Reichert, Austria). Nicol prisms crossed at 80°, and magnifications of 468 for the mosaic and 156 for the larger textures were used. Of the 41 coke samples, the sample VSZ/C did not contain anisometric textures, and could not be used in the estimation of the effective sizes.

The upper-size values of isometric textures (except Dc) were defined close to a geometric sequence as if to be the argument values of a logarithmic distribution. Not surprisingly, this was confirmed. The cumulative texture contents related to the isometric textures, shown as full square marks in Fig. 2, were found consistent with the log-normal distribution (LNDF) in all the coke samples. This was proved by the Kolmogorov's test on 5% significance level. The upper effective size of an anisometric or Dc texture, shown as full circle in Fig. 2, could be determined as the intersection of the distribution function and the respective texture content.

The six upper effective sizes of the Fm, Fc, Dc, FD, As and Al textures, common to 40 pitch coke samples, and the LNDF mean and dispersion pairs (GMS and GSD) of the cokes, that is $6 + 40 \times 2 = 86$ parameters, were estimated simultaneously from the total of 379 texture content values. The non-linear least-squares regression procedure that respected the lower and upper cut points of the individual sample distributions, was written for BMDP Statistical Software, CA.

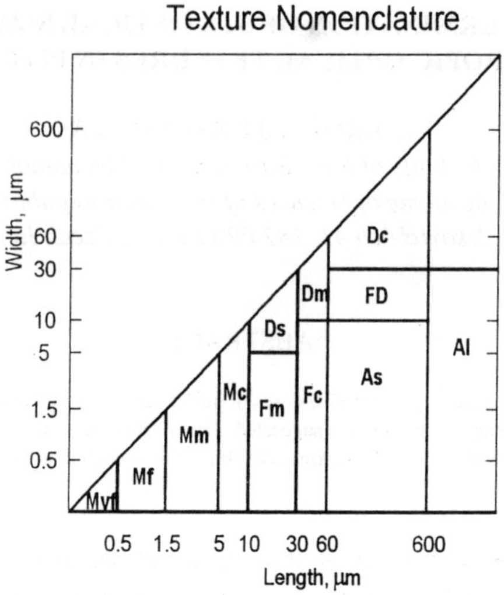

Fig. 1. Texture size definitions: M - mosaic, D - domains, F - flow, A - acicular, vf - very fine, f - fine, m - medium, s - small/short, l - long.

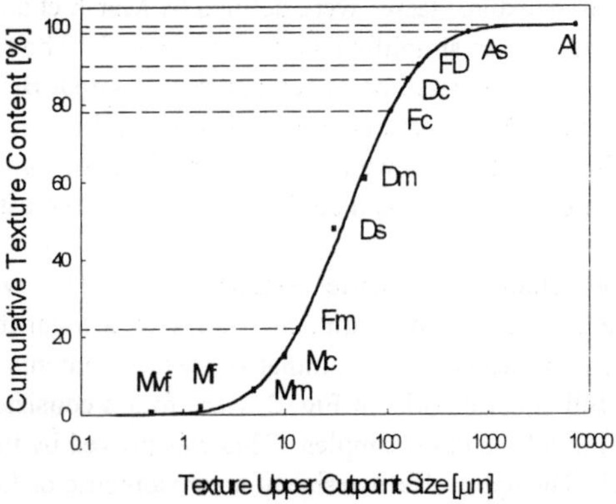

Fig. 2. Principle of the texture effective size estimation (petroleum coke Borzesti, Romania).

The definition of an order of the textures by increasing anisotropy was the critical part of the estimation design. The natural choice would be the order by increasing length as the major attribute, and increasing anisometry (that is decreasing width) as the minor attribute. However, the position of the Fm texture was unacceptable in the order ... Mc, Ds, Fm, Dm.... Many of the estimated GMS values were unreasonably high (in four cases even above the Al size), the As size converged under 600 μm, and controversial data outliers were signaled.

Respecting the requirement of the LNDF continuousness and the values of texture indexes suggested by Oya et al.[3], who suggested the value of the Fm texture index to be the same as of the Mc texture, the texture order shown in Fig. 2 was accepted. This removed the inconsistencies, and significantly better fit of the data was reached. The residual standard deviation from the set of 40 LNDF's with the 293 (= 379 - 86) degrees of freedom was 3.3%. This value did not evidently exceed the experimental error. The results of the estimation are summarized in Table 1.

Table 1. Comparison of Quantitative and Optical Texture Characterization

	Defined or estimated effective size			Texture index			
Texture	Upper value, µm	Accuracy, µm	Geom. mean, µm	GMS_i, µm	OI[1]	IOT[3]	OTI[3]
Iso	*(0.2)*		—	0	1	0	0
Mvf	*0.5*		*0.27*	0.3	—	—	0
Mf	*1.5*		*0.87*	0.9	2	1	1
Mm	*5.0*		*2.74*	2.7	3	3	3
Mc	*10.0*		*7.1*	7	4	7	7
Fm	13.7	0.4	11.7	12	—	7	20
Ds	*30.0*		20.3	20	—	20	10
Dm	*60.0*		*42.4*	40	—	20	12
Fc	98.9	3.6	77	80	5	20	20
Dc	161.5	8.5	126	130	7	30	15
FD	278	22.3	212	220	6	30	20
As	706	122.0	443	500	—	—	20
Al	7332	$>10^4$	2275	2500	—	—	30

The reason for the low accuracy (asymptotic standard deviations) of the As and Al size estimates is clearly seen in Fig. 2. By principle, the coke samples with the right-limited distribution contributed to the right-limit estimate with a value within the last size interval, thus decreasing the estimated value. Therefore, the geometric means calculated from the estimated upper sizes of the textures Fc, Dc, FD, As and Al, were rounded upward to obtain reasonable values of the "texture indexes" GMS_i.

The geometric mean size GMS as a measure of the coke anisotropy, can be calculated from Eq. (1). In addition, the geometric size dispersion–GSD– as a measure of the coke texture distribution width, can be evaluated from Eq. (2). The x_i is the point count ratio of the I-th texture ($\Sigma x_i = 1$), and the GMS_i has a value from Table 1. It should be stressed that, provided the sample texture distribution is close to a log-normal one, these two parameters are the estimates of the complete set of parameters describing the distribution. In analogy with the mean ± standard deviation of a normal distribution, the understanding of the anisotropy parameters is $\ln(GMS) \pm \ln(GSD)$. However, interpretation of the anisotropy value in linear size units (µm) is more transparent, and seems to be preferred. The dispersion parameter GSD is a special form of a variation coefficient. It tells, that 68 % of the observed texture counts in the sample hit the texture effective size interval from GMS/GSD to GMS*GSD.

$$GMS = \exp\left[\sum_i x_i \ln(GMS_i)\right] \quad (1)$$

$$GSD = \exp\left[\sqrt{\sum_i x_i (\ln(GMS_i))^2 - (\sum_i x_i \ln(GMS_i))^2}\right] \quad (2)$$

The ability of the GMS and GSD parameters to characterize optical properties and discriminate between cokes is demonstrated in the "coke-map" in Fig. 3. The parameters were calculated from the experimental data using the formulas of Eqs. 1 and 2. The obtained pattern was nearly identical to that f cokes plotted in the space of the same parameters that had resulted from the non-linear estimation, except minor differences in GSD. Owing to the properties of normal distribution, the suggested simple method of the parameter's estimation by direct calculation from the optical texture content data, is satisfactorily robust.

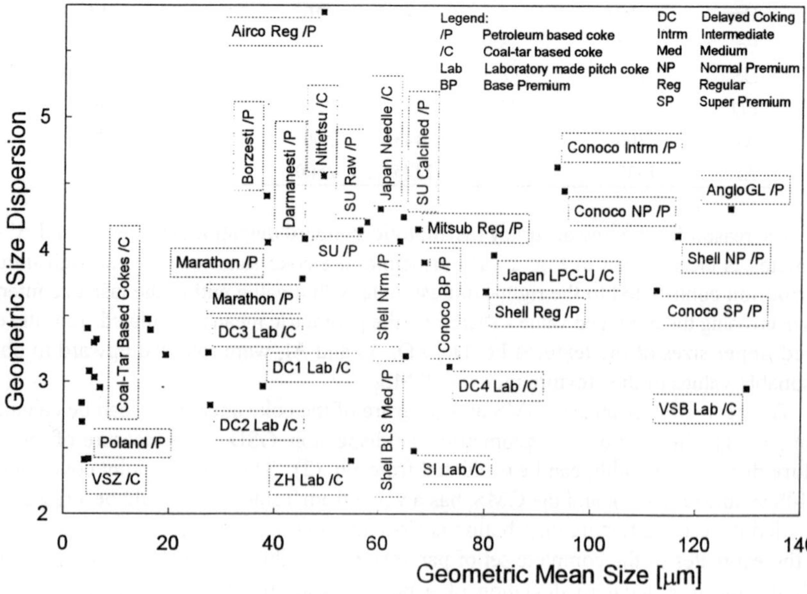

Fig. 3. Comparison of optical properties of pitch cokes.

Fig. 3 shows that, in the range of the available samples, all the regular coal-tar pitch (CTP) based cokes have the GMS value under 20 μm. The ordinary "normal" petroleum pitch R (PP) based cokes have the GMS between 20 and 60 μm. The best both commercial and laboratory coal-tar pitch based cokes fell into the GMS range between 60 and 90 μm, where the normal to premium petroleum pitch based cokes are found as well. All the commercial high anisotropy cokes with GMS above 100 μm were PP based. An exception is the CTP based semi-coke "VSB" (prepared in the laboratory of Technical University in Ostrava). However, it was the only coke in the set known to be processed to a temperature under 800K, while all the other cokes were, as believed, treated to a temperature over 1200K. The reason for the relatively low size dispersions of all the laboratory-prepared CTP based cokes is unclear.

There is a speculation by the authors that dispersion of texture size plays a role in mechanical and/or temperature dependent properties of pitch coke based products. However, no systematic effort was made in this direction.

References

1. A. Grint and H. Marsh, *Fuel* **60**(6) 513 (1981).
2. Z. Qian and H. Marsh, *Fuel* **63**(11) 1588 (1984).
3. A. Oya, Z. Quin and H. Marsh, *Fuel* **62**(3) 274 (1983).

INFLUENCE OF SULFUR ON THE FORMATION OF VAPOR-GROWN CARBON FIBERS PRODUCED ON A SUBSTRATE USING DIFFERENT IRON CATALYST PRECURSORS

P. SERP and J. L. FIGUEIREDO
Laboratório de Catálise e Materiais, Faculdade de Engenharia da Universidade do Porto, Portugal

C. A. BERNARDO
Departamento de Engenharia de Polimeros, Universidade do Minho, Guimarães, Portugal

ABSTRACT

Vapor-grown carbon fibers (VGCF) were produced from a CH_4-H_2 or a CH_4-H_2-H_2S mixture on a surface of a reconstituted graphite support (Grafoil®). The catalyst used consists of iron particles germinated from different precursors. We have investigated the influence of the nature of the anion present in the precursor on the VGCF yield using as starting materials: $Fe(NO_3)_3.9H_2O$ (denoted 1), $FeCl_3.6H_2O$ (denoted 2), and $Fe(SO_4).7H_2O$ (denoted 3). We have noticed that the presence of sulfur in precursor 3 leads to a higher yield of carbon fibers. Additionally, the utilization of a neutral organometallic complex such as $[Fe_3(CO)_{12}]$ (denoted 4) also increases the yield of carbon fibers compared to the results obtained starting from 1 and 2 when sulfur is not present. The carbon fibers so obtained have been characterized by scanning electron microscopy (SEM) and X-ray photoelectron spectroscopy (XPS) as well as by measurement of their tensile strength and modulus.

1. Introduction

Commercially available carbon fibers are generally produced from polyacrilonitrile or pitch by carbonization and graphitization. Vapor-grown carbon fibers are prepared via the decomposition of a gaseous carbon source in the presence of dihydrogen. This process is characterized by the utilization of catalyst particles (e.g., iron) and involves the initial creation of carbon nanofibers. The temperatures usually used range between 950 and 1200°C. This type of material, in spite of the encouraging mechanical, electrical, and thermal properties[1–4], as well as a potentially low cost of production, is not yet mass produced.

Recently, several processes describing the production of this kind of fibers were reported in the literature[5–10]. Among them, we have to distinguish the continuous processes[5–7] that allow the growth of very short and thin fibers (some micrometers in length and hundreds of nanometers in diameter) and the batch processes[8–10] where the fiber grows on a substrate. In this later case, longer contact time with the gaseous phase leads to fibers with several centimeters in length and some micrometers in diameter.

Different carbon sources have been used with success, such as methane[11], benzene[12], 1-methylnaphtalene[13], or carbon monoxide-carbon dioxide mixtures[14], and in many cases it appears that adding sulfur to the catalyst promotes the formation of VGCF[15,16]. The most usual sulfur sources are thiophene and hydrogen sulfide. They are introduced in the gas phase during the whole run. Using the substrate seeding method, Kunzru et al.[15] and Kato et al.[8] obtained negligible amounts of VGCF without a sulfur source in the feed. On the contrary, Tibbetts et al.[9] and Gadelle et al.[10], using a quite similar procedure, reported the production of carbon fibers without the addition of sulfur.

The objective of this study was to investigate the effect of the presence of sulfur on the VGCF yield, using a process that allows the production of carbon fibers without any sulfur source. Another objective was the assessment of the effect of the nature of the anion present in the iron catalyst precursor on the yield and the morphology of the fibers.

2. Experimental Procedure

The experimental apparatus used in this study is schematically shown in Fig. 1. The reactor, a mullite tube having 25 mm i.d. and 1 m length, was placed vertically and externally heated by an electric furnace connected to a temperature programmer. Reactants were dihydrogen (99.99999% purity), methane (99.999% purity) as the carbon source, and hydrogen sulfide as the sulfur source. In these experiments, gas flows were adjusted by mass flow controllers for H_2 and CH_4 and by a permeation system (Metronics®) for H_2S. The reactants were introduced at the top of the reactor without any particular treatment (deoxygenating, preheating). Samples of the exhaust gas could be collected with a gas syringe and analyzed with a gas chromatograph equipped with a 15 m capillary column of DB-5ms and a FID detector.

The substrate (6 x 7 cm), a reconstituted graphite support (Grafoil®), was impregnated by spraying with 3 cm^3 of an ethanolic solution (0.025M) of the iron precursor, dried in an oven at 100°C for ten minutes, and then placed in the mullite reactor. The conditions for fiber growth are shown in Fig. 2. The support was heated from room temperature to 950°C under an atmosphere of 70% H_2–30% CH_4 (catalyst reduction) or 70% H_2–30% CH_4-H_2S (molar ratio S/Fe = 0.5). The total flow rate was 150 $cm^3.min.^{-1}$. Then the temperature was raised to 1150°C, this second stage leading to the production of very thin fibers (diameter $\leq 1\mu m$) which can reach several centimeters in length, depending on the precursor used. Finally a thickening stage was carried out at 1150°C in order to obtain fibers that are easier to handle. Noticeably, the last stage does not need a flow rate as high as the second one.

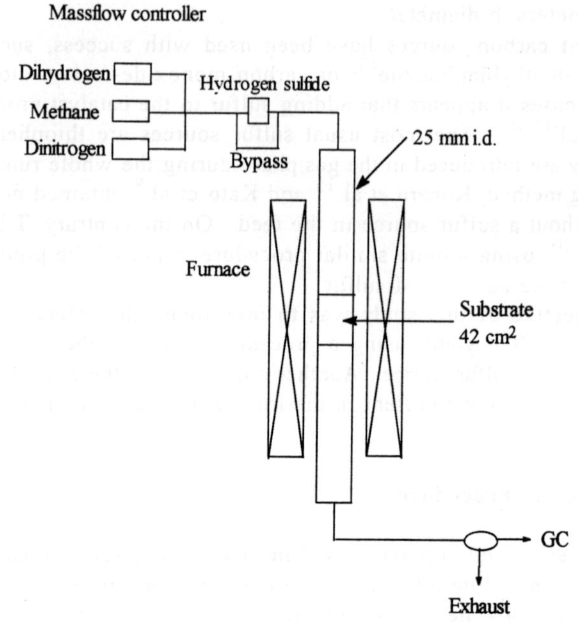

Fig. 1. Representation of the reactor.

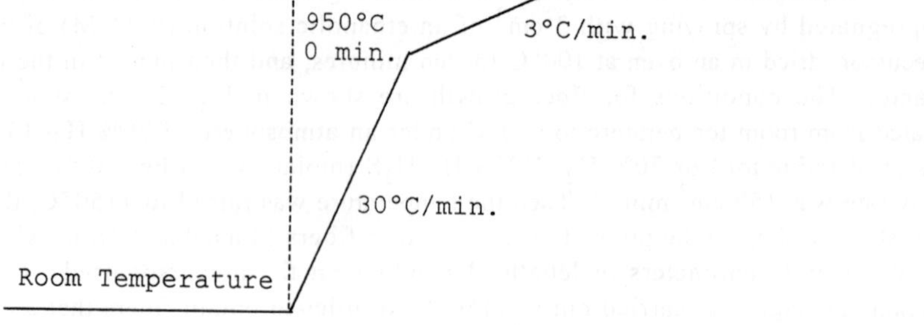

Fig. 2. Temperature conditions for fibers growth.

The morphology of the as-grown carbon fibers was determined by SEM with a JEOL LSM 630IF microscope. The diameters of the fibers were measured by using the laser beam diffraction test (32 fibers per run). X-ray photoelectron spectra were recorded without any Ar^+ ion bombardment on an ESCALAB 200A-VG Scientific spectrometer. Elemental analyses were made on a Carlo Erba EA1108 elemental analyzer. The dinitrogen adsorption isotherms (77 K) were performed on a Coulter Omnisorp 100CX apparatus. Tensile test experiments were performed on an Instron 1122 tensile testing machine. Depending on the fibers' lengths, the measurements reported herein were made with one or two gauge lengths, 6 and 10 mm, and the fibers were strained at 50 $\mu m.min.^{-1}$.

3. Results and Discussion

3.1. VGCF Growth

Representative results obtained starting from precursors 1 to 4 are listed in Table 1. When no sulfur is present in the gas phase, we notice that the masses of deposited carbon (mainly carbon fibers with some additional soot) are very similar when we start from precursors 1 and 2. Precursor 4 has been used in order to obtain very pure iron particles. Indeed, carbonyl ligands are very labile and the decomposition-reduction process of this kind of complex leads to the production of small and pure aggregates of metal[17]. This could be an explanation for the better yield that we have obtained starting from precursor 4 (as compared to 1 and 2). Moreover, we can see from Table 2 that the fibers obtained starting from 4 are significantly longer than those obtained by the use of 1 or 2. This latter result is in good agreement with a smaller metallic particle size as evidenced by Gadelle et al.[18].

Table 1. Influence of the Precursor and the Presence of H_2S on the Yield of VGCF

Precursor	Presence of H_sS [fiber type]	Mass of deposited carbon (mg)
(1) $Fe(NO_3)_3.9H_2O$	no [FC1]	152
	yes [FC1S]	250
(2) $FeCl_3.6H_2O$	no [FC2]	170
	yes [FC2S]	225
(3) $Fe(SO_4).7H_2O$	no [FC3]	275
	yes [FC3S]	277
(4) $[Fe_3(CO)_{12}]$	no [FC4]	209
	yes [FC4S]	255

We believe that the results obtained starting from precursor 3 cannot be explained just by a particle size effect. They are also due to the presence of some sulfur in the iron particles even after the reduction stage. Thus, a temperature-programmed reduction experiment performed under pure dihydrogen on precursor 3 confirmed that the mass loss did not correspond to a complete reduction of the metal. Furthermore, semi-quantitative XPS analysis of the surface of the substrate, performed after the reduction step, showed the presence of sulfur when starting from precursor 3 (atomic ratio Fe/S: 30/1). No sulfur had been detected by XPS after the reduction step of precursor 1. The promoting effect of sulfur on the VGCF yield has also been observed in the case of precursors 1, 2, and 4 when H_2S is added to the gas phase (see results in Table 1). A more pronounced effect is noticed on precursor 1. No special effect of sulfur on the yield is remarked when we used the sulfate salt 3 as starting material. In this substrate seeding method, and in our experimental conditions, the promoting effect of sulfur is noted provided the sulfur source is introduced during the pretreatment of the catalyst, e.g., before 950°C. Indeed, no significant amount of fibers is formed if hydrogen sulfide is introduced through the entire duration of a run.

3.2. VGCF Characterization

Table 2 displays some characteristics of the VGCF which were located on the first centimeters of our support. In this case, the thickening stage was 120 minutes.

Table 2. Morphological and Mechanical Characterization of the VGCF

Fiber type	Average diam. (μm)	Average length (mm)	$\sigma^{(a)}$ (GPa)	Modulus$^{(b)}$ (GPa)	BET area ($m^2.g^{-1}$)
FC1	15 ± 3	9	0.87 ± 0.10	141	5.7
FC1S	18.5 ± 4	20	1.46 ± 0.20	155 (270)	3.9
FC2	21 ± 7	12	0.65 ± 0.11	80	2.6
FC2S	20 ± 7	22	0.71 ± 0.10	90	2.3
FC3	16 ± 4	21	1.22 ± 0.22	99 (184)	2.3
FC3S	21 ± 7	20	1.32 ± 0.24	193 (274)	6.2
FC4	11 ± 3	21	0.98 ± 0.16	139 (170)	1.8
FC4S	12 ± 4	26	0.88 ± 0.18	124 (162)	2.6

(a) σ is given with a confidence interval of 95%.
(b) The value in parenthesis is the corrected modulus.

The diameters and lengths that we report are representative of the longest fibers. In fact, when we have observed (by SEM) the surface of the support after removal of these fibers, we were able to see very short fibers of some micrometers length (see Fig. 8).

We can notice in Table 2 that sulfur influences mainly the length of the fibers (except for 3) and to a lesser extent, the diameter. Thus, we believe that the promoting role of sulfur occurs before the thickening stage of the reaction (1150°C). Indeed, during an experiment stopped at 1150°C (0 min. of thickening), using precursor 1 on one part of the support and 3 on the other part, we were already able to observe fibers of some centimeters length on the precursor 3 side, whereas the fibers on the precursor 1 side were only of some millimeters length. The diameters of these fibers were almost the same: 1μm. The thickening stage that followed does not affect the length of the fibers. A typical SEM micrograph of fibers obtained with precursor 3 and 60 min. thickening is shown in Fig. 3.

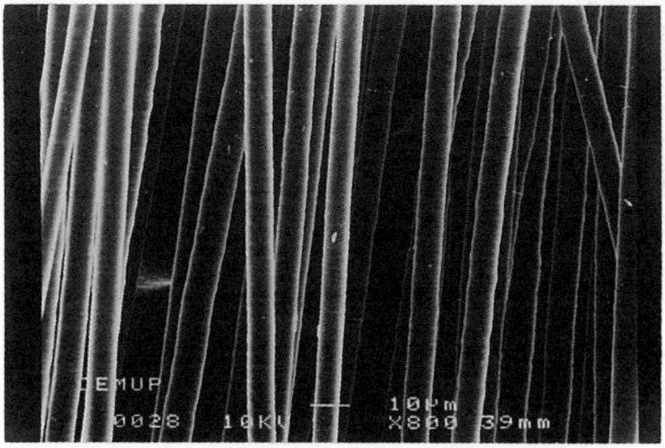

Fig. 3. Typical SEM micrograph of VGCF obtained with precursor 3 (thickening stage 60 min.), x 800.

The BET surface areas reported in Table 2 were measured with all the fibers of each sample (thickening stage: 120 min.). In this situation, it was quite difficult to have an exact idea of the mean diameter of the VGCF. But in any case, this diameter will be larger than one micrometer. Also, all the values of the BET area are larger than the geometric area ($\approx 1 m^2.g^{-1}$), implying a certain porosity, as far as gaseous adsorption is concerned. From these values, it is difficult to correlate the BET surface with the presence of sulfur in the gas phase.

The tensile strengths (stress at break) and moduli reported were measured for VGCF with mean diameters ranging from 11 to 21 μm and have to be considered as preliminary results. In fact, due to the length of our fibers, the moduli of the VGCF were measured with only two gauge lengths. The values that we have obtained are lower than those reported by Tibbetts et al.[19] for un-heat treated, 7.5 μm diameter VGCF. This result could be explained by a dependence of these mechanical properties both on the diameter of the fibers and on the rate at which the pyrolytic carbon is deposited by chemical vapor deposition. This rate determines the degree of crystalline perfection and the flaw population of the fibers, and hence, the ultimate tensile strength. The diameter of the fiber also influences the tensile strength. This is shown in Fig. 4 for the case of precursor 3. A constant decrease of the tensile strength is observed when the mean diameter increases from $\sigma = 1.47$ GPa for $\phi = 10\ \mu m$ to $\sigma = 0.43$ GPa for $\phi = 41\ \mu m$.

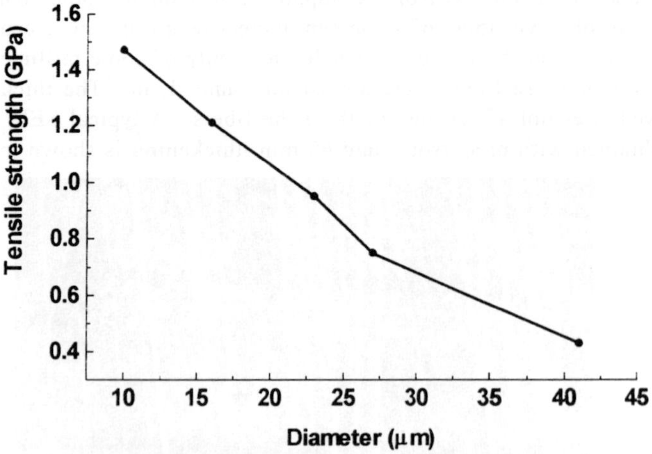

Fig. 4. Tensile strength as a function of the diameter (fiber FC3).

In spite of these observations, it appears that some differences exist in the mechanical properties according to the starting material and to the presence of sulfur. The influence of sulfur on the mechanical properties of the fibers is remarkable in the case of precursor 1. It is worth mentioning that it is also in this case that sulfur has the more pronounced effect on the yield. In a recent publication, Madroñero et al.[20] have proposed that the diameter of the inner core of the VGCF is an important factor for the mechanical properties of this kind of fiber. According to Madroñero, the higher hydrogen content in this part of the fiber (compared to the outer CVD coating) is responsible for its greater hardness. A treatment at high temperature, enhancing the overall perfection of the fibers, should thus increase their overall mechanical properties.

Table 3 displays some XPS and elemental analysis results which allow the comparison between the superficial (XPS) and bulk (elemental analysis) chemical composition of the VGCF. We have particularly focused our attention on the oxygenated functional groups. As XPS does not detect the hydrogen atom, we have recalculated the different percentages on the basis of the common quantified elements (C, N, S, O). These results show a clear difference between the surface and the bulk as far as the O/C ratio is concerned. This ratio changes from almost 1% on the surface of the VGCF to 0.1% in the bulk. This could indicate that the VGCF are prone to superficial oxidation when they are in contact with the atmosphere. Another explanation could be an oxidation occurring at high temperature during the cooling of the reactor under a dinitrogen atmosphere (99.999% purity). The O1s peaks are in all cases centered at 534.1 ± 0.2eV and 532.1 ± 0.2eV after deconvolution. These values are not consistent with adsorbed water (O1s: 536eV). The first peak, at 534.1eV, corresponds to functions such as alcohol, phenol, or C-OH of acid group. The second peak, at 532.1eV, can be attributed to ketones, COC or C=O in a carboxylic group. The C1s peak is centered at 284.5eV, whatever the type of VGCF, and the width at half height (FWHM) is 1.32eV. This last result reflects the high degree of graphitization of the VGCF surface, as shown on by Takahagi et al.[21], who reported FWHM values of 1.25eV for highly oriented graphite and 2eV for evaporated carbon. Furthermore, as in the work of Katsuki et al.[22], no sulfur was detected analytically in the VGCF, either on the surface (XPS, SEM-EDXA) or in the bulk (elemental analysis).

Table 3. Superficial and Bulk Chemical Composition of the Fibers

	XPS analysis (% at.)				Elemental analysis (% at.)
Fiber type	C	N	S	O	O
FC1	98.3	–	–	1.7	0.12
FC1-S	99.1	–	–	0.9	0.11
FC2	98.7	–	–	1.3	0.09
FC2-S	98.9	–	–	1.1	0.11
FC3	97.4	–	–	2.5	0.09
FC3-S	98.9	–	–	1.1	0.1
FC4	98.3	–	–	1.7	0.18
FC4-S	98.9	–	–	1.1	0.13
PITCH	87.6	–	0.3	12.1	0.10
PAN	78.6	0.13	0.5	18.5	0.42

Compared to commercial PAN (Amoco) or pitch fibers, these results reflect the high degree of purity of the VGCF as well as a high degree of graphitization (C1s FWHM of PAN: 1.82eV; C1s FWHM of pitch: 1.41eV). Fig. 5 shows a general spectrum for both VGCF (FC3) and PAN carbon fibers.

3.3. SEM Observation of the Support and Gas Phase Analysis

It is generally accepted that in order to reach some centimeters in length, a vapor-liquid-solid model is necessary to explain the growth process of VGCF. Indeed, the diffusion of carbon in a solid material at the fiber tip cannot permit such a growth. Two different possible models are proposed in the literature. First,

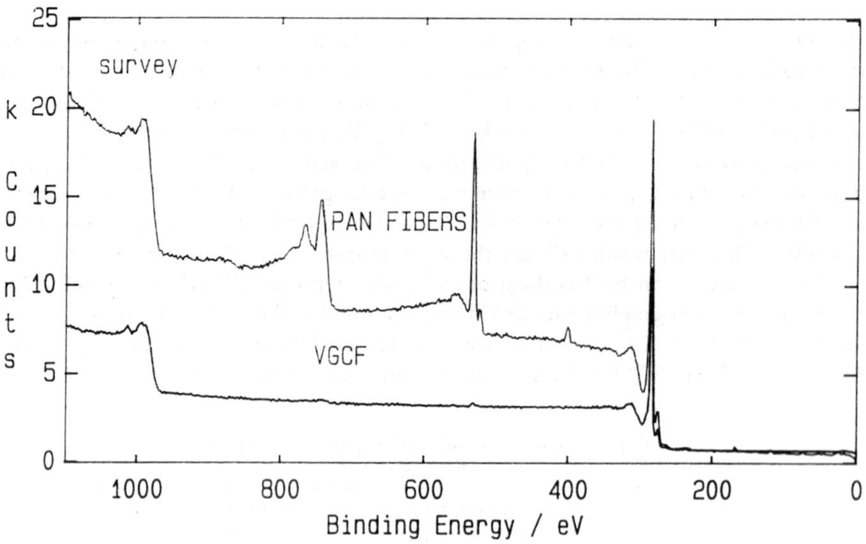

Fig. 5. General XPS spectra of VGCF (FC3) and PAN carbon fibers.

Gadelle et al.[23,24] proposed that the non-solid drop consists of an iron-carbon eutectic, the particle size influencing the melting point of the particles. In a recent investigation on the role of sulfur in VGCF production, Tibbetts et al.[16] suggested that the liquefaction of the particle increases the rate of filament nucleation. Although Tibbetts's work was relative to a continuous process with the production of short filaments, it supports the model of Gadelle. In this model, the metallic particle is located at the tips of the fibers. Very recently, a second model has been proposed by Madroñero[25] for the production of long VGCF in which the non-solid drop consists of a carbon material: coronene ($C_{24}H_{12}$, melting point 434 to 436°C). In this later model, the role of the metallic particle is to initiate the growth process by a discharge of dihydrogen that produces a molten drop of coronene. In this model, the metallic particle remains on the support.

Under our experimental conditions, a simple way to establish which mechanism occurs could be to check for the presence of the metallic particle at the tips of the filaments. Figs. 6 through 8 display the aspect of our support (SEM observation) after removal of the longest fibers (at different temperatures and with the use of different precursors). In the great majority of our observations, no iron particles can be observed at the tip of the filaments.

This could be explained by a thin carbonaceous coating on the seeds. In a recent publication, Endo et al.[26] report that they have observed pyrolitic carbon nanotubes (PCNT) coexisting with nanometer-scale VGCF during the early stage of the reaction of formation of VGCF. PCNT appear to grow as ultra-thin graphene

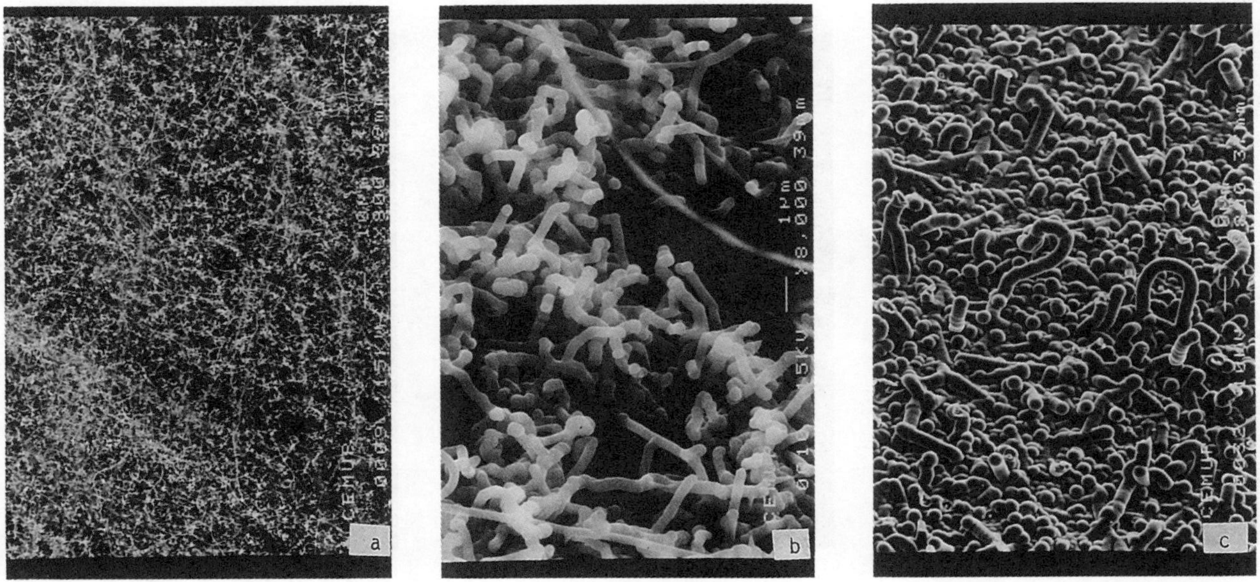

Fig. 6. Aspect of the support after removal of the longest fibers (precursor 1), (a) 1100 °C (0 min.), x 800; (b) the same magnified, x 8000; (c) 1150 °C (60 min.), x 800.

Fig. 7. Aspect of the support after removal of the longest fibers (precursor 3), (a) 1100°C (0 min.), x 800; (b) the same magnified, x 8000; (c) 1150°C (60 min.), x 800.

tubes, and catalytic particles are not observed at the tips of these tubes. Thus, it seems that several mechanisms could occur at the same time. Under our experimental conditions, it is probable that the mechanism that does not imply a metallic particle at the tip of the filament predominates. Figs. 6 through 8 also allow us to observe some differences on the morphology of the substrate according to the use of precursors 1 or 3. In particular, some important aggregates appear in the case of precursor 3 around 1100°C (Fig. 7b). The nature of this phenomenon has not been elucidated. Even at 1150°C after 60 minutes of thickening, some differences remained, with a germination that looks more advanced in the case of precursor 3 (Fig. 7c). We can also see that when H_2S is used with precursor 1 (Fig. 8), the morphology of the support at 1150°C (60 min. thickening) is close to that observed when 3 is used as a starting material (Fig. 7c).

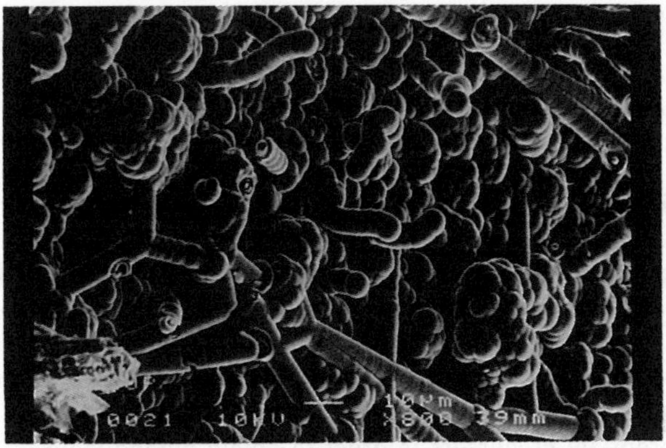

Fig. 8. Aspect of the support after removal of the longest fibers (precursor 1 with H_2S), 1150°C (60 min.), x 800.

Finally, the analysis of the gas phase at the exit of the reactor by gas chromatography has allowed us to identify ethylene, acetylene, ethane, and traces of propene as the light products formed during a run. The iron particles present on the support catalyze the decomposition of methane, as shown in Fig. 9. The presence of sulfur does not affect this phenomenon.

It is worth noting that the distribution of the products is different when iron is present on the support. The amounts of ethylene, acetylene, and ethane were 60%, 13%, and 27%, respectively, for a blank experiment (no iron on the support) and 67%, 23%, and 10% in the case of precursor 3. The presence of iron favors the formation of the unsaturated compounds at the expense of ethane, even in an H_2-enriched atmosphere.

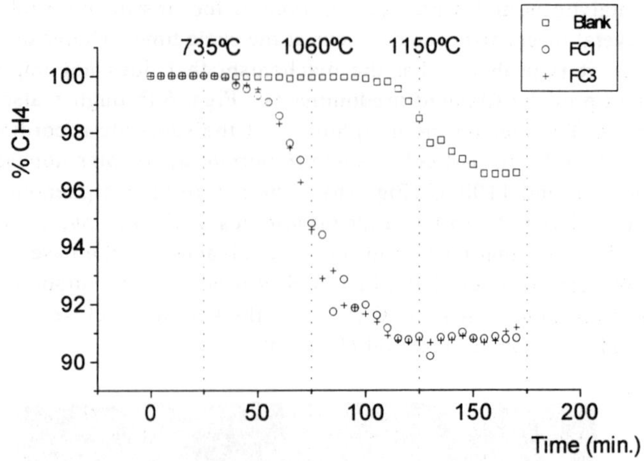

Fig. 9. Percentage of methane in the gas phase as a function of the reaction time for a blank experiment (□), precursor 1 (o) and precursor 3(+).

4. Conclusions

- We have demonstrated the promoting effect of sulfur on VGCF production, using a batch process that allowed us to prepare carbon fibers without addition of sulfur.
- It appears that the choice of the iron precursor can have an influence on the VGCF yield. Sulfur-containing catalysts or carbonyl complexes seem to be the more appropriate precursors in order to obtain a good yield.
- The as-grown carbon fibers were characterized morphologically and mechanically, and some of their surface properties were determined. According to the precursor, an influence of sulfur on the morphological as well as on the mechanical properties can be noticed. We believe that the promoting role of sulfur occurs before the thickening stage of the process and acts mainly on the VGCF's inner core.

5. Acknowledgments

This work was supported by the European Economic Community through the Human Capital and Mobility program (contract number CHRXCT940457). The authors are indebted to Professor D. D. Edie for providing the pitch fibers and to Union Carbide for supplying the Grafoil®. Acknowledgments are also made to Dr. Carlos Sá for the performance of XPS and SEM-EDXA measurements and to Mrs. Conceição Paiva for helping in the mechanical characterization of the VGCF.

6. References

1. G. G. Tibbetts, *Carbon* **27** (1989) 745.
2. G. G. Tibbetts, Vapor Grown Carbon Fibers, in *Carbon Fibers, Filaments and Composites*, eds. J. L. Figueiredo, C. A. Bernardo, R. T. K. Baker, and K. J. Hüttinger (Kluwer Academic Publishers, Dordrecht, 1990) 87.
3. G. G. Tibbetts, G. L. Doll, D. W. Gorkiewicz, J. L. Moleski, T. A. Penny, C. J. Dash, and M. J. Balogh, *Carbon* **31** (1993) 1039.
4. R. L. Jacobsen, T. M. Tritt, J. R. Guth, A. C. Ehrlich, and D. J. Gillespie, *Carbon* **33** (1995) 1217.
5. G. G. Tibbetts, D. W. Gorkiewicz, and R. L. Alig, *Carbon* **30** (1993) 809.
6. M. Ishioka, T. Okada, K. Matsubara, and M. Endo, *Carbon* **30** (1992) 865.
7. T. Kato, K. Kusakabe, and S. Morooka, *J. Mater. Sci. Lett.* **11** (1992) 674.
8. T. Kato, K. Haruta, K. Kusakabe, and S. Morooka, *Carbon* **30** (1992) 989.
9. G. G. Tibbetts, *Carbon* **30** (1992) 399.
10. F. Benissad, P. Gadelle, M. Coulon and L. Bonnetain, *Carbon* **26** (1988) 61.
11. M. Coulon, N. Kandani, L. Bonnetain and J. Maire, *French Patent* **2564** (1984) 110.
12. T. Koyama, *Carbon* **10** (1972) 757.
13. T. Kato, T. Matsomoto, T. Saito, J. I. Hayashi, K. Kusakabe and S. Morooka, *Carbon* **31** (1993) 937.
14. M. Ishioka, T. Okada and K. Matsubara, *Carbon* **31** (1993) 699.
15. M. Jayasankar, R. Chand, S. K. Gupta and D. Kunzru, *Carbon* **33** (1995) 253.
16. G. G. Tibbetts, C. A. Bernardo, D. W. Gorkiewicz and R. L. Alig, *Carbon* **32** (1994) 569, and references therein.
17. Y. I. Yermakov, *J. Mol. Catal.* **21** (1983) 35.
18. E. Benissad, P. Gadelle, M. Coulon and L. Bonnetain, *Carbon* **27** (1989) 585.
19. G. G. Tibbetts and C. P. Beetz Jr., *J. Phys. D* **20** (1987) 292.
20. A. Madroñero and M. Verdú, *Carbon* **33** (1995) 247.
21. T. Takahagi, K. Matsunaga, M. Egashira and S. Kawasumi, *Carbon* **26** (1988) 389.
22. H. Katsuki, K. Matsunaga, M. Egashira and S. Kawasumi, *Carbon* **19** (1981) 150.
23. F. Benissad, P. Gadelle, M. Coulon and L. Bonnetain, *Carbon* **26** (1988) 425.
24. F. Benissad and P. Gadelle, *Carbon* **31** (1993) 21.
25. A. Madroñero, *J. Mater. Sci.*, **30** (1995) 2061.
26. M. Endo, K. Takeuchi, K. Kobori, K. Takahashi, H. W. Kroto and A. Sarkar, *Carbon* **33** (1995) 873.

STUDY ON STRUCTURE AND PERFORMANCE OF GRAPHITE FIBER FROM MESOPHASE PITCH

Z. SHEN, H. GUO, H. JIN, R. HUANG,
J. DU, Y. LI, and S. YAN
Institute of Carbon Fiber and Composites
Beijing University of Chemical Technology
Beijing 100029, China

ABSTRACT

Two kinds of mesophase pitch with different anisotropic content were prepared by two-stage heat condensation. Through melt spinning, oxidation, carbonization, and graphitization, we obtained graphite fibers (GF) with tensile strength (σ) of 2.80 to 3.20 GPa and modulus (E) of 580 to 710 GPa. In the meantime, the changes of composition and structure as a result of heat treatment were studied using an element analyzer, SEM, and X-ray diffractometer. The results show that spinning temperature has an obvious effect on the properties of GF. From the results of SEM, we can see that GF exhibit the cross-section of onion-radial structure and high performance under optimal spinning conditions. The proper temperature of spinning is higher by about 90 to 110°C than the softening point of mesophase pitch. The anisotropic contents are higher in mesophase pitch, the best oxygen contents and oxidation temperature are lower in oxidation process, and the performance of its production is better. D_{002} of GF gradually decreased, while L_c and L_a increased with the increasing of carbonization temperature.

1. Introduction

The high tensile strength and modulus graphite fiber derived from mesophase pitch is a new kind of special engineering material. It has taken an important place in material industries for its distinctive characteristics[1,2]. Because of its rich resource and broad market requirements, the development of mesophase pitch graphite fiber in our country will become more important.

Two kinds of mesophase pitch with different anisotropic content were prepared by two-stage heat condensation[3]. Through melt spinning, oxidation, carbonization, and graphitization, we obtained graphite fibers with tensile strength (σ) 2.80 to 3.20 GPa and modulus (E) 580 to 710 GPa.

In the meantime, by element analyzer, SEM, and X-ray diffractometer, the changes of the fibers during heat treatment were studied.

2. Experimental

Raw material: two kinds of mesophase pitch, one with anisotropic content ≥ 80% (No. 4-X-31) and one with anisotropic content ≥ 95% (No. 4-X-47).

Spinning: temperature: 340 to 375°C; pressure: 0.001 to 0.01 MPa.

Oxidation: flow rate of air, 1.0m^3/h; heating rate, 3°C/min; final temperature of oxidation, 280 to 360°C; load, 9.00 Kg/cm^2.

Carbonization and graphitization: carried out under N_2 between 1000 and 2500°C.

Analysis and testing: The element analysis was carried out with Carloeria's EA-1106 element analyzer. The microstructure of carbon and graphite fibers was investigated by Cambridge S-250 MK11 SEM and Japanese Science and Motor Co. Ltd. D/Max-rB X-ray diffractometer. The mechanical performance of carbon and graphite fibers were determined by an Instron 1211 system.

3. Results and Discussion

3.1. Influence of Melt Spinning Conditions on Structure and Performance of GF

From the results of SEM (Fig. 1), we can see that GF with lower spinning temperature (340°C) has a radial transverse structure, whereas GF with higher spinning temperature (375°C) has the cross-section of onion-radial structure. The performance of the former is lower than that of the latter. Their performances are respectively tensile strength (σ) 2.78 GPa and 2.88 GPa; tensile modulus (E) 433.9 GPa and 584.3 GPa.

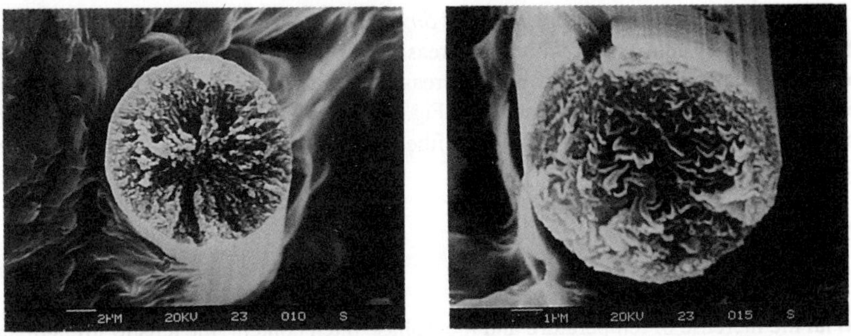

Fig. 1. SEM of mesophase pitch graphite fibers (No. 31). Left, low temperature (340°C); right, high temperature (375°C).

3.2. Influence of Oxidation Process Conditions on the Properties of CF and GF

3.2.1. Influence of Final Temperature on the Performance of CF and GF

Under constant air flow rate and heating rate, the oxidation of mesophase pitch fiber was carried out. Table 1 gives the results. The tensile strength and modulus of CF and GF varied with the increase of temperature. Among them there is an optimum value of tensile strength and modulus (see Fig. 2). The oxygen content and final temperature are the key factors for the preparation of high performance of CF and GF. Particularly interesting is the fact that the results show that the oxygen up-take of pitch fiber with low anisotropic content (No. 31) was faster than pitch fiber with high anisotropic content (No. 47). So the best oxygen content of No. 31 is 11.8%, whereas the best oxygen content of No. 47 is 7.18%. Therefore, we concluded that at the same temperature, the activity of oxidation of pitch fiber with low anisotropic content is higher.

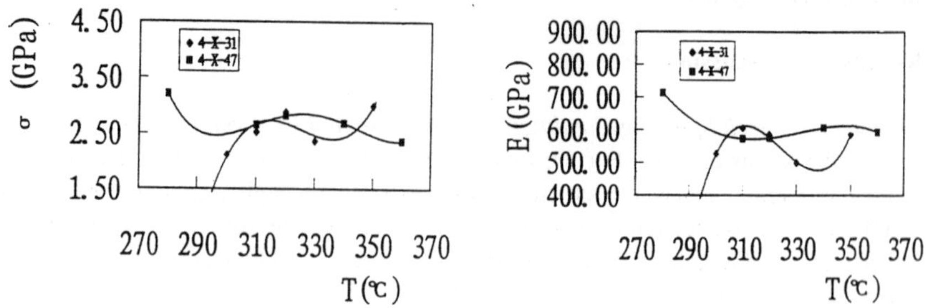

Fig. 2. The effect of various oxidation temperatures on σ and E of mesophase graphite fiber.

3.2.2. Variation of Element Composition and Structure During Oxidation

It was shown (Table 1) that with the increase of oxidative temperature, the contents of carbon and hydrogen in the pitch fibers decrease, the oxygen content increased, and the degree of oxidation gradually increased (see Fig. 3). Besides, although the variation trend was similar, the oxidation activities of pitch fibers with different anisotropic content were different.

3.3. Carbonization and Graphitization

3.3.1. Variation of Mechanical Properties

With the increase of temperature and mesophase pitch content, the tensile strength and modulus increased gradually (Table 2). This shows that mesophase pitch fiber can orient and graphitize more easily than isotropic pitch fiber or PAN-based fiber, for the tensile strength of isotropic pitch fiber and PAN-based fiber decreased with the increase of graphite temperature (see Fig. 4).

3.3.2. Variation of Structure

The results of X-ray diffraction are shown in Table 2. It can be seen that with the increase of temperature, the interlayer distance d_{002} decreased gradually, L_c and L_a increased gradually (Fig. 5). This is shown that the orientation along the fiber axis becomes better and the structure of the pitch fiber varied from turbostatic carbon layers to the ordered basal plane graphite structure. The interlayer spacing d_{002} of the graphite fiber (3.391 to 3.395A) is near the d_{002} of ideal graphite (3.354A).

Table 1. The Influence of Stabilization Temperature on Element Composition of Fibers and Performance of Obtained CF and GF

sample			Element Composition						Performance of GF and CF				
			C%	H%	O%	N%	O/C	H/C	CF		GF		
									d(um)	σ (Gpa)	d(um)	σ (GPa)	E(GPa)
4-X-31	mesophase pitch		94.33	4.40	0.70	0	0.0056	0.56					
	pitch fiber		94.72	4.79	0.69	<0.3	0.0054	0.61					
	stability fiber	300℃	87.14	3.66	8.24	<0.3	0.0710	0.50	7.18	1.89	10.48	2.11	527.5
		310℃	84.55	3.57	11.18	<0.3	0.0990	0.51	8.11	2.11	7.81	2.52	605.2
		320℃	80.83	3.43	14.15	<0.3	0.1310	0.51	7.54	1.89	9.33	2.88	584.3
		330℃	78.67	3.18	15.05	<0.3	0.1430	0.48	6.36	2.33	5.92	2.36	500.1
		350℃	76.22	3.01	16.54	<0.3	0.1630	0.47	5.91	1.73	6.14	2.99	586.6
	carbon fier		98.28	<0.3	0	<0.3	0	<0.036					
	graphite fiber		99.67	<0.3	0	<0.3	0	<0.036					
4-X-47	mesophase pitch		94.21	4.14	1.08	0	0.0086	0.53					
	pitch fiber		94.90	4.60	0	<0.3	0	0.58					
	stability fiber	280℃	84.46	4.02	7.18	<0.3	0.0610	0.54	7.84	1.65	7.41	3.20	713.6
		310℃	88.40	3.61	9.95	<0.3	0.0870	0.51	9.82	2.03	8.09	2.65	572.6
		320℃	84.10	3.60	11.03	<0.3	0.0980	0.51	11.51	1.71	10.04	2.81	574.6
		340℃	81.97	3.26	12.84	<0.3	0.1170	0.48	10.69	1.71	9.36	2.68	605.9
		360℃	78.53	2.99	15.68	<0.3	0.1500	0.46	9.23	1.68	8.33	2.36	593.1
	carbon fiber		98.94	<0.3	0	<0.3	0	<0.036					

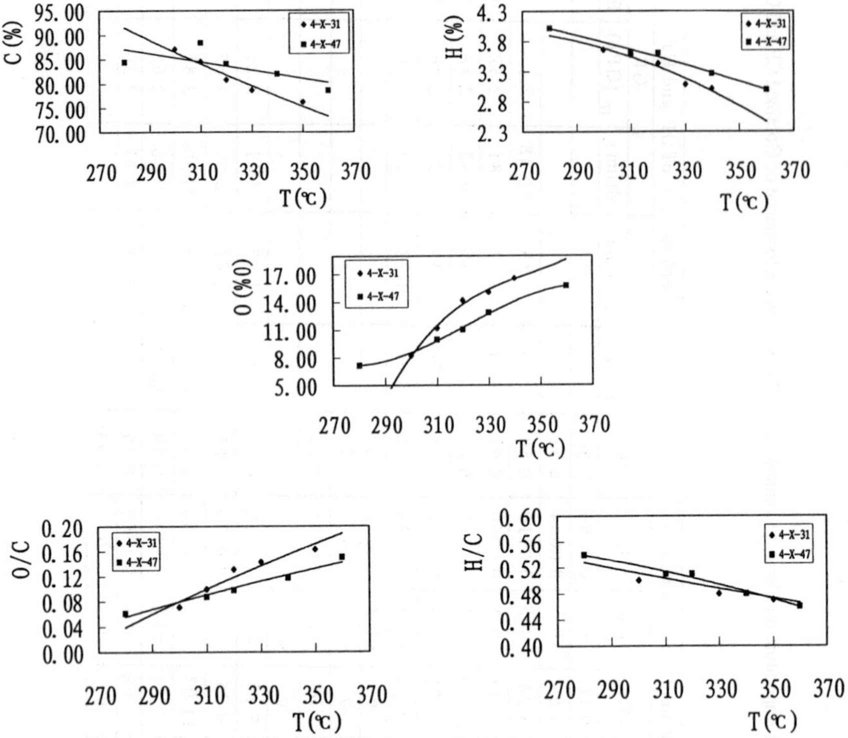

Fig. 3. Element change of fiber of various oxidation temperature.

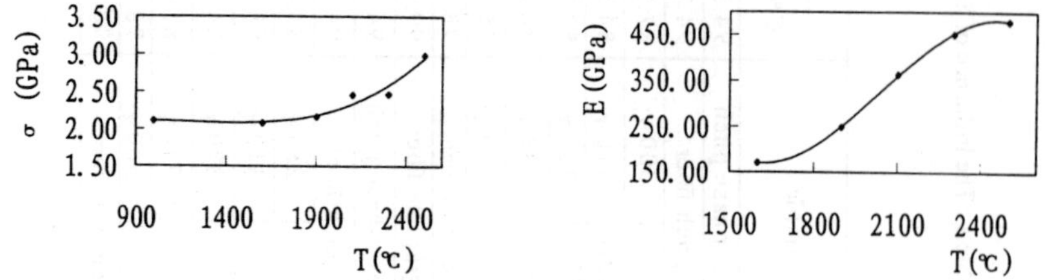

Fig. 4. Change of graphite fibers' σ and E in various graphitized temperature.

Table 2. Performance and Crystalline Parameters of Mesophase Pitch Graphite Fiber

sample	temp. °C	the performance of GF				the crystallite parameter of GF			
		d(μ m)	σ (GPa)	ε (%)	E (GPa)	d_{002} Å	d_{001} Å	Lc Å	La Å
4-X-31	1000	8.11	2.11			3.567		40.249	
4-X-31	1600	8.34	2.08	1.24	171.2				
4-X-31	1900	8.54	2.17	0.84	249.8	3.437	2.114	106.774	181.328
4-X-31	2100	6.51	2.46	0.68	363.2	3.423	2.121	151.623	268.877
4-X-31	2300	6.32	2.47	0.54	451.5				
4-X-31	2500	7.52	2.99	0.59	479.0	3.395	2.125	198.059	295.425
4-X-37	2500	7.84	3.20	0.55	713.6	3.391	2.126	187.045	302.849

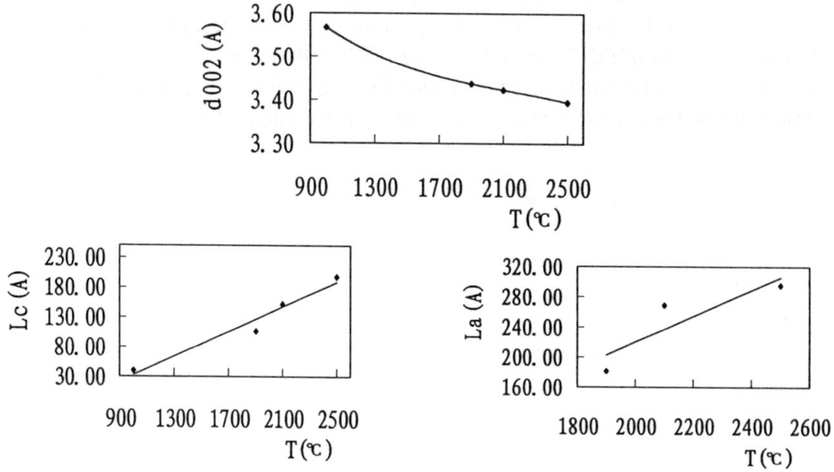

Fig. 5. Change of structural parameters of graphite fiber during the graphitization process.

4. Conclusions

- Two kinds of mesophase pitch with different anisotropic content (≥ 80% and ≥ 95%) and different softening points (264°C and 269°C) were prepared by two-stage heat condensation. These two mesophase pitches were spun into fiber by melt spinning. Then through oxidation, carbonization, and graphitization (2500°C), we obtained mesophase pitch graphite fiber with tensile strength (σ) 2.80 to 3.20 GPa and modulus

(E) 580 to 710 GPa. The GF performance with high mesophase content is better than that of low mesophase content.

- The process condition of melt spinning was low pressure and high temperature. The proper temperature of spinning is about 90 to 110°C higher than the softening point.
- The best oxygen contents of No. 31 and No. 47 are 11.81% and 7.18%, respectively, with oxidation temperatures of 310°C and 280°C.
- With the increase of temperature during the process of carbonization and graphitization, d_{002} decreased gradually, and L_c and L_a increased. The values of d_{002} are 3.391A and 3.395A, which are near the value of ideal graphite interlayer distance.

5. References

1. J. B. Barr, S. R. Didchank, J. C. Lewis and L. S. Singer, High Modulus Carbon Fibers from Pitch Precursor, *Appl. Polym. Symp.* **29** (1976).
2. L. S. Singer, Carbon Fibers from Mesophase Pitch, *Fuel* **60** (1981) 839.
3. Y. D. Park and I. Mochida, A Two-stage Preparation of Mesophase Pitch from the Vacuum Residue of FCC Decant Oil, *Carbon* **27** (1989) 925.
4. Z. Shen and R. Qin, Study on Stabilization Process of Petroleum Derived Pitch Fiber, *36th International SAMPE Symposium and Exhibition* **36** (1991) 1109.

EVALUATION OF CARBONACEOUS FIBERS

M. KALOČ, P. DVOŘÁK*, and V. ROUBÍČEK**
Department of Ironmaking and Coking
**Department of Mineral Raw Materials and Geoinformatics*
***Department of Chemistry*
VŠB - Technical University of Ostrava, Czech Republic

ABSTRACT

When evaluating carbonaceous fibers, it is necessary to consider the methods that are employed in the evaluation of materials with similar characteristics. Based on this concept, the methods of evaluation used in the fields of coal matter and coke can be applied to other carbonaceous materials. Ordinary reflected light is used primarily in assessing the homogeneity of fibers, contents of impurities, surface quality, and the occurrences of fissures, cavities, and fragments. It is possible to evaluate the shape in the longitudinal and transversal sections under ordinary reflected light. Polarized light enables the origin of optical anisotropy, its shape, size, and pattern to be assessed. Using knowledge acquired in this manner, the quality of fibers can be evaluated and appropriate improvements can be pursued.

1. Introduction

The term "fiber" is related to polycrystalline or monocrystalline solids in which the ratio of the minimum length to the maximum cross-section is more than 10:1. Fibers have a more specific structure and more perfect atomic structure than non-fiber mono- or polycrystals, and are characterized by higher mechanical properties. It is known that carbonaceous fiber materials have different forms. They can be produced in the shape of threads, cables, felts, etc. The final form of a carbonaceous material is determined by the form and type of the original raw material.

A polyacrylonitrile (PAN) fiber is considered to be a very suitable fiber precursor, whereas fibers based on black-coal tar, coal-tar pitch, or oil products, with their variable properties, rank among the less valuable materials. However, the development of technological procedures has proved that even these less valuable raw materials are able, under appropriate regimes, to form fibers that meet required parameters.

If fibers produced from pitch are to be used for preparing composites, especially C-C composites, it is necessary to observe their surface properties, which are given by the degree of roughness, the shape of the fracture surface, etc.

In this paper, some knowledge is stated that was obtained via microscopical study of carbonaceous fibers that were prepared in the Research Institute of Chemical Fibers in Svit.

The research of carbonaceous fibers utilizes methods known from the field of research of materials containing a considerable share of carbon, such as coal and coke. A precondition of proper evaluation is to obtain a specimen that was prepared in a manner that corresponds to the evaluation methodology used. When observing by the optical microscope, a polished section is used, which is evaluated in reflected normal and polarized light, or fractured specimens are observed by the scanning microscope.

Under normal reflected light, especially the homogeneity of fibers, impurity of contents, surface quality, and the occurrences of fissures, cavities, and fragments can be judged. It is possible to judge shapes both in the longitudinal and cross-sectional directions. Homogeneity of the fracture surface, origin of inclusions, and contents of pores and impurities are evaluated.

The homogeneity of fibers can be understood in terms of changes in composition, impurity content, and other microstructural imperfections that could influence the quality of fibers, above all, their flexibility and strength.

The surface quality is given by a character of the outer part of the fiber which can be smooth, rough, warty, or wavy. This feature is connected with the course of the fiber-forming process and also with the quality of the properties of the original matter, e.g., with the contents of impurities.

The occurrence of fissures can be observed either on the outer faces or in the polished sections by the scanning microscope. In some cases, such fissures are then oriented in various directions, namely, longitudinal, transversal, and oblique. These fissures are a manifestation of the mass character and its strength characteristics. A result of these observations is the ability to predict the behavior of an appropriate fiber to certain mechanical stress. It is connected with the question of flexibility and strength of the fiber, similar to the above-mentioned homogeneity.

Fragments can occur in specimens which show a great number of fissures. This is a feature that is influenced by the character of the solid, especially by its strength, flexibility, and, mainly, brittleness. This can be caused by fibers that are not sufficiently thermally processed or are processed in an unsuitable temperature area.

Cavities can be an indication of utilization of an erroneous procedure during the processing of the fiber-forming material, unsuitable temperature range or speed of fiber-forming, when a partial foaming could cause pore formation. Such pores adversely influence the strength of the fiber.

Evaluation of sections helps assess the quality and degree of wear of the spinning nozzle as, according to this factor, a quality of the spun material (e.g., viscosity) can be evaluated. The size of the section can change as a result of undesired impurities.

In polarized light, the behavior of planes can be evaluated, whether the planes are optically isotropic or anisotropic. In the case of an anisotropic plane, the shape, arrangement, and size of anisotropic areas can be judged. According to the pattern, we can differentiate between mosaic, fibrous, granular, pencil, and planar anisotropy. According to the arrangement, flow, ordered, disordered, and other anisotropies can be distinguished. It is possible to define fine-grained, middle-, and coarse-grained anisotropy according to the size of the observed areas. Anisotropy is more or less related to a degree of alteration of the spun material. It can be judged that the most suitable type will be fibers that are isotropic.

Anisotropic parts have been mostly found at cavities in the fibers. Moreover, various degrees of alteration of the development of so-called mesophases have been observed; however, they are not always obvious due to their small dimensions.

All of the stated properties can be observed in the polished section be means of optical microscopy. This technique requires polished specimens of good quality, since only a small amount of material is examined. A certain difference in qualitative evaluation exists thanks to using the scanning microscope. The fibers are thus strongly marked. It is possible to judge very well their surface properties, plane homogeneity, and the occurrences of pores and cavities.

2. Conclusion

Knowledge achieved using optical and electron microscopy techniques makes it possible to evaluate the quality of fibers and to gain, in this manner, underlying data to delineate a further way to improve the quality.

CARBON FIBERS FROM PAN FIBERS STABILIZED UNDER FLUIDIZED BED

R. B. MATHUR, V. V. JANAKI, and O. P. BAHL
Carbon Technology Unit
National Physical Laboratory
New-Delhi 110012, India

ABSTRACT

Polyacrylonitrile (PAN) fibers have been stabilized under graphite powder fluidized bed in order to suppress the exotherm associated with the stabilization process. It has been possible to stabilize the fibers up to temperatures of greater than 300°C with a much faster heating rate as compared to the conventional process. A special reactor of Pyrex glass was designed for creating fluidized bed. Experiments were conducted under different fluidized conditions. Finally, the carbon fibers produced by using the new technique show an improvement of 50% in the tensile strength of the resulting carbon fibers as compared to ones produced by following the conventional process.

1. Introduction

PAN fibers have been found to be a highly successful precursor for producing high performance carbon fibers. Production of carbon fibers from PAN involves two major steps:

(1) low temperature thermal stabilization up to 300°C
(2) carbonization at temperatures greater than 1000°C.

Stabilization is comparatively the most crucial step of the two which involves converting PAN to an infusible and non-flammable structure. PAN fibers, when heated to 180°C and above, undergo thermal degradation which causes cyclization, also known as oxidation, of the structure[1–3]. This cyclization is exothermic in nature and causes excessive weight loss and chain scission or even fusion or burning of fibers during the process. This exotherm must be controlled to produce good quality carbon fibers using PAN precursor. Gassie, Hautz, McGuchan and Hay[2–6] have carried out detailed studies to understand the exothermic behavior of PAN using calorimetric techniques. Several authors[7,8] have taken recourse to such DSC curves for carrying out proper thermal stabilization of PAN fibers before carbonization as it gives vital information regarding heat flow, energy of activation, cyclization reaction start temperature and final temperature of oxidation. These parameters depend largely on the physical and chemical characteristics of PAN fibers, rate of heating, and the medium of stabilization such as air, oxygen, sulphur dioxide, ammonia vapors, etc.[9-11] A broader exothermic peak is observed in the pyrolysis of PAN in air than in inert atmosphere[12]. Thus these media modify the exotherm by varying the mechanism of the various cyclization reactions during stabilization. The over-heating of the precursor can be prevented by lowering the heat treatment temperature and decreasing the heating rate during the stabilization treatment. Since PAN fibers are thermal insulators, the exothermic heat gets localized and therefore the rise in temperature is quite sharp. The only alternative is to keep the heating rate very slow or the stabilization temperature quite low, about 220°C.

In the present work we have utilized a novel way to control the exotherm build-up by carrying out stabilization of PAN fibers, which are poor thermal conductors, in a thermally conducting medium which remains in contact with the PAN fiber surface during the entire period of thermal stabilization. It is supposed that such medium should immediately conduct away the exothermic heat without chemically reacting with the fibers and should therefore prevent the localized temperature rise at the fiber surface. It is believed that such stabilized fibers should have a lesser number of defects comparatively and should therefore become precursor for high-performance carbon fibers. Natural flaky graphite powder, purity > 99.5%, has been chosen as the thermal conducting medium or heat sink. A special reactor made up of Pyrex glass was designed to produce a fluidized bed of the graphite powder around the PAN fiber tow. Sufficient nitrogen flow is maintained so that the graphite powder forms a fluidized bed during thermal stabilization. Carbon fibers have been prepared using these stabilized PAN fibers as well as from those stabilized under conventional conditions. Mechanical properties of the carbon fibers stabilized under the two conditions have been compared.

2. Experimental

Studies have been carried out on a special variety of PAN fibers obtained from M/S Courtaulds, UK. A single fiber tow contains 6000 manofilaments of 1.2 d'tex each. The fiber tow was spread to about 1 cm on a Pyrex glass frame which was kept in a fluidized bed reactor made up of Pyrex glass and shown in Fig. 1. Fine mesh at the bottom of the reactor supported the graphite powder bed which was sufficient to cover the complete sample and the glass frame while under dynamic conditions.

The reactor was placed inside a tubular furnace such that the glass frame containing the fiber tow was in the constant temperature zone. The fluidized bed was created by passing nitrogen gas, and oxidation of the fibers was carried out by maintaining a supply of dry air at the rate of 5 LPM during the entire course of thermal stabilization. The samples were then carbonized in a quartz reactor under

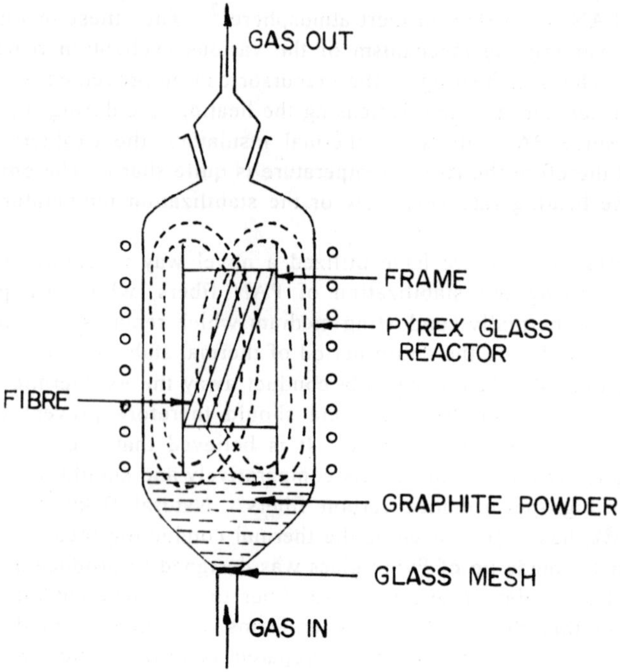

Fig. 1. Fluidized bed reactor.

nitrogen atmosphere up to the HTT of 1300°C. The carbonization step was completed in 45 minutes. Experiments were repeated under identical conditions without using a fluidized bed, i.e., the conventional process, by cutting off the supply of nitrogen and passing only air inside the reactor.

Mechanical properties of the carbon fiber samples have been measured on the single filaments of gauge lengths of 2 cm on an Instron 1122 universal testing machine, maintaining a cross-head speed of 0.05 cm/min. The results are reported on the average of at least 20 samples in each case, and the coefficient of variation in the tensile strength values has also been reported in each case.

3. Results and Discussion

Thermal stabilization of the fibers both under fluidized bed as well as non-fluidized bed has been carried out under following conditions:

- keeping the final temperature of stabilization same and changing the heating rates
- keeping the heating rate constant and changing the final temperature of stabilization.

Figure 2 compares the tensile strength of carbon fibers stabilized under fluidized bed (FB fibers) and ones prepared under non-fluidized bed conditions (NFB), using different heating rates and keeping the fiber isothermally for 1 hr. at 270°C. It is observed that the FB fibers possess higher tensile strength (TS) for all the heating rates with a maximum value of 2.93 GPa for FB as compared to 2.36 GPa for NFB fiber. One finds that there is a maxima of strength at a 7°C/min heating rate which corresponds to the optimum stabilization conditions. Since the experiments have been conducted under identical processing parameters, e.g., temperature, time, tension per unit filament, etc., the improvement in the tensile strength can only be due to the improved structure of FB fibers having a lesser number of surface defects. If the heating rate is increased to 10°C/min under normal stabilization conditions, the structure of the fiber is destroyed so much that upon carbonization it becomes extremely brittle and unable to be handled. The effect is less severe in the case of FB fibers. Quite interestingly, one finds from Table 1 that the coefficient of variation is always less for FB fibers than for NFB fibers.

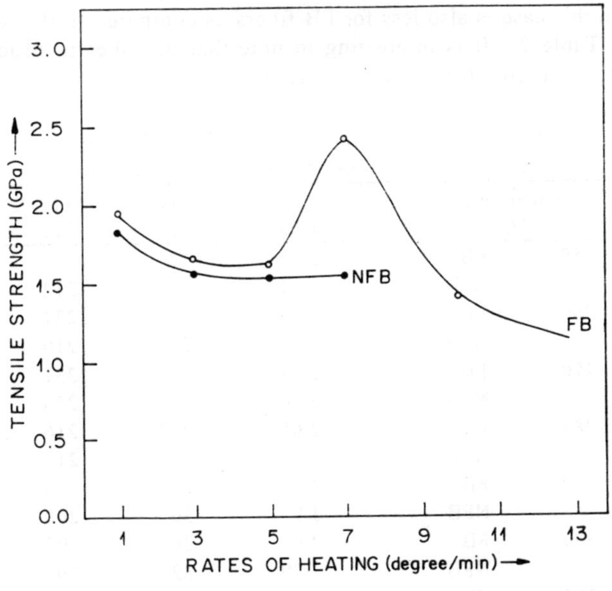

Fig. 2. Tensile strength of carbon fibers for different rates of stabilization.

Table 1. Mechanical Properties of FB and NFB Stabilized Carbon Fibers, Stabilized under Different Heating Rates

Heating Rate °C/min		TS GPa	COV %	YM GPa
1	FB	2.0	24	198
	NFB	1.87	25	205
3	FB	1.65	20	182
	NFB	1.59	31	189
5	FB	1.64	17	200
	NFB	1.58	31	196
7	FB	2.43	16	208
	NFB	1.57	24	187
10	FB	1.49	34	163
	NFB	—	—	—

Figure 3 shows the variation in the tensile strength of the carbon fibers produced from PAN fibers heated isothermally for one hour at different stabilization temperatures, i.e., 270, 280, 290, 300, and 325°C, respectively, under FB and NFB conditions. It is seen that in the case of FB, the tensile strength is always higher as compared to NFB. As the temperature increases, the percentage increase of TS increases. The TS shows an increase of 77% up to 325°C. The coefficient of variation in this case is also less for FB fibers as compared to that of NFB fibers as shown in Table 2. It is interesting to note that the fiber is structurally stable under fluidized bed conditions even at 325°C.

Table 2. Mechanical Properties of FB and NFB Stabilized Carbon Fibers, Stabilized at Different Temperatures

Temperature °C		TS GPa	COV %	YM GPa
250	FB	2.36	22	210
	NFB	2.41	19	212
260	FB	2.34	18	221
	NFB	2.38	24	210
270	FB	2.93	18	231
	NFB	2.36	24	226
280	FB	2.65	23	216
	NFB	1.97	24	210
290	FB	2.65	28	230
	NFB	1.92	30	208
300	FB	2.65	26	207
	NFB	1.79	32	190
325	FB	2.43	16	208
	NFB	1.57	24	187

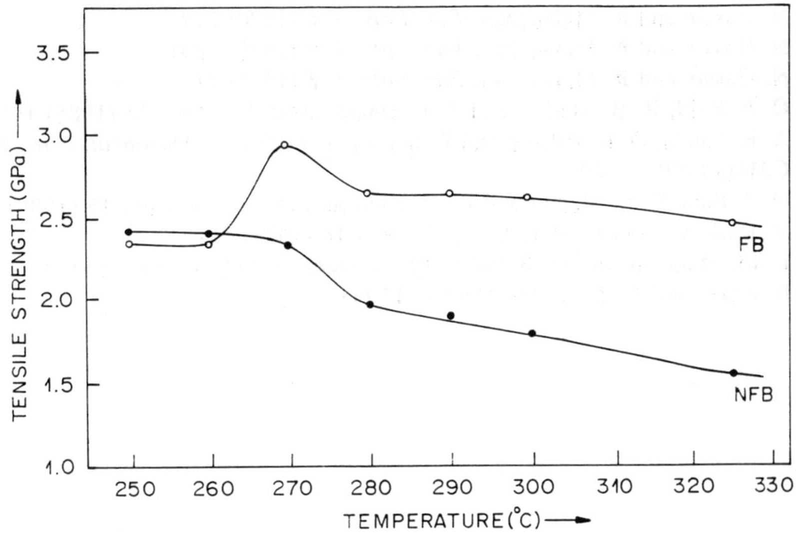

Fig. 3. Tensile strength of carbon fibers for different stabilization temperatures.

4. Conclusion

Fluidized bed stabilized fibers possess better structure and therefore the resulting carbon fibers show better mechanical properties compared to non-fluidized bed ones. It is possible to keep the heating rates higher and the final stabilization temperature quite high. There is no influence of the reaction conditions on the Youngs modulus of the resulting carbon fibers.

5. Acknowledgments

The authors are grateful to Dr. E. S. R. Gopal for his encouragement and kind permission to publish these results. One of the authors (V. V. Janaki) is thankful to CSIR for providing a Research Fellowship.

6. References

1. J. H. G. Cowie, *Polymers: Physics & Chemistry of Modern Materials* (International Textbook Ltd., London, 1973).
2. N. Gassie and Hay, *J. Polym. Sci.* **56** (1962) 189.

3. R. C. Hautz, *Textile Res. J.* **20** (1950) 786.
4. N. Gassie and R. McGuchan, *Eur. Poly. J.* **6** (1970) 277.
5. N. Gassie and R. McGuchan, *Eur. Poly. J.* **7** (1971) 1891.
6. N. Gassie and R. McGuchan, *Eur. Poly. J.* **8** (1972) 257.
7. O. P. Bahl, R. B. Mathur and T. L. Dhami, *Mat. Sci. Eng.* **73** (1985) 105.
8. A. K. Gupta, D. K. Paliwal and Puspa Bajaj, *JMS-REV. Maroncol. Che., Phys.* **C31(1)** (1991) 1-89.
9. O. P. Bahl, R. B. Mathur and K. D. Kundra, *Fib. Sci. Technol.* **13** (1980) 155.
10. V. Raskovic and S. Marinkovic, *Carbon* **16** (1978) 351.
11. L. M. Manocha and O. P. Bahl, *Agnew Makromol Chem.* **64** (1977) 115.
12. E. Fitzer and D. J. Muller, *Carbon* **13** (1975) 63.

SOME PROPERTIES OF COAL TAR BINDERS

D. KOŠT'ÁL and M. JANÍK
DEZA Co. Ltd., 757 Valašské Meziříčí, Czech Republic

M. KALOČ
Faculty of Metallurgy and Material Engineering
Technical University of Mining and Metallurgy
tř. 17. listopadu, 708 33 Ostrava - Poruba, Czech Republic

ABSTRACT

The properties and behavior of carbon binders are defined by parameters such as softening point, viscosity, surface tension, penetration ability, mesophase formation and structural changes that occur during thermal treatment. In laboratory conditions all parameters noted above were studied, especially their dependence on the group and chemical composition. Each material was characterized according to the content of toluene insolubles [TI] and quinoline insolubles [QI]. The extent of mesophase formation and transformation of the group composition under heating was used to quantify this dependence. The mesophase forming tendency of a particular group is different according to the higher distillation fractions contained in the coal tar. These processes were studied under microscope. The coke forming and solid remainder yield of the binder is affected both by heating rate and by their chemical structure. The reaction velocity constants and other kinetic parameters were found. Our hypothesis that the softening point depends on the group composition was validated and could be formulated as a logarithmic dependence. This dependence was shown to apply to the various pitch mixtures.

1. Binder Pitch

Binder pitch must fill the gaps between coke grains and at the same time must penetrate into the coke grains capillars (Fig. 1). For this purpose we search the optimum analytical values (Table 1).

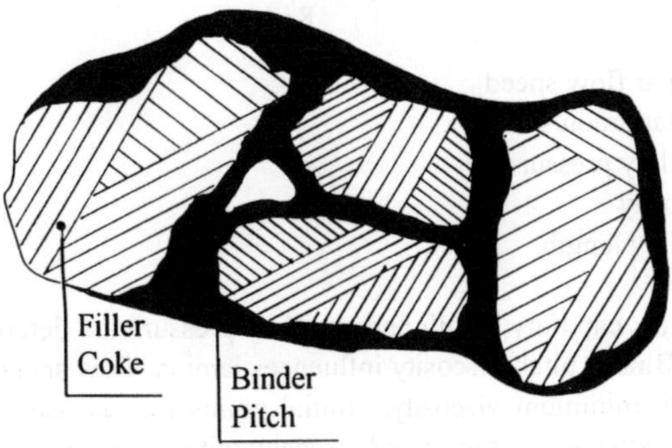

Fig. 1. Grains of coke connection.

Table 1. Results of Pitch Sample 80 M.Q.I., January to May 1993.

Ash content		0.1901 + 0.0224 wt.%
Softening point:	Kraemer-Sarnow	84.0181 + 1.3491°C
	Mettler	103.5912 + 1.0994°C
Coking value		55.2750 + 0.6655 wt.%
Toluene insoluble		28.5737 + 1.6831 wt.%
Quinoline insoluble		9.4533 + 0.7748 wt.%
Sulfur content		0.4726 + 0.0100 wt.%
Distillation range 0 – 360°C		5.2840 + 0.3297 wt.%

Binder pitch evaluation assumptions:

1. The physical properties of binder pitch are determined by its chemical composition and its structure.
2. Binder pitch is a solution of high molecular tars contained in low-molecular oils with plastification effect.
3. Chemical and structural properties of binder pitch can be approximately evaluated according to high-molecular tars and low-molecular oils content.
4. Binder pitch coking ability depends on its degree of aromaticity.
5. Binder pitch sintering properties are conditioned on the base of plastification and coking properties.

We will now discuss some approaches to the problem.

2. Viscosity

Laminar flow speed into the capillars pores is controlled by Poiseuille equation:

$$u = \frac{r^2 P}{8\,\mu l}$$

where: u = laminar flow speed
r = capillary radius
P = capillary pressure
μ = viscosity
l = capillary length.

Capillary radius, capillary length and capillary pressure are determined by the quality of coke material. Binder pitch viscosity influences laminar flow speed in coke. Our main purpose is to reach minimum viscosity. Initial points for viscosity calculation are both content of TI high-molecular particles and viscosity initial value, i.e., softening point which corresponds to viscosity value $3.5 \bullet 10^7$ mPa.s. as shows the following equation:

$$\log(\nu + 0.8) = 0.8 + 0.4 \log TI + (5.8 - 0.3\ TI).[\log(S.P. + 273) - \log TI]$$

ν, viscosity	cSt
TI, content of toluene insoluble	wt.%
S.P., softening point Kraemer-Sarnow	°C
K, temperature	K

Properties of high-molecular fraction (cut) are characterized by TI and QI content. We recognize two types of QI, primary and secondary QI, where the primary QI are spherically formed with higher molecular weight which is made by connecting aromatic hydrocarbons to the large discotic molecules (Fig. 2). . From the reason of binding pitch use primary QI are more suitable. Relation between pitch viscosity and M_n average molecular weight can be expressed by the following Staudinger-Sakurada-Houwing equation:

$$\eta = k \cdot M_n^a$$

where: k = 1.41

exponent a = 1/10, depending on molecules form, for rigid spherical molecules this mentioned exponent comes up to 0 and for planar molecules this one comes up to 0.5.

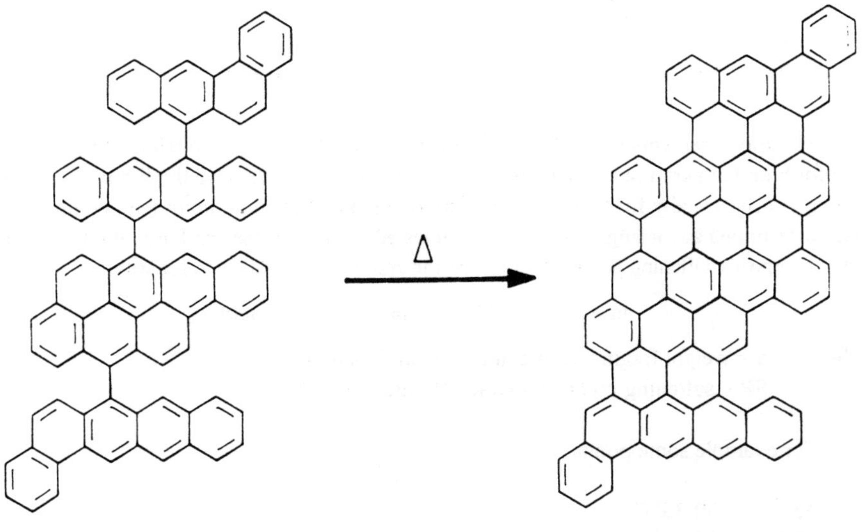

Fig. 2. Ring closure reaction yielding planar aromatic systems from oligo-aryls.

3. Softening Point

Coal tar is a colloidal system. It is a mixture of crystalline materials and high-molecular components which prevents crystallization, this solid phase transforms to a liquid by softening, not by fusion. Softening point depends not only on individual compound content but also on the methods used. Some methods are based on irreversible deformation by precise mercury amount or by ball or rod formed defined particles and others are formed on the base of deformation by their own weight at increased temperature (Table 2).

Table 2. Softening Point Methods

Method	Reference	Difference against Kraemer-Sarnow
Kraemer-Sarnow	-	-
Wendriner	Schumann L:Kokerei-teer und Rohbenbol, Verlag von Ferdinand Enke in Stuttgart, 1940	+ 3 ÷ + 5
Ring and ball	-	+ 12 ÷ + 15
Ring and rod	Russian method	- 10
Cube in Air	ASTM:D 2319-66	+ 16 ÷ + 22
Cube in Water	ASTM: D 61 - 68	?
Mettler	-	+ 18 ÷ + 20
Boetius	-	- 4 ÷ - 5

The measurements by the Mettler's apparatus are based on the pitch melting principle and for high temperature measurements–up to 360°C– has been used the Boetius apparatus which is used usually for melting point determination. Softening point temperatures have been determined according to the pitch particles edge fused in the capillary. Pitch softening point composes softening points of its extraction fractions according to logarithmic equation:

$$\ln SP_{mix} = a_1 \ln SP_1 + a_2 \ln SP_2 + \ldots + a_n \ln SP_n$$

where: a = weight fragments of extraction cuts/fractions/
SP = softening points of extraction cuts /fractions/.

For example some points:

TS	30-35°C
TI-PyI	185°C
PyI-QI	265°C

QI 600-1100°C
Carbon 3650°C

The softening temperature range for QI materials is very wide due to the variability of molecule shape. It is clear from equation and fig. 3

$$SP = K.QI + 45.5$$

where: K < 4.2 for high-condensated QI
K > 4.2 for low-condensated QI

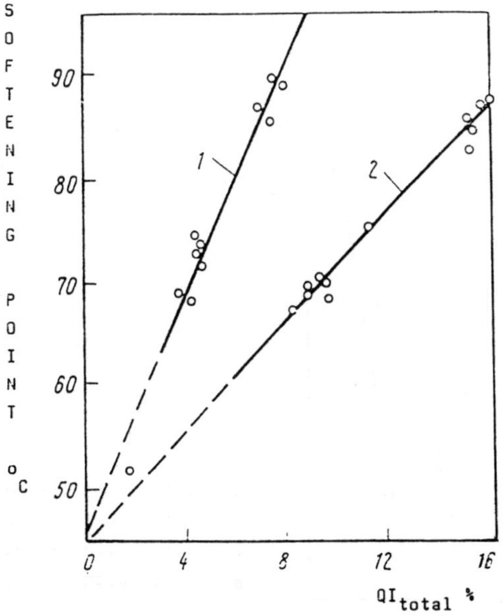

Fig. 3. Dependence of softening point on QI content. Line 1 shows pitch with 4% QI prim, line 2 shows pitch with 10% QI prim.

Pitches with primary QI content are more suitable for use as binders due to their ease of plastification.

4. Insoluble Resins

Solvent fractionation is the most important method for pitch characterization. The main insoluble resins are TI and QI which are coke precursors, additionally QI content has a special function at plugging capillars in the coke grains. It is possible to say that without QI content, pitch can be totally soaked into coke. The size and form of the QI particles are also very important. Preferred QI/TI ratio is 0.25 because formed pitch from these above-mentioned fractions influence anodes porosity (Fig. 4). Particle sizes are different: primary QI have a dominant size 1 micrometer; secondary QI size is bigger, usually at least 8 micrometers and can reach hundreds of micrometers and as a big mesophase spherules are situated in the low layer as a new physical phase. QI's are very useful for us and these ones are formed during coal coking in the gas-stream area of coke oven (Fig. 5). QI quality depends on coal chemical structure, type of the coking chamber, coking rate, total coking temperature and blowing air and the combustion products in the coking chamber. Primary QI are formed on the preheated wall and in the gas collector.

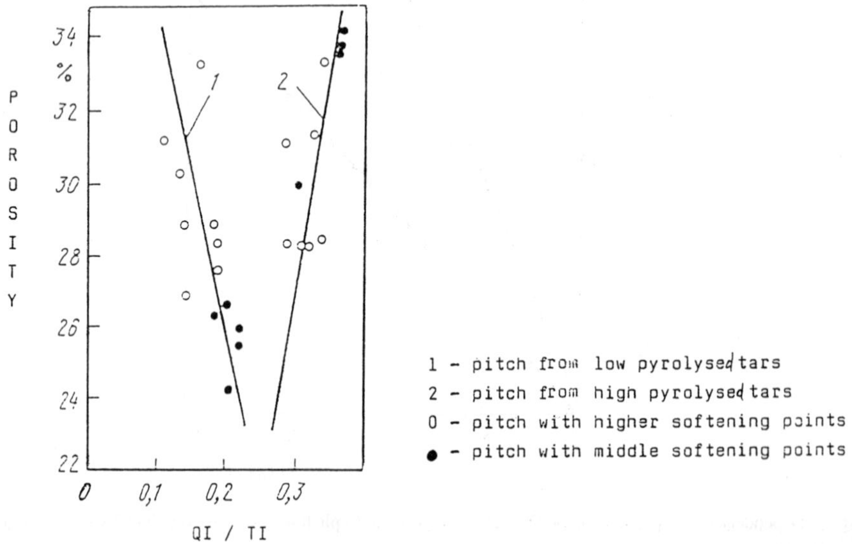

$$\text{coking value \%} = 1.26\ QI_{prim} + 0.54\ QI_{sek.} + 1.23\ \beta - \text{resin} + 0.25\ TS$$

Fig. 4. Porosity of the anode mass in proportion to QI/TI.

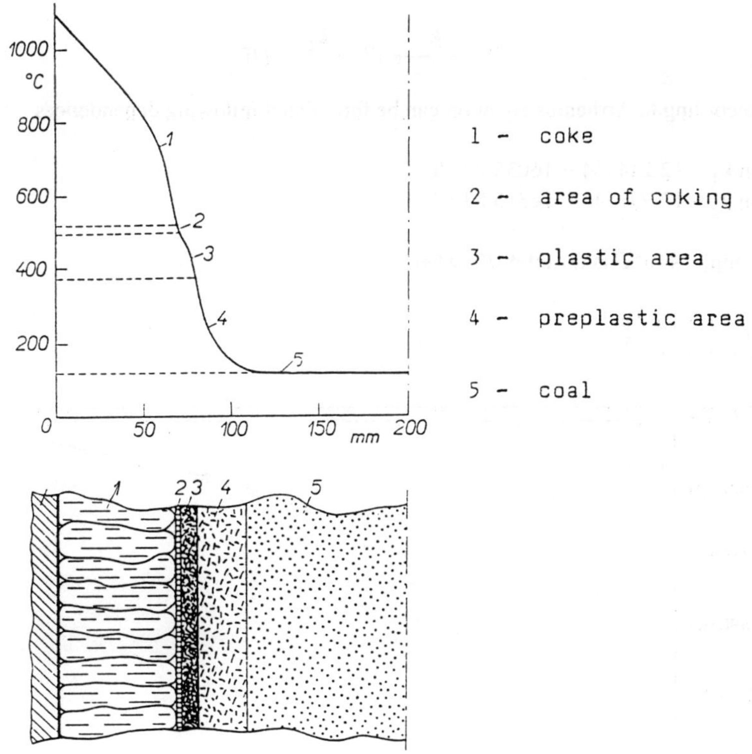

Fig. 5. Temperature spread in the coking chamber.

Aromatizaton temperature is above 800°C; proper aromatic compounds disproportionation between gas and high-temperature resin starts at 1000°C. It is needful to obtain B-resin content min 20%. But the question is, how high level can be obtained? B-resins are derived from toluene solubles (TS) by heat polycondensation, according to this scheme:

$$TS \xrightarrow{k_1} TI$$

where K_1 is rate constant of first degree reaction (Fig. 6), for pitch 70 at the temperature 410°C.

Simultaneously QI are formed and for this reason B-resin content is limited. These reactions are consecutive and can be expressed by following scheme:

$$TS \xrightarrow{k_1} TI \xrightarrow{k_2} QI$$

According to Arrhenius equation can be formulated following dependences

$$\text{Ln}\, k_1 = 12.244084 - 16035.23\,/\,T$$
$$\text{Ln}\, k_2 = 12.991544 - 16315.83\,/\,T$$

For example, 410°C is for pitch 70, where:

$$k_1 = 1.2\ 10^{-5}\ s^{-1}$$
$$k_2 = 1.85\ 10^{-5}.$$

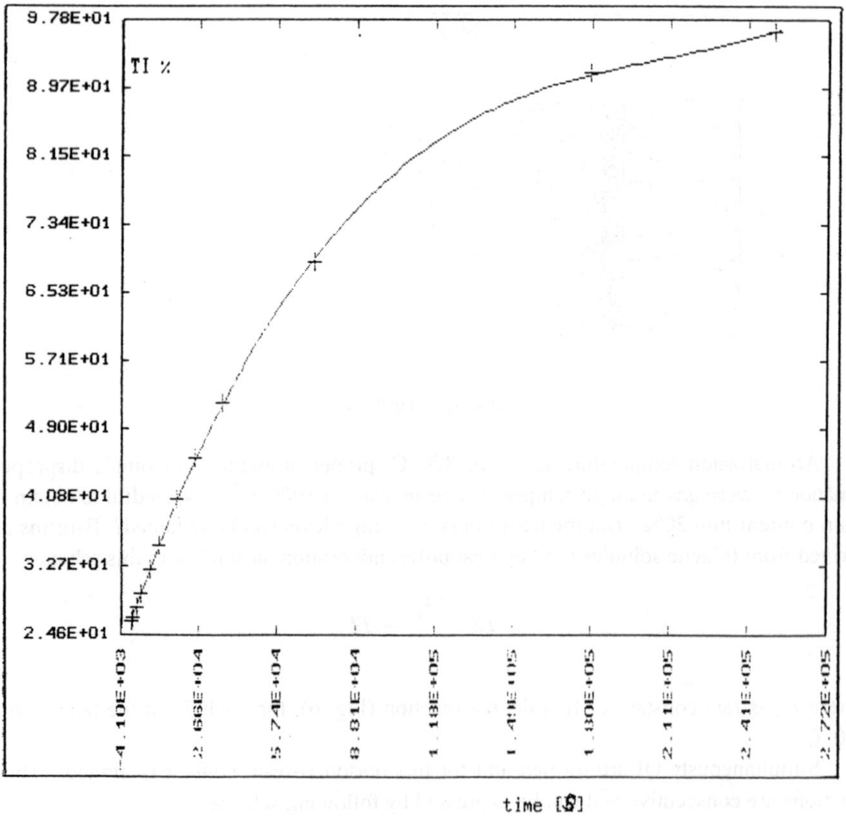

Fig. 6. TI increase in the pitch at 410°C.

During destillation of the tar at 390 – 400°C, changing TI and QI according to equations

$$TI_{pitch} = TI_{tar} \,.\, 2 + 12.6$$
$$QI_{pitch} = QI_{tar} \,.\, 2 + 2.5$$

First reaction is slow and thus is controlling. Secondary QI originates during heating and, following consecutive reactions, forms mesophase before finally transforming into coke.

5. Coking Values

Pitch transfer ability to coke depends on pitch aromaticity degree. This phenomena we express according to C_{at}/H_{at} ratio.

For example:

biphenyl	1.20
naphthalene	1.25
terphenyl	1.33
pyrene	1.60
pitch	1.77
sec. QI	2.40
prim. QI	5.20

We perform this phenomena determination as coking value determination by pitch sample pyrolysis. The results of this method depend on precise conditions of analyses and we used the Alcan method. Sample (1g) was pyrolysed at 550°C for 2.5 hours in the crucible under coke layer. Coke values calculation have been explored from values of primary and secondary QI, B-resins and toluene solubles (Fig. 4). For primary QI determination two methods have been used. The first method uses softening point values. Low slope of primary QI softening point has been observed. The second method determines primary QI on the base of hydrogen content. Primary QI contain 1.5% H, versus secondary QI with 3.5% H. This titled method is more precise but much more laborious. The primary and secondary QI content in Deza's pitch is showed on sample determination:

(17.5.1993, sample 80)

Kraemer-Sarnow softening point	83.8°C
QI hydrogen content	1.61 wt.%
Total QI content	7.8 wt.%
Primary QI content from H content	7.34 wt.%
Primary QI content from softening point	6.0 wt.%
Alcan coking value	54.92 wt.%
TI content	27.6 wt.%
Viscosity at 160°C	729 $mPa.s^{-1}$

6. Literature

1. H. G. Franck and O. Wegener (1958). *Brennstoff-Chemie*, Bd. 39, Nr. 13/14, p. 195.
2. M. Sakai, M. Yoshihara and M. Inagaki (1981). *Carbon,* **vol. 19,** pp. 83-87.
3. V. I. Šuvalov and V. V. Močalov (1983). *Koks I chimija,* Nr. 36-38.
4. M. H. Wagner, I. Leticia, S. Giannini and G. Wilhelmi (1986). *Carbon.*

Fabrication of Carbon Artifacts, Fibers and C-C Composites

THE RELATIONSHIP BETWEEN SURFACE TENSION, VISCOSITY, PENETRATION ABILITY AND GROUP COMPOSITION OF COAL TAR PITCHES

M. KALOČ, M. ŠVIČEK, and E. LINZER
Faculty of Metallurgy and Material Engineering
Technical University of Mining and Metallurgy
tř. 17, listopadu, 708 33 Ostrava - Poruba, Czech Republic

ABSTRACT

Information about the important properties of coal tar pitches as binders for composites and other industrial carbons is contained in this contribution.

The surface tension, as measured by the lying drop method and the maximal pressure of a gas, was used to model the penetration ability of binder samples that contained various contents of QI and TI matter. Very significant is thermostability index, which was also measured and compared to the viscosity parameter. The experimental results show that the relations between the mentioned physical-chemical properties of pitches are difficult to establish because more factors influence the binders' behavior than have been considered here.

1. General

With regard to a specific character of the production of carbonaceous materials, binders play an important role which influences properties of the carbonaceous artifact.

Characteristic features of the binders are as follows[1]:

- rheological properties which promote good wettability of the warp at working temperatures
- ability to form a film around the warp
- ability to be absorbed on the warp surface
- ability to deform together with the warp and to exhibit good compressibility.

Wettability depends upon viscosity, group composition and temperature of the binder. Wetting intensity of the warp is related to the type of material forming the warp, especially important is the structure of the surface.

These issues have been discussed, along with other matters, in two studies[2,3].

2. Effect of Temperature

In the course of heating to working temperatures coal tar pitch undergoes pyrolysis, which results in alterations in its chemical composition, above all the C_{at}/H_{at} ratio increases.

Through this process, the matter makes a transition to a highly polyaromatic state and subsequently to coke. Schematically, the whole complex of changes can be seen in Fig. 1.

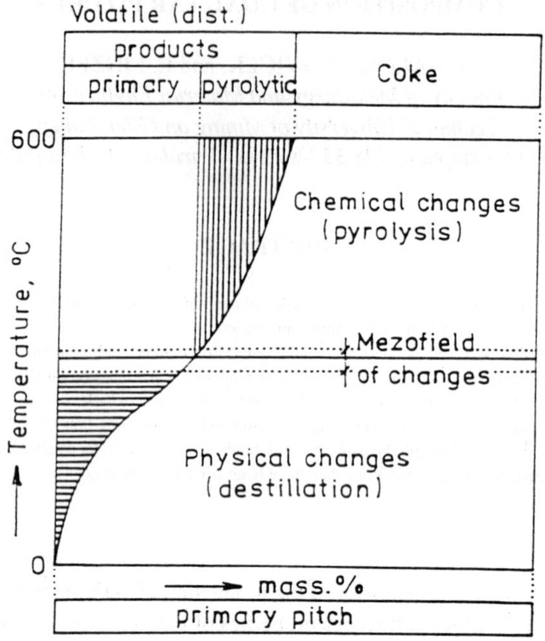

Fig. 1. Changes in coal tar pitch.

2.1. *Solubility Changes*

By using suitable solvents, pitch can be divided into components of various molecular weights and structures. Toluene, pyridine and chinoline are usually utilized that decompose the original matter into the various fractions: toluene soluble (ToS); pyridine fraction (FPy), soluble in pyridine but insoluble in toluene (ToI-PyS); chinoline fraction (FCh), soluble in chinoline but insoluble in pyridine (PyI-QS); and the share insoluble in chinoline (QI). Owing to action of temperature, material initially soluble in toluene can become insoluble in toluene, pyridine and in chinoline. Newly developed low-molecular substances, both original and formed, are simultaneously formed.

These solubility changes result from an increase in molecular weights, as presented in Fig. 2.

Based on another of our studies[4], it is possible to state that the control reaction of the whole process is conversion of ToS → ToI → FPy. The rate of FPy forming is minimal; greater rates occur with the origin of FCh and QI.

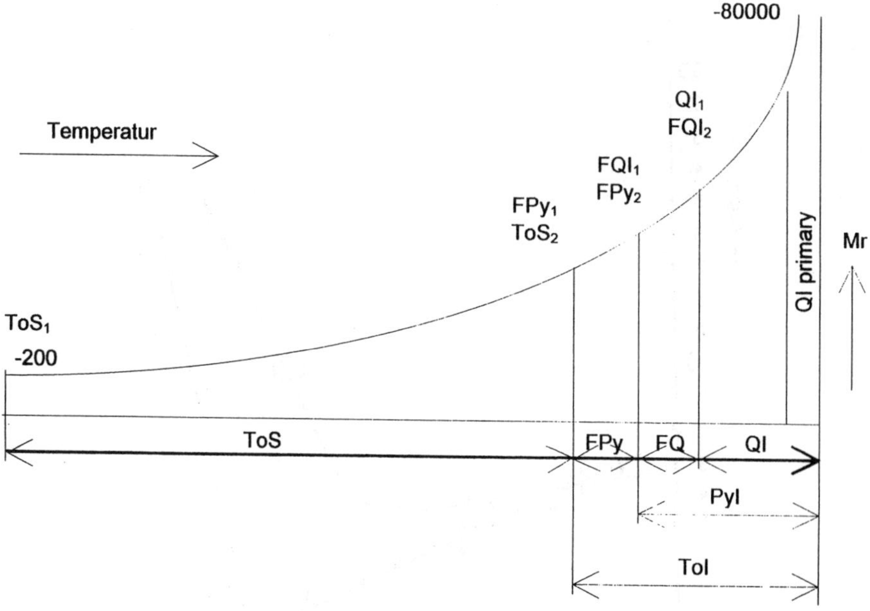

Fig. 2. Solubility changes.

Pyrolytical transformations can also be monitored by changes in the softening point. This subject is discussed in the literature[5].

2.2. *Viscosity Changes*

The effect of pyrolysis on viscosity can be studied using a vibration viscosimeter. For details see reference 6 and 7.

We shall limit ourselves to statement that viscosity is influenced most by the QI, especially QI of the secondary type as proved by data in Fig. 3, which shows the temperature dependence of viscosity in the Frenkel relation:

$$\log \eta = \frac{E_\eta}{2{,}303RT} + \log B\eta \qquad \textbf{(1)}$$

where B is a constant.

The point of inflexion indicates, with high probability, a moment of the origin of a new phase in the matter studied[8].

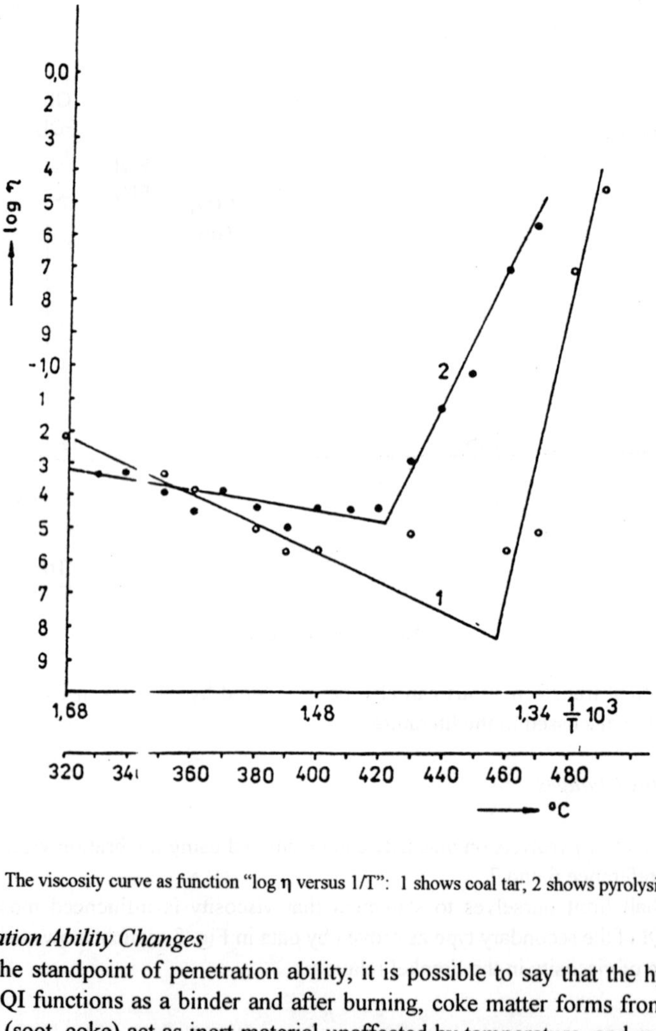

Fig. 3. The viscosity curve as function "log η versus 1/T": 1 shows coal tar; 2 shows pyrolysis oil.

2.3. Penetration Ability Changes

From the standpoint of penetration ability, it is possible to say that the hydrocarbon component QI functions as a binder and after burning, coke matter forms from it. Other components (soot, coke) act as inert material unaffected by temperature, and remain in the coke. These components affect the strength, thickness and mass of the formed binding coke. However, in the original state, the QI influences the penetration of pitch into the warp because it deposits on its surface and forms thus a filtration layer that makes the penetration of the ToI component of pitch into the underlying rocks difficult. The QI content of the binders is therefore important and seems to sensitively influence the utilization of this binder.

For the aim of studying penetration ability, an apparatus was constructed, a schematic of which is presented in Fig. 4. In total, seven types of pitch were investigated, the properties of which are shown in Table 1. The penetration depth of ToI material into the underlying cloth material was evaluated microscopically; the results obtained are illustrated schematically in Fig. 5.

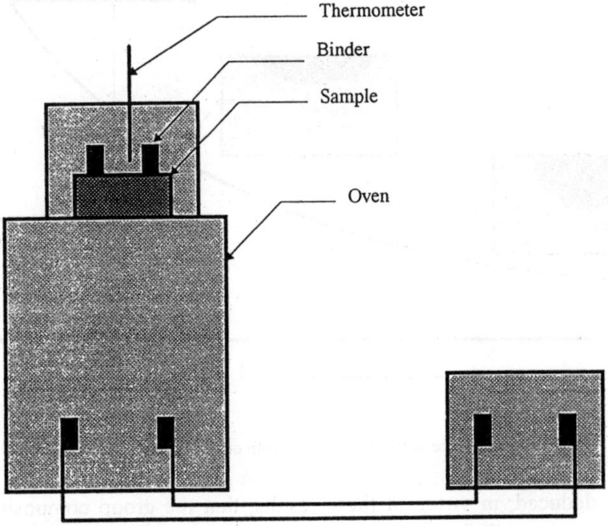

Fig. 4. Schematic of apparatus for measuring penetration ability.

Table 1. Properties of Pitch Samples

Pitch properties	Cp 900	Cp 1200	SM70	SM182	SM188	SM183	SM187
Softening point (°C)	88.4	115.1	87.8	107.4	110.7	124.2	142.0
ToI (% mass.)	13.1	14.59	25.8	32.4	39.9	42.9	50.9
QI (% mass.)	0.26	0.31	8.1	9.3	13.6	15.6	19.3
Viscosity 160°C (mPa.s)	/	/	190	727	1412	4803	48501
Volatile matter (% mass.)	/	/	15.1	13.7	10.1	8.2	7.3

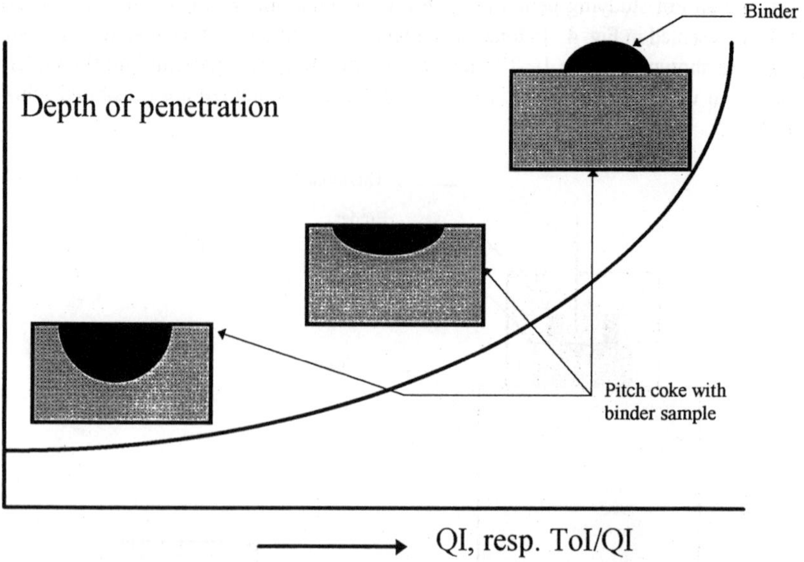

Fig. 5. Dependence of the penetration depth on QI, resp. ToI/QI content.

It can be deduced, in virtue of these results, that the group composition expressed especially by parameters ToI and QI as well as their mutual ratio markedly influence the penetration characteristics of the pitch.

3. References

1. M. Kaloč, *Průmyslový uhlík - ES VŠB Ostrava* (1992) 197 p.
2. M. Kaloč and M. Janík, *Hutnické listy* **5** (1982) 248.
3. M. Kaloč, D. Košťál and M. Janík, *Ext. Abstr. Carbon 94* (1994) 194.
4. D. Košťál, M. Janík and M. Kaloč, *Chemický průmysl* **39/64** (1989) 32.
5. D. Košťál, M. Janík and M. Kaloč, *Chemický průmysl* **41/66** (1991) 91.
6. E. Linzer and J. Kaluža, *Abstracts of Carbon and Carbonaceous Composite Materials: Structure Property Relationships*, (1995).
7. D. Košťál and M. Kaloč, *Abstracts of Carbon and Carbonaceous Composite Materials: Structure Property Relationships*, (1995).
8. M. Kaloč, M. Janík and Z. Adolf, *Hutnické listy* **34** (1979) 779.

THE MEASUREMENT OF DYNAMIC VISCOSITY OF COAL TAR USING THE VIBRATION VISCOMETER

E. LINZER and J. KALUŽA
VSB-Technical University of Ostrava
708 33 Ostrava-Poruba, Czech Republic

ABSTRACT

With respect to the usability of coal tar for the preparation of composite materials, an understanding of viscous characteristics is necessary. Besides the running property, which is a rather technological parameter embracing soaking attributes of the underlying material used and viscosity of the melted state, the number of characterization methods is relatively limited. From the techniques that are available, the rotation and the vibration method are the most appropriate for measurements on coal tar. In our department we use a vibration viscometer for measuring dynamic viscosity. Growing requirements regarding sensitivity and reproducibility of measurements led to changes in the existing construction of the viscometer. The main differences involved the resolution damping oscillate of vibration measuring body and bringing the system body-drive coil to a state of electromechanical resonance. In this paper, we present the results of dynamic viscosity measurements of coal tar used as a binding agent for industrial carbon.

1. Introduction

The optimal uses of coal tars are conditioned by the knowledge of the temperature dependence of their physical properties.

Important properties are the temperature of softening, density, surfaces tension and viscositiy. An important technological parameter is the running property, which combines the soaking attributes of the underlying material with the viscosity of the melted tar.

The dynamic viscosity is expressed by Newton's equation:

$$\eta \bullet \frac{dv}{dn} = \tau \qquad (1)$$

The properties of the liquid phase are a reflection of the resistance of the environment to the mutual moving of neighboring layers within the volume of liquid. This situation is conditioned by the principle of the experimental method used to measure the temperature dependent dynamic viscosity.

2. Experimental and Results

The number of methods available to measure the behavior of the tar in the melted state is very limited. Methods based on rotation or vibration can be applied for low to medium viscosities, and for the region of high viscosities the penetration method can be used.

The Faculty of Metallurgy and Materials Engineering, VŠB-Technical University of Ostrava, are using the vibration viscometer for the measurement of dynamic viscosity. The arrangement of the equipment as shown in Fig. 1 is mainly utilized for studying the viscosity of the melted metallurgical slags[1,2] at the interval of the viscosities 0–7 [Pa s].

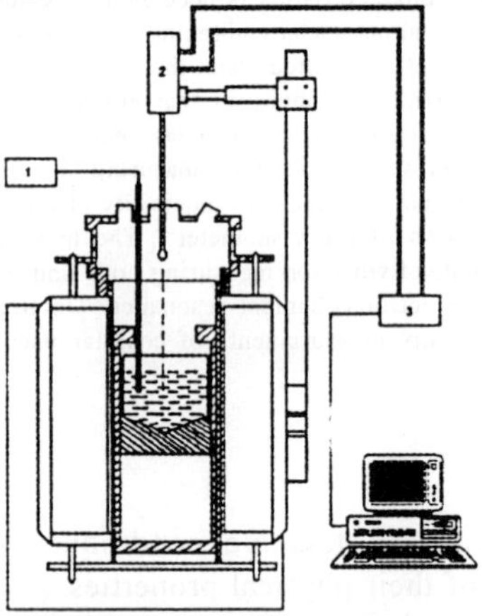

Fig. 1. The equipment for measuring viscosity using the vibration method: position 1, temperature measurement; position 2, driving coil and measuring transducer, position 3, electrical circuits.

In order to improve the sensitivity and the reproducibility of the measurement the vibration equipment was redesigned. The main difference involves the measurement of the natural frequency of oscillation of the measuring body. It consists of bringing the system and the body-drive coil to a state of the electro-mechanical resonance. In this way the exactness of the measuring was increased and the range of the measuring dynamic viscosity was extended.

Before taking any measurements the viscometer was calibrated. This was done by determining the mutual damping of the vibration body and the dynamic viscosity of the liquid in which the body is plunged (Fig. 2).

In this work, the vibration viscometer was used to determine the temperature dependent dynamic viscosity of "hard" coal tar such as that used in the preparation of industrial carbon.

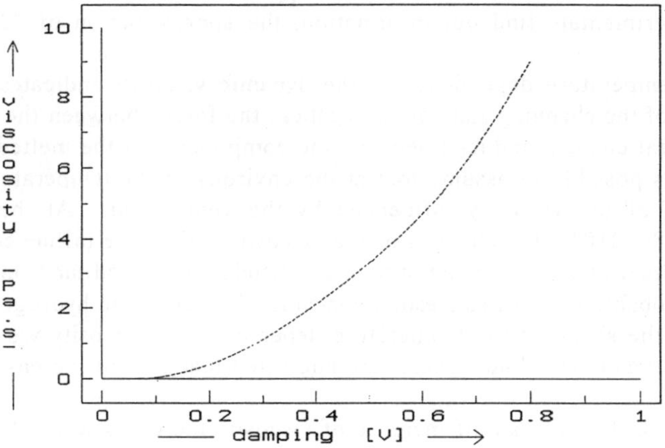

Fig. 2. The calibration's graph of the vibration viscometer.

The opposite arrangement of the measuring apparatus (Fig. 1) the pot furnace with fluently regulation of the temperature was used at the measuring. The temperature was measured by Hg-thermometer.

The melting temperature of the tar was 300°C. The dynamic viscosity was measured with the decreasing temperature–at the velocity of cooling 2°C/min, the values of damping oscillation was subtracted from the corresponding value of the dynamic viscosity. The obtained temperature dependence is shown in Fig. 3.

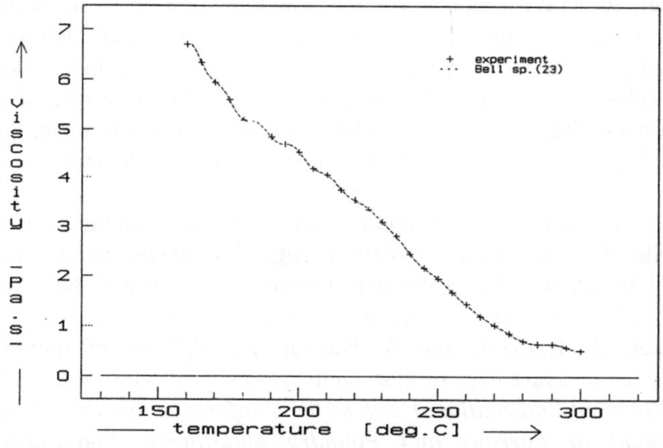

Fig. 3 . The temperature dependence of the dynamic viscosity coal tar.

The experimental measured point was approximated by Bell's cubic spline. As all 28 experimentals find out information, the approximation of 23 splines was used.

The temperature dependence of the dynamic viscosity indicates the possible influences of the changing state of aggregation, the forces between the particles and the structural changes induced between the components of the melted tar.

This is possible to assume that at the environs of the temperature 300°C are the course of the viscosity influenced by the venting tar. At the temperature interval » 280–210°C the changes of the viscosity reflect the influence of temperature on mutual affecting among structural particles of melted tar from temperature » 235°C probable with the suggestion nucleation higger cyclic hydrogen carbonates.

From the shape of the temperature dependence of viscosity we assumed that between 210 and 160°C new phases are nucleated that exhibit increasing molecular mass.

We note that the rate of increase of viscosity increases when the temperature decreases from » 180°C.

3. Conclusion

In conclusion we confirm that the resonant vibration method for measuring viscosity is a useful variant of the vibration viscometer. The results can be interpreted to indicate both the structure changes and the changes between the forces between the particles in the melted tar.

4. References

1. E. Linzer, R. Krayzel and A. Buchal, The contribution to study of the heterogeneous melts, in Symposium III. International conference "*Modeling the metallurgical processes with application at industry*." Komorní Lhotka (1993) 93.
2. R. Krayzel and E. Linzer, The increasing exactness measuring of the viscosity blast-furnace slag, in Symposium III. International conference "*Modeling the metallurgical processes with application at industry*." Komorní Lhotka (1993) 90.
3. J. Kaluža and E. Linzer, The modification at the other elaboration the principle of the vibration viscometer, in *Proceedings of International Scientific Conference of VŠB-Technical University of Ostrava*, Session No. 20 - Automation of Technological Processes, Czech Republic (1995) 20.
4. E. Linzer, R. Krayzel, and A. Buchal, Possibilities of increasing of the accuracy for measurement of slag melts viscosity by vibration viscosimeter, in *Proceedings of International Scientific Conference on the Occasion of the Department of Ferrous and Foundry Metallurgy*, Metallurgical Faculty Technical University Košice, Slovakia-Tálle (1994) 161-166.

CARBON-CARBON COMPOSITES BASED ON A POLYIMIDE MATRIX ITA

V. E. YUDIN, M. YA. GOYKHMAN, A. V. GRIBANOV, G. N. GUBANOVA, and V. V. KUDRYAVTSEV
Institute of Macromolecular Compounds, Russian Academy of Sciences, V.O. Bolshoy pr., 31, St. Petersburg 199004, Russia

K. BALÍK and P. GLOGAR
Institute of Rock Structure and Mechanics, Academy of Sciences of the Czech Republic V. Holešovičkách 41, 182 09 Prague 8, Czech Republic

ABSTRACT

It is shown that polyimide matrix ITA in carbon fiber reinforced composite can have a maximum carbon yield up to 80% after thermolysis at 1000°C in vacuum. Chemical synthesis of ITA is based on the polycondensation of acetyl derivatives of diamine with dianhydrides with following cross-linking at 300–350°C. Carbon-carbon composites based on ITA binders have a monolithic structure with high shear modulus of 8–9 GPa. It is possible to suggest that high carbon yield of ITA is due to the monolithic structure of the carbon-carbon composite, which results from co-carbonization between the fiber and the matrix by the free-radical mechanism.

1. Introduction

Polyimides (PI), because of their high thermal stability, have recently attracted increasing attention as promising materials for carbonization and obtaining carbon-carbon (C/C) composites. In 1987 E. Fitzer[1] mentioned PI matrix QX-13 as a possible candidate for C/C composites. The reasons for using PI as matrix precursor for C/C may be as follows:

- The products of PI thermolysis or combustion are no more toxic than those of wood. The products are CO, CO_2, H_2O, and N_2.
- High thermal stability and high carbon yield up to 70%[2].
- It is possible to carbonize PI films without any cracks and pores and to obtain a material with high graphitizability[3,4].

In the present work, PI with different molecular structures were synthesized, particularly taking into account their compatibility with carbon fibers, and the carbonization behavior of these polymers in carbon fiber reinforced plastics (CFRP) were investigated. The carbon yield of the PI matrix in C/C composites is discussed from the possible process of chemical bonding between the fiber and the matrix.

2. Experimental

Some new types of PI resins named ITA, IPO, PEI, and IPM (Fig. 1) have been developed at the Institute of Macromolecular Compounds (Russia). These PI resins are analogues of such well-known binders as QX-13, PMR-15, and LARC-TPI, but can differ from these commercially available polymers in other chemical structures of the dianhydride and diamine components.

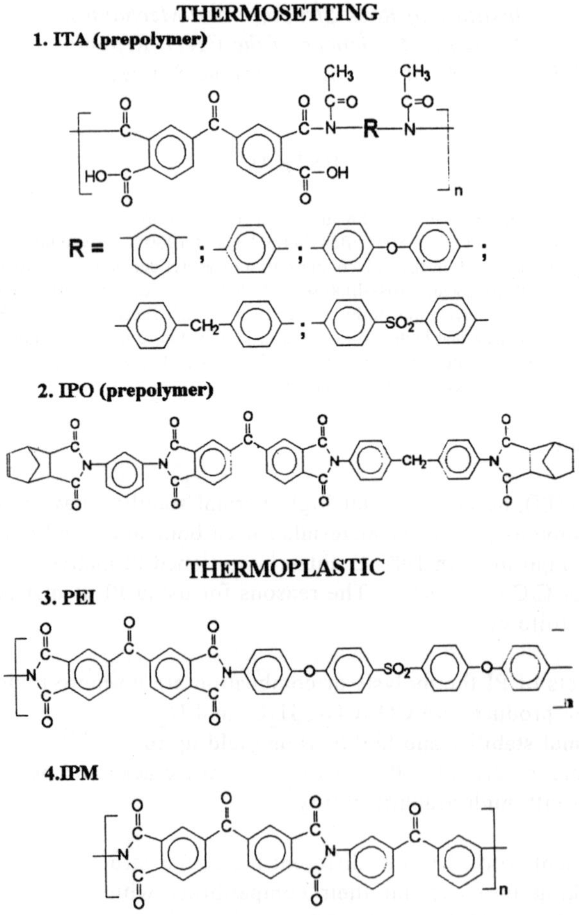

Fig. 1. Molecular structure of the polyimides synthesized for CFRP.

Surface treated commercial carbon fiber (ELUR, Russia) is used for the C/C composite. The main mechanical properties of ELUR fibers are as follows: the tensile strength at 10 mm length is about 2.2 GPa and the elasticity modulus is about 200 GPa.

The thermomechanical properties of "green" carbon fiber reinforced plastics (CFRP) based on ELUR and different PI resins are represented in Table 1. Fiber volume fractions V_f of these CFRP differ and depend on the melt viscosity of the binders in their initial state before curing. For such thermosetting binders as ITA and IPO, V_f is high and equals 58–60% because they are based on oligomer's technology. For PEI and IPM thermoplastics with high melt viscosity V_f is lower and is not more than 50–53%.

Table 1. Thermomechanical Properties of CFRP based on the ELUR Fiber and PI Matrices before Thermolysis

Binder type	V_f (%)	σ_b (MPa)	T_g (°C)	G_{1C} (J/m^2)	T_o (°C)	Carbon yield (%)
ITA	58	1270	370	230	570	80
IPO	60	1320	350	170	400	55
PEI	48	1050	230	950	530	60
IPM	53	1170	255	630	500	60

In accordance with V_f, the bending strength σ_b of CFRP is higher for CFRP based on ITA and IPO thermosetting resins. In contrast to bending strength, the interlaminate toughness G_{1C} of CFRP based on PEI and IPM thermoplastics is about 2–4 times higher than that for analogous CFRP based on thermosetting resins. Glass transition temperature T_g of binders for CFRP depends mainly on the chemical structure and cross-linking of the PI and, of course, is higher for thermosetting resins.

The temperature at which weight loss begins T_o depends on the chemical structure of PI and is highest for ITA (Table 1). The carbon yields of IPO, PEI, and IPM are typical for PI and lie in the range of 55–60% after thermolysis up to 1000°C at vacuum (10^{-2} bar). The highest carbon yield is for ITA in carbon plastics and attains 80%. This result makes the ITA resin a good candidate for C/C composites.

3. Results and Discussion

The process of CFRP thermolysis based on PI is shown in Fig. 2. It can be seen that for ITA binders, the weight loss of matrix ITA in CFRP is lower than that for other PI matrices, including PEI, IPM, and IPO. The matrix weight loss in the composite was calculated using the equation:

$$\frac{\Delta M_m}{M_m} = \frac{\Delta M_c}{M_c} \times \left(1 + \frac{\rho_f}{\rho_m} \times \frac{V_f}{V_m}\right) \tag{1}$$

where $\Delta M_c/M_c$ is the weight loss of CFRP, ρ_f and ρ_m are the densities of the fiber and the matrix, respectively, and V_f and V_m are the volume fractions of the fiber and the matrix in CFRP. For carbon fiber ELUR, ρ_f is equal to 1.7g/cm^3, and for ITA, ρ_m is equal to 1.3g/cm^3.

The comparative analysis (Fig. 2) of shear modulus change during thermolysis of CFRP shows that for CFRP based on ITA, it increases during thermolysis and the value of the shear modulus is approximately 8–9 GPa at 1000°C. The shear modulus of CFRP based on PEI decreases and this decrease during thermolysis is

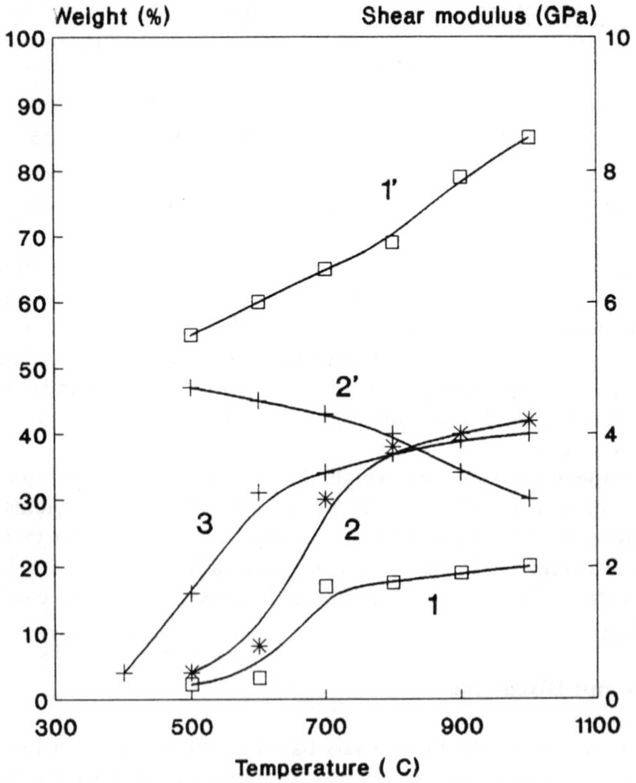

Fig. 2. Changes of weight loss (1–3) of PI matrices according to Eq. 1 and shear modulus (1', 3') of CFRP based on PI matrices with heat treatment temperature on the polymers: 1,1' - ITA, 2,2' - PEI; 3 - IPO.

quite typical of most CFRP based on thermoplastic PI and thermosetting (phenolic, epoxy, IPO, etc.) resins. Thus, it is possible to assume that during the thermolysis of CFRP based on ITA, the monolithic structural type of C/C composites takes place. The ratio of the Young's modulus E_{11} to the shear modulus G_{12} is 120 GPa/5.4 GPa ≈ 22; after thermolysis at 1000°C, E_{11}/G_{12} is 90 GPa/8.5 GPa ≈ 11 for C/C composites based on ITA.

The monolithic type of structure of C/C composites based on ITA also follows from microscopical investigation. Fig. 3a shows the failure surface of the C/C composite based on ITA after bending strength testing. The failure is quite brittle with crack propagation transverse to fiber direction without fiber pullout. Detailed investigation of the failure surface of C/C composite by using SEM (Fig. 3b) shows a very good matrix/fiber bonding with almost no gaps around the fiber circumference.

Acoustic emission (AE) analysis of CFRP based on ITA before and after thermolysis at 1000°C demonstrates a different pattern of their failure during loading along the fiber axis (Figs. 4a, b). For CFRP before thermolysis, the typical pattern is that powerful peaks of AE signals occur from the beginning of sample loading up to its catastrophic failure (Fig. 4a). This fact indicates the brittleness of the matrix and low delamination toughness of CFRP based on ITA. After thermolysis, the failure pattern of this C/C composite (Fig. 4b) differs from that of "green" CFRP. One crack at the beginning of sample loading can be sufficient for its catastrophic failure. Thus, the high level of matrix/fiber adhesion in the C/C composite does not prevent the spread of brittle cracking.

It is possible to suggest that the reason for high carbon yield in C/C composites based on ITA is the specific interaction between the fiber and the matrix, which in turn follows from the monolithic structure of C/C composites.

In fact, in comparison to the general scheme of PI synthesis, the preparation of ITA oligomers and polymers is unusual (Fig. 5). In the case of ITA, acetyl derivatives of diamine can interact with dianhydride up to oligoimides formation. At a temperature of more than 300°C, these oligoimides can cross-link by the free-radical mechanism.

Simultaneously in the temperature range of 300–350°C and higher, as can be seen from mass spectrometry data (Fig. 6), some gases (CO, CO_2, and H_2O) evaporate from the surface of the ELUR fibers. These gases are the products of the decomposition of functional groups (hydroxalic oxide, dianhydride, etc.) on the fiber surface. After heating to 300°C, the occurrence of free radicals on the fiber surface is possible and their existence time is 1–10 min. This time may be sufficient for cross-linking between the fiber and the matrix by the free-radical mechanism at 300–350°C. Thus, the conditions of curing of the ITA binder and previous heat treatment of fiber can profoundly affect the mechanical properties of CFRP. Indeed, after the heat treatment of ELUR fiber at 250°C for 2 h and the removal of active functional groups from the fiber surface, the bending strength of

A. Optical microscope magnification 10x

B. Scanning electron microscopy magnification 2500x

Fig. 3. Optical microscope (a) and scanning electron microscope (b) photographs of the fracture surface of the C/C composites based on ITA after thermolysis in vacuum. Sample is loaded in flexure.

A

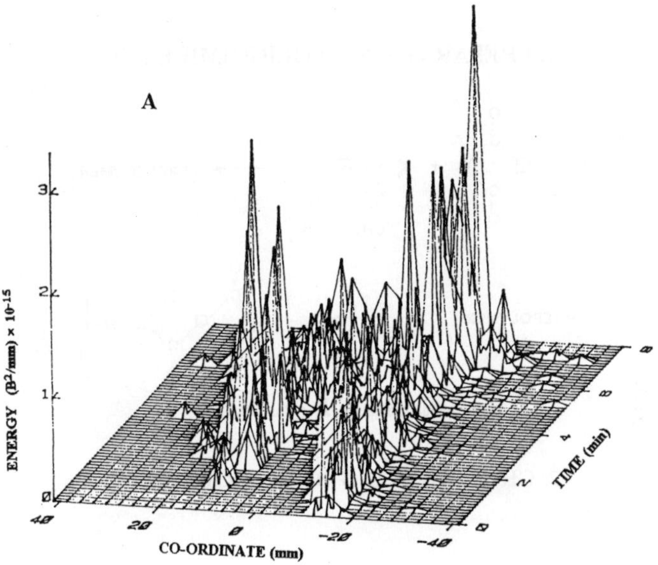

B

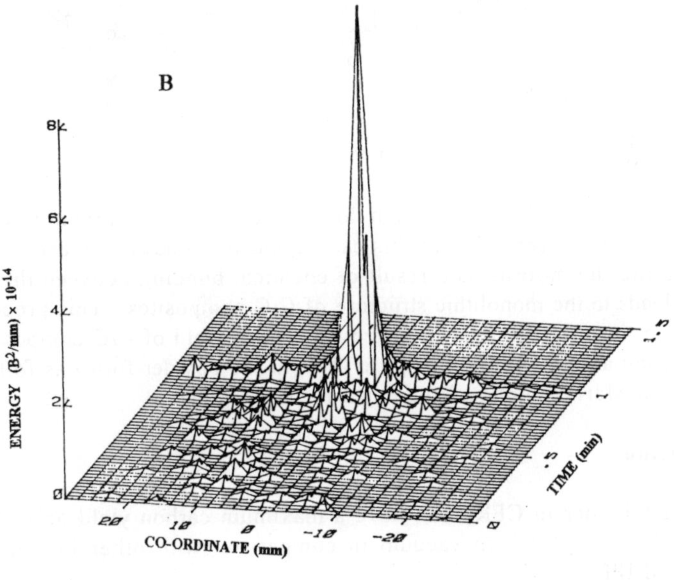

Fig. 4. AE failure's patterns of CFRP based on ITA before thermolysis (a) and C/C composite based on ITA after thermolysis at 1000°C (b). Samples are loaded along fiber direction.

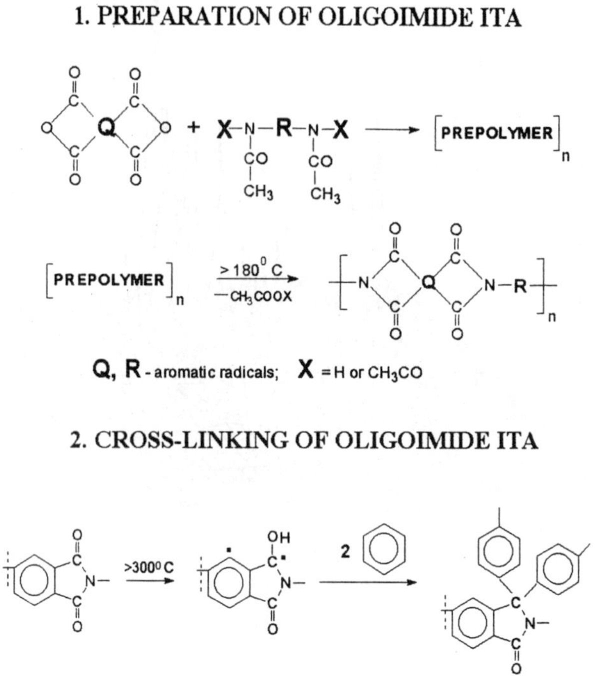

Fig. 5. General reaction scheme for ITA polyimide.

CFRP based on ITA becomes much lower than that for carbon fiber without temperature treatment (Fig. 7). In this case, good fiber/matrix adhesion which can increase during thermolysis as a result of chemical bonding between the fiber and the matrix leads to the monolithic structure of C/C composites. This process of co-carbonization, in turn, can lead to improved carbon yield of C/C composites based on ITA in comparison with that of pure ITA (in the powder form) as follows from Markovic and Marinkovic[5] and the experiments (Fig. 8).

4. Conclusion

- The ITA binder in CFRP can have a maximum carbon yield up to 80% after thermolysis at 1000° in vacuum in comparison with other PI binders: IPO, PEI, and IPM.
- C/C composites based on ITA binders have a monolithic structure with high shear modulus of 8–9 GPa but low toughness.

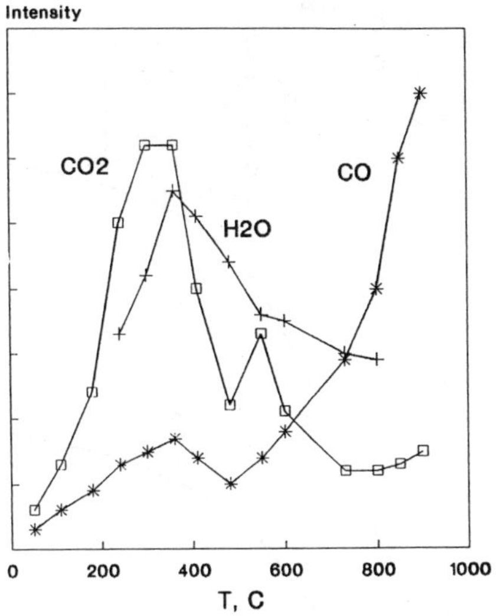

Fig. 6. Changes of decomposition products measured by mass spectrograph with heat treatment temperature.

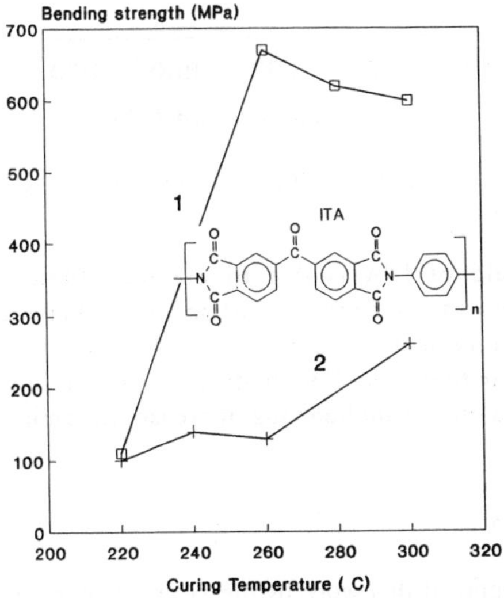

Fig. 7. Curing temperature dependence of bending strength for ITA polyimide composites based on untreated (1) and heat treated at 350°C/2 h carbon fiber ELUR (2).

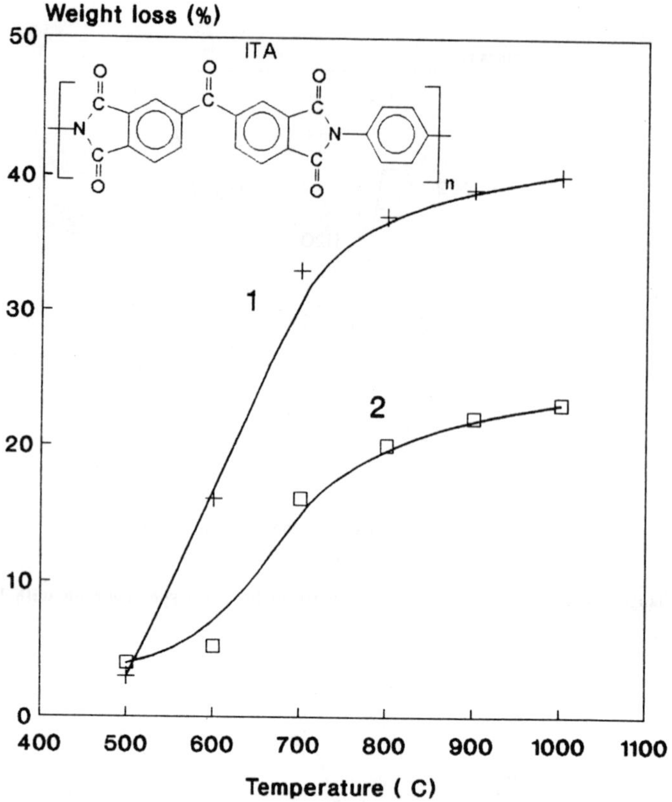

Fig. 8. Thermolysis of pure ITA (1) and ITA in CFRP (2) at vacuum. Weight loss of ITA in CFRP (2) is calculated by using Eq. 1.

- High carbon yield of ITA is due to the monolithic structure of the C/C composite which results from co-carbonization between the fiber and the matrix by the free-radical mechanism.
- It is necessary to optimize the fiber/matrix interaction in C/C composites based on ITA for attaining simultaneous increase in carbon yield and fracture toughness.

5. Acknowledgments

The financial support of this work by the Grant Agency of the Czech Republic, grant no. 104/94/1789, is gratefully acknowledged.

6. References

1. E. Fitzer, *Carbon* **25** (1987) 163.
2. M. I. Bessonov, M. M. Koton, V. V. Kudryavtsev and L. Laius, *Polyimides–Thermally Stable Polymers* (Consultants Bureau, Plenum Publishing Corp., New York, 1987) 199.
3. M. Inagaki, T. Ibuki and T. Takeichi, *J. of Appl. Polym. Sci.* **4** (1992) 521.
4. M. Inagaki. Y. Hishiyama and Y. Kaburagi, in *Proc. Int. Conf. Carbon '94*, Granada, Spain (1994) 546.
5. V. Markovic and S. Marinkovic, *Carbon* **18** (1980) 329.

SYNTHESIS OF SILICON CARBIDE WHISKERS FROM RICE HUSK + CARBONACEOUS BINDER MIXTURE

V. RAMAN, O. P. BAHL, and V. K. PARASHAR
Carbon Technology Unit
National Physical Laboratory, New Delhi -110012, India

ABSTRACT

Silicon carbide whiskers were synthesized from burnt rice husks + phenolic resin, raw rice husks + phenolic resin and burnt rice husks + pitch mixtures. The mixtures of rice husks and carbonaceous binders were pyrolyzed at 1400°C in argon atmosphere to get a mixture of silicon carbide whiskers, silicon carbide particulate, free silica and carbon. The removal of silica and carbon by chemical methods yielded silicon carbide whiskers and particulates. The characterization of the products obtained at 1000°C, 1400°C and at 1500°C and of the precursors by scanning electron microscopy (SEM), X-ray diffraction studies and also by Fourier infrared spectroscopy (FTIR) is reported. The possible reasons for the difference in morphologies and the advantage of using an external carbon source are discussed.

1. Introduction

Silicon carbide whiskers (SiC_w) are widely employed as reinforcing materials for the fabrication of ceramic and metal matrix composites due to their high strength[1,2]. Silicon carbide whiskers which also serve as high temperature resistant materials are generally synthesized by chemical vapor deposition or by the carbothermal reduction of silica or silicon containing compounds[3,4]. Pyrolysis of rice husks, an agrowaste, also yields SiC along with silicon carbide particulates (SiC_p). This process has received much attention since it is highly economical. Several reports have been published on the synthesis of SiC from rice husk and also on the effect of various parameters on the process[5–10]. The present investigation reports the synthesis of SiC_w from rice husk without employing catalysts at relatively low temperature (1400°C).

2. Materials and Method

The rice husk used in the present study was obtained from West Bengal and was found to contain 15% silica. The rice husk as such was burnt at 700–800°C in a tubular furnace fitted with silica tube with one end closed. The burnt rice husk was analyzed for carbon and silica content and then mixed with excess of carbonaceous binders like phenolic resin or pitch. Fifty grams of the mixture were pyrolyzed at 1400°C in a Glowbar furnace fitted with alumina tube under a flowing argon atmosphere. A heating rate of 175°C/h was used up to 1100°C and 50°C/h from 1100–1400°C. The samples were held at 1400°C for two hours and were cooled in flowing argon.

The pyrolyzed products were heated at 700°C and then treated with hydrofluoric acid for the estimation and elimination of excess carbon and silica present in these samples.

A few samples were prepared by directly pyrolyzing raw rice husk mixed with phenolic resin at 1400°C and also by carbonizing the burnt rice husk and phenolic resin mixture at 1000°C before subjecting it to carbothermal reduction. The carbonized product was analyzed for carbon and silica content before pyrolyzing at 1400°C.

The pyrolyzed products were characterized by scanning electron microscopy (SEM), (Jeol 35CF using a 25 KV accelerating voltage) and the X-ray patterns were obtained using a Siemens D-500 X-ray diffractometer. The Fourier transformed infrared spectra of the samples were recorded using a Nicolet spectrometer with KBr pellets.

3. Results and Discussion

The chemical analysis of the pyrolyzed products showed that SiC contents differ depending upon the precursor materials used and processing conditions employed. The total yield varied from 5 to 10% in these samples. The yield of SiC_w from directly pyrolyzed rice husk + phenolic resin mixture and burnt rice husk + pitch mixture was found to be small when compared to the SiC_w yield obtained with other precursors employed in the present investigation.

The X-ray diffraction patterns of the pyrolyzed products are given in Fig. 1. The X-ray studies showed that the pyrolyzed products have excess carbon and silica. The diffraction patterns of the samples derived from burnt rice husk + pitch and directly pyrolyzed rice husk + phenolic resin mixtures indicated that these samples have a relatively slaller content of SiC_w, consistent with the results of the chemical analysis.

The infrared spectra of the pyrolyzed product synthesized at 1000°C, 1500°C and of the precursor material [burnt rice husk + phenolic resin] are given in Fig. 2. The sample pyrolyzed at 1500°C shows a strong absorption peak around 800 cm^{-1} assignable to Si-C linkage. This peak could not be assigned to Si-O linkage since characteristic peaks of the Si-O bond were not observed around 1100 cm^{-1} and 450 cm^{-1}. This may be due to the absence of bead like whiskers in the sample subjected to infrared studies[3]. The sample pyrolyzed at 1000°C (a mixture of silica and carbon) showed peaks attributable to SiO_2 at 1100,450 and also at 800 cm^{-1}. The IR spectrum of the precursor material (burnt rice husk + phenolic resin) exhibited peaks around 1500–1620 cm^{-1} due to the dissubstitution of the benzene ring present in phenolic resin. The strong peaks at 1100 cm^{-1} are characteristic of Si-O-Si bonding. The strong IR absorptions at 470 cm^{-1} and 800 cm^{-1} are attributed to O-Si-O linkage and also to the CH out-of-plane mode of the aromatic ring present in the phenolic resin. The peak due to silica at 800 cm^{-1} is generally of medium intensity and hence the strong peak observed at 800 cm^{-1} should be due to the overlapping of absorptions characteristic of aromatic rings and silica.

SEM photographs are presented in Fig. 3. It is observed that the whiskers were well grown in all the samples. This may be attributed to the excess of carbon present in the

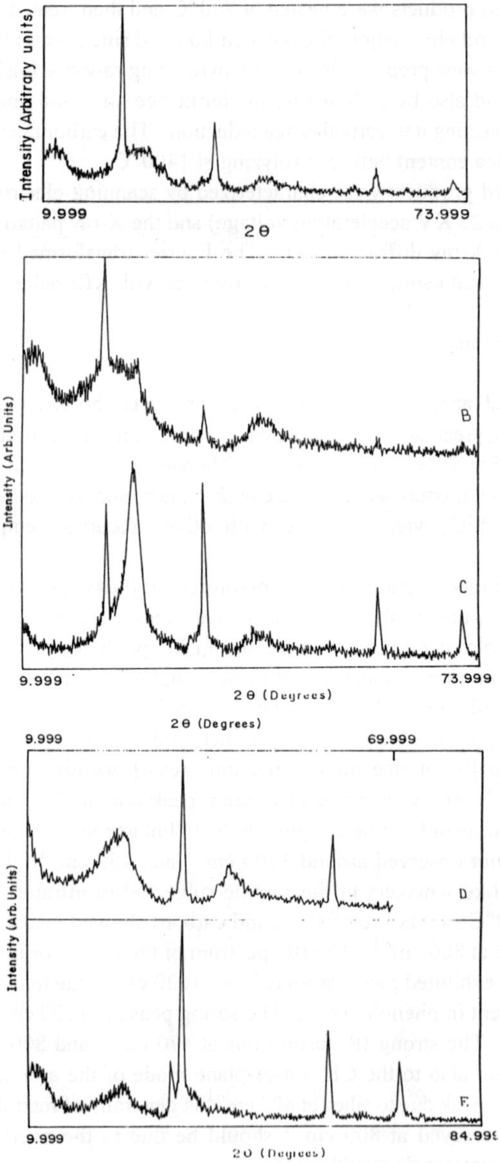

Fig. 1. X-ray diffraction patterns of A. burnt rice husk + phenolic resin pyrolyzed at 1400°C in argon, B. rice husk + phenolic resin pyrolyzed at 1400°C in argon, C. burnt rice husk + pitch pyrolyzed at 1400°C in argon, D. burnt rice husk + phenolic resin carbonized at 1000°C and pyrolyzed at 1500°C in argon, E. burnt rice husk + phenolic resin pyrolyzed at 1400°C, heated in air and HF treated.

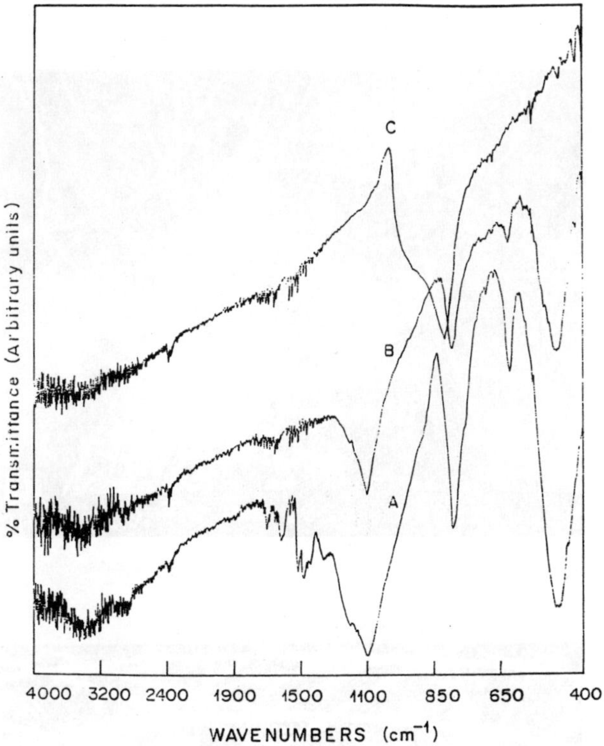

Fig. 2. FTIR spectra of A. burnt rice husk + phenolic resin, B. burnt rice husk + phenolic resin carbonized at 1000°C in argon, C. burnt rice husk + phenolic resin carbonized and pyrolyzed at 1500°C in argon.

precursors which provided a carbon substrate for the growth of whiskers. It was observed by earlier workers that the needle like whiskers grow around graphite containers used for whisker growth[3,8]. It was also observed that when a mullite container was used in the experiment for whisker growth, only bead like whiskers containing silica on the necks and surface were formed[3]. It is noticed from Fig. 3 that SiC samples obtained by precarbonization of burnt rice husk + phenolic resin mixture and by subsequent carbothermal reduction of the carbonized product at 1500°C showed only a few whiskers. This is in agreement with the earlier observation that whiskers grow at lower temperatures, with the temperatures differing depending on the process parameters[7]. Close observation of Fig. 3 shows that samples synthesized by employing pitch as the carbon source yielded bead like whiskers in addition to needle like whiskers. This may be attributed to the incomplete carbothermal reduction of silica at some sites. This observation explains the low yield of SiC_w with burnt rice husk +

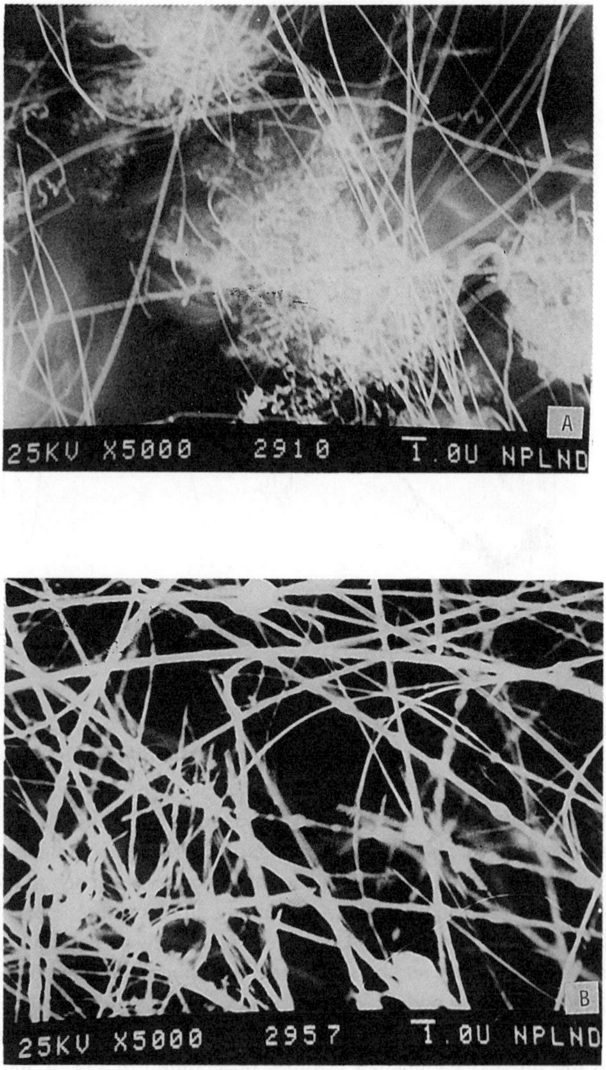

Fig. 3. SEM micrographs of A. burnt rice husk + phenolic resin pyrolyzed at 1400 °C, B. burnt rice husk + pitch pyrolyzed at 1400 °C in argon.

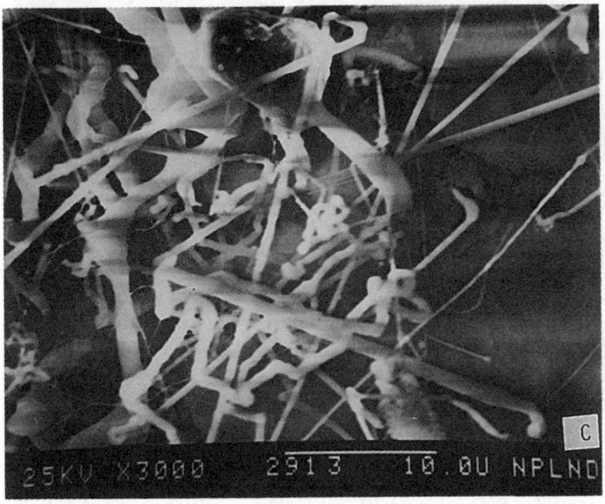

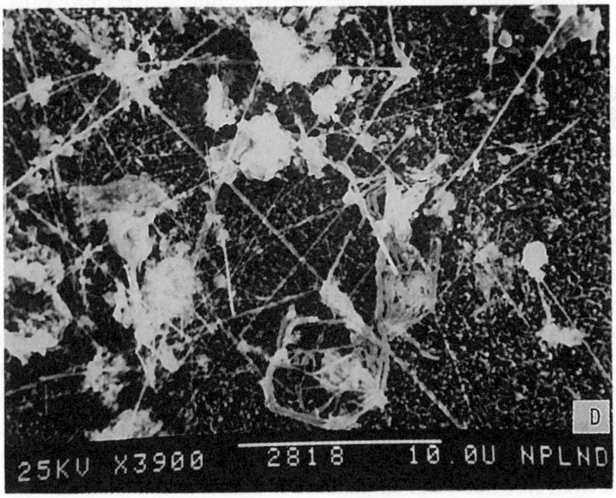

Fig. 3, cont'd. SEM micrographs of C. rice husk + phenolic resin pyrolyzed at 1400°C in argon, D. burnt rice husk + phenolic resin pyrolyzed at 1500°C in argon.

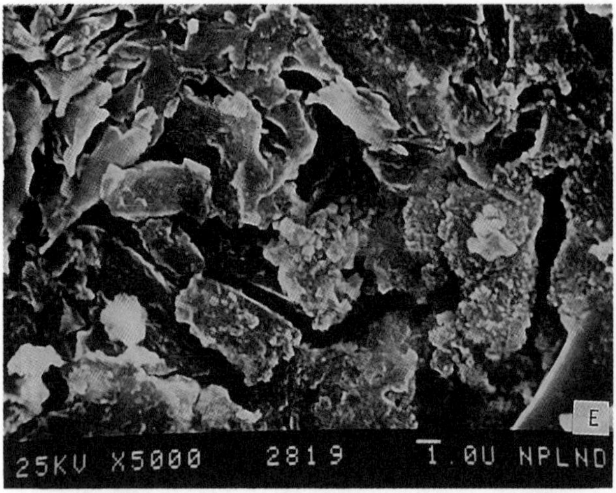

Fig. 3, con't. SEM micrograph of E. burnt rice husk + phenolic resin carbonized at 1000°C in argon.

pitch precursor material. SiC prepared by directly pyrolyzing rice husk + phenolic resin mixture showed bundles of thin whiskers having long as well as bent whiskers. This is due to the products of decomposition of rice husk and phenolic resin during pyrolysis, namely silica and carbon, which underwent carbothermal reduction above 1300°C to produce SiC_w. The whiskers obtained in the above pyrolyzed samples are not due to fibrous silica since SEM of carbonized products does not reveal any fiber (Fig. 3).

The formation of SiC_w and SiC_p can be explained by the following reactions which take place during carbothermal reduction:

$$SiO_2(s) + C(s) \rightarrow SiO(g) + CO(g) \quad (1)$$
$$SiO(g) + 3CO \rightarrow SiC(s) + 2CO_2(g) \quad (2)$$
$$3SiO(g) + CO(g) \rightarrow SiC(s) + 2SiO_2(s) \quad (3)$$
$$CO_2(g) + C(s) \rightarrow 2CO(g) \quad (4)$$

The over all reaction is written as:

$$SiO_2(g) + 3C(s) \rightarrow SiC(s) + 2CO(g).$$

It should be pointed out here that the rate controlling step for the formation of SiC is the evolution of SiO and its subsequent reaction with CO. The production of CO is important in the synthesis of SiC. The reduction of CO_2 by C (equation 4) aids the production of SiC_w and in our investigation, this process was assisted by the addition of phenolic resin or pitch which yields around 50% carbon when heated in inert atmosphere. It is also possible that the carbon produced from rice husk and carbonaceous binder is activated carbon which reduces CO more effectively than non-activayed carbon. The direct pyrolysis of rice husk + phenolic resin produces carbon in situ which in turn reacts with SiO to produce SiC_w. This explains the better yield of SiC_w in the case of directly pyrolyzed raw rice husk + phenolic resin mixture.

The formation of SiC_w was noticed only at temperatures above 1500°C even when carbon-black was used as a carbon source in addition to the carbon from the rice husk without employing additional catalysts[11]. The earlier observation by Ray et al. reported that SiC_w formation did not take place at 1400°C even after the addition of a costly catalyst, and the yield of SiC_p formed at 1400°C was found to be small[7]. Our experiments yielded SiC_w at 1400°C; this can only be attributed to the addition of carbonaceous binders that serve as an additional carbon source. The charred rice husk prepared by Krishnarao et al. contained around 55% carbon which agrees with the value of 50% carbon obtained by chemical analysis of the burnt rice husk in this study. It was observed that neither the charred rice husks + catalysts nor the mixture of charred rice husks + carbon black yielded SiC_w on pyrolysis at 1400°C. The results of the present investigation show that addition of cabonaceous binders to the raw and to the burnt rice husks helped in producing SiC_w at 1400°C. It is inferred by our results that the carbon derived from rice husk alone is insufficient to produce SiC_w below 1500°C and the nature of the additional carbon source affects the SiC_w yield at 1400°C.

4. Conclusion

These observations show that the nature of carbon obtained from phenolic resin and pitch during pyrolysis helps in the carbon-thermal reduction of silica to yield SiC_w at relatively lower temperatures.

5. Acknowledgements

The authors thank Dr. E. S. R.Goapl, Director, NPL, for encouragement and also for his permission to publish these results. Thanks are due to Mr. S. U. M. Rao and Ms. U. Dhawan for their help in characterization studies.

6. References

1. T. Ohji, Y. Goto and A. Tsuge, *J. Am. Ceram. Soc.* **74** (1991) 739.
2. J. M. Chiu and D. D. L. Chung, *J. Mat. Sci.* **25** (1991) 2583.
3 J. M. Saito, S. Nagashima and A. Kato, *J. Mat. Sci. Lett.* **11** (1992) 373.
4. H. Wang and G. S. Fischman, *J. Am. Ceram. Soc.* **74** (1991) 1519.
5. J. C. Lee and I. B. Cutler, *Am. Ceram. Soc. Bull.* **54** (1975) 195.
6. N. K. Sharma, W. S. Williams and A. Zangvil, *J. Am. Ceram. Soc.* **67** (1984) 715.
7. A. K. Ray, G. Mohanty and A. Ghose, *J. Mat. Sci. Lett.* **10** (1991) 227.
8. R.V. Krishnarao, M. M. Godkhindi, P. G. Mukunda and M. Chakraborthy, *J. Am. Ceram Soc.* **74** (1991) 2869.
9. R. V. Krishnarao, M. M. Godkhindi and M. Chakraborthy, *J. Mat. Sci.* **27** (1992) 1227.
10. R. V. Krishnarao, M. M. Godkhindi, M. Chakraborthy and P. G. Mukunda, *J. Mat. Sci.* **29** (1994) 2741.
11. R. V. Krishnarao, *J. Mat. Sci.* **12** (1993) 1268.

PITCH PRE-TREATMENTS TO IMPROVE THE STRUCTURE AND PROPERTIES OF C/C COMPOSITES

R. MENÉNDEZ, J. BERMEJO, A. FIGUEIRAS,
J. J. FERNÁNDEZ, and M. GRANDA,
Instituto Nacional del Carbón (CSIC),
Apartado 73, 33080 Oviedo, Spain

J. BONHOMME and J. BELZUNCE
Instituto Tecnológico de Materiales,
Llanera - Asturias, Spain

ABSTRACT

This work is concerned with the use of commercial pitches, with different specifications, in the preparation of uni-directional C/C composites, and the study of the influence of pitch composition on the properties of such composites. In a second stage, it is also studied the optimization of pitch utilization by subjecting the pitches to different pre-treatments in order to promote the polymerization of the components of pitch which could improve carbon yield and density of the subsequent carbon materials. This paper reports on the effects of air-blowing and carbon black additions on the properties of resultant composites. The composites are characterized using SEM and optical microscopy in terms of fiber/matrix interactions, porosities and optical textures of the matrices. Mechanical properties of the uni-directional composites are discussed from the structure of the composite and the chemical composition of pitch precursors.

1. Introduction

Carbon fiber-reinforced carbon composites are advanced materials with industrial applications resulting from their unique properties, which depend not only on the fiber properties but also on those of the matrix. C/C composites have a wide application as refractory and structural materials, and more recently the brake industry has developed methods to use carbon materials as disc brake materials. Nevertheless, C/C composite manufacture processing and design methodology are still underdeveloped.

The matrix plays an important role in the final properties and performance of C/C composites. Considering that matrix microstructure is determined by the physicochemical properties of the matrix precursor, the importance of an adequate development of the matrix is obvious. Matrix precursors for C/C composites should meet the following requirements[1]:

- high carbon yield, even when carbonized at atmospheric pressure,
- good fluidity and efficient wetting of the carbon substrate,

- ability to produce matrices with low porosity, high strength and resistance to oxidation at high temperatures.

In recent years, pitch has proven to be an acceptable material as a matrix precursor for the preparation of C/C composites, because of its low price and excellent graphitizability on heat treatment. Properties of derived carbons are controlled by chemical and physical properties of pitch and its behavior during pyrolysis treatment. However, commercial pitches have some disadvantages for the preparation of high-density materials, as their low carbon yield and the development of porosity after carbonization lead to poor physical and mechanical properties of the final material. To diminish these effects, several impregnation and carbonization cycles are necessary, with the subsequent rise in price of the product.

It is possible to optimize pitch utilization by subjecting them to different pretreatments in order to promote the polymerization of the components of pitch which could improve carbon yield and density of the subsequent carbons. In this end, pitch oxidation by air-blowing promises to be very effective. Molecular cross-linking at low temperatures, induced by oxygen functionality[2–4], increases the molecular weight of some molecules, preventing their distillation and removal during the carbonization stage. Carbon black addition[5] followed by thermal treatment can also be beneficial, as this produces a reduction in volumetric expansion on carbonization of pitches due to an enhancement of gas removal, and the retention of the light components within the pitch with a subsequent increase in carbon yield.

It is important to develop a good understanding of those properties of pitch which have a significant impact upon the operational properties of composite materials. Hence, a program of research concerning the preparation of C/C composites from coal derived products is ongoing at *Instituto Nacional del Carbón*, with financial support from the Spanish Education Ministry and the European Commission.

This paper is concerned with the use of commercial pitches, of different specifications, and the same type of fiber, in the preparation of uni-directional C/C composites, and the study of the influence of pitch composition on the structure and properties of resultant composites. In a second stage, different treatments (air-blowing and carbon black addition) are applied to pitch in order to improve its properties as a matrix precursor. Structure and properties of composites obtained from modified pitches are compared with those from untreated pitches.

2. Composites from Commercial Pitches

2.1. Experimental

2.1.1 Materials

Four commercial pitches of different specifications were selected. Table 1 summarizes their main characteristics, in terms of elemental analysis, softening point, carbon yield and toluene and quinoline solubility.

Table 1. Properties of the Pitches

Pitch	Elemental analysis (wt.%)					SP[1] (°C)	TI[2] (wt.%)	QI[3] (wt.%)
	C	H	N	S	O			
CTP1	92.3	4.1	1.2	0.7	1.8	75.1	35.7	13.0
CTP2	91.3	4.6	1.2	0.9	1.9	58.7	20.8	3.0
LP	91.1	5.2	1.6	0.5	1.6	169.1	44.5	0.5
A240	90.9	5.5	0.2	2.6	0.7	82.1	3.9	0.0

[1] Softening point, DMA
[2] Toluene insolubles
[3] Quinoline insolubles

CTP1 and CTP2 are coal-tar pitches, a binder and an impregnating pitch, respectively, supplied by Química del Nalón, LP is a pitch obtained from the liquefaction of a bituminous coal, supplied by British Coal, and A240 is a commercial petroleum pitch from Ashland. The carbon fiber used (AS4-12K, Hercules Aerospace) is a continuous, high-strength, high-strain, PAN-based type, with a filament diameter of 7 μm. This fiber has been surface-treated and sized to improve its interlaminar shear properties. Table 2 summarizes typical properties of the fiber.

Table 2. Hercules PAN-based AS4 Carbon Fiber Properties

Typical fiber properties	SI units
Tensile strength	4071 MPa
Tensile modulus	248 GPa
Ultimate elongation	1.65%
Carbon content	94.0%
Density	1.80 g cm^{-3}
Filament diameter	7 μm

The four pitches were characterized by FTIR, extrography and gas chromatography of their toluene-soluble fraction following procedures already described[6]. Pitch evolution during pyrolysis was monitored by hot-stage microscopy, and approximate temperatures of pitch softening, initiation of mesophase formation and hardening to semi-coke were determined.

2.1.2. *Composites Preparation*

In order to minimize the number of variables for the evaluation of materials, the same processing conditions were applied for the preparation of all composites.

A flow diagram of the laboratory procedure is shown in Fig. 1. In the first step, plate-shaped unidirectional composites were prepared by liquid impregnation of the fiber with a THF solution of the pitch in a 1:1 concentration. Fig. 2 shows a schematic diagram of the winding machine employed to prepare the prepreg laminates. Initially, the prepreg is heated at 60°C under vacuum, to remove the THF. Then, the temperature is increased to a value slightly higher than the SP of the pitch and maintained overnight (about 10 h). During this process the prepreg is placed between two metallic laminates in order to prevent deformation. Laminates of 170 mm × 70 mm were cut and molded to about 4 mm thickness. Molded laminates, 150 mm × 45 mm, were pyrolyzed at a heating rate of 5°C min^{-1} to 450°C and molded under a pressure of 1.5 MPa, before being pyrolyzed at a heating rate of 1°C min^{-1} to 750°C, under a nitrogen pressure of 10 MPa. In a subsequent pyrolysis, laminates of 150 mm × 45 mm were heated to 1200°C under argon at atmospheric pressure.

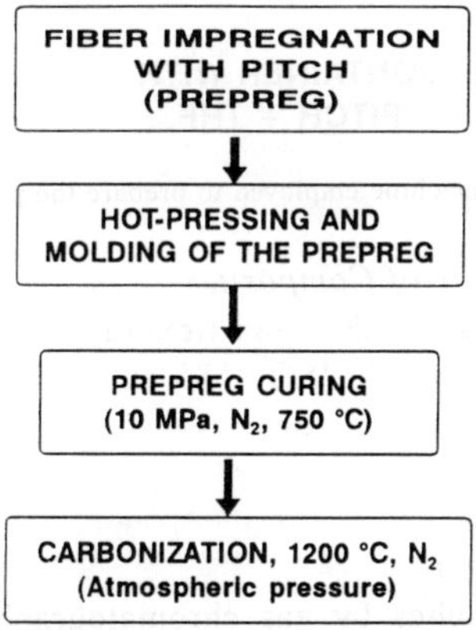

Fig. 1. Flow diagram of the laboratory process for the preparation of C/C composites.

2.1.3. *Microstructural Characterization of Composites*

Samples of composites for microscopical examinations were prepared as previously described[6]. Microscopy analysis was conducted using a Leitz Laborlux 12 Pol microscope with a ×100 objective in conjunction with a Swift automatic point-counter. The microscope included a one-wave retarder plate to record the optical texture of the anisotropic matrix. The analysis of each composite involved the recognition of fibers, matrix and pores (voids, intramatrix cracks and fiber/matrix interface cracks) located at the intersection of the graticulate cross-hair.

Quantification was achieved by recording 500 individual determinations per composite using an interpoint and interline distance of 0.5 mm. The number of counts recorded was converted to a proportion by volume of the total. The optical texture of the matrix[7] was also characterized. The orientation of constituent lamellar structures of the matrix with respect to the fibers on the surface of the composites was qualitatively observed by SEM, after etching with a chromic acid solution for 4 h[(8)].

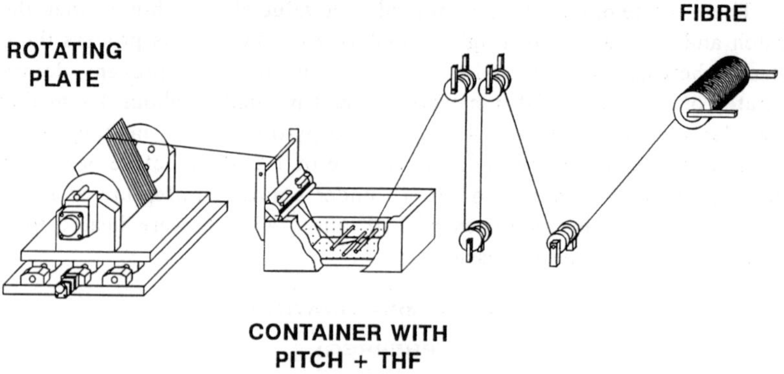

Fig. 2. Winding machine employed to prepare the prepreg laminates.

2.1.4. Mechanical Properties of Composites

Interlaminar shear strength (ILSS) and flexural strength (FS) tests were carried out on the composites according to ISO 4585 and ASTM D2344-84 standards.

2.2. Results

2.2.1. Pitch Characterization

Characterization of pitches by gas chromatography of their toluene-soluble fractions (TS) shows differences in the composition of this fraction from each pitch, as observed in Fig. 3. Basically, CTP1 and CTP2 have the same composition, as they are both coal-tar pitches, but with a different distribution of components. CTP2 has a higher proportion of light compounds than CTP1. A240 pitch gives a chromatogram with a fingerprint typical of petroleum pitches with a large number of compounds, different from those of coal-tar pitches. Finally, LP differs in the number and retention times of the main peaks, the aromatic compounds commonly present in coal tar pitches being absent.

FTIR spectra of the four pitches are shown in Fig. 4. The main differences are in the absorbances (Abs.) at 3050 (aromatic C-H), 2950–2800 (aliphatic C-H), 1600

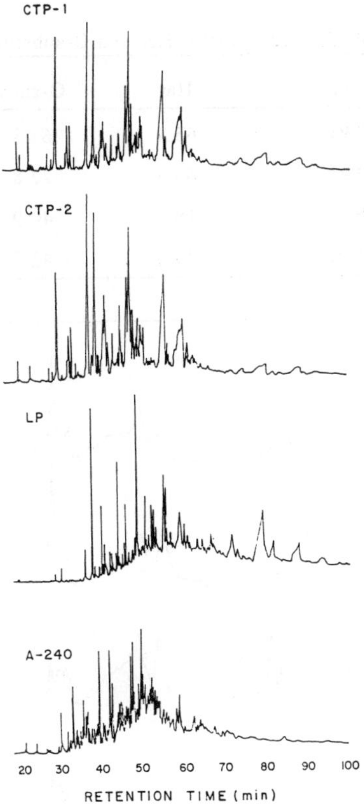

Fig. 3. Gas chromatograms of the TS fraction of pitches.

(aromatic C=C) and 900–750 cm^{-1} (aromatic substitution). As expected, coal-tar pitches contain a higher percentage of aromatic hydrogen (Har) and also a higher C/H atomic ratio than the other two. The prevalent components of the two coal-tar pitches are those with ortho-substituted rings. Values of the percentage of Har (Abs. 3050/Abs. 3050–2920 cm^{-1}) and percentage of ortho-substitution (Abs. 750 cm^{-1}/Abs. 850–750 cm^{-1}) obtained from the spectra, are shown in Table 3, and were calculated according to Guillén et al.[9] LP shows a strong band at 1600 cm^{-1} which is indicative of a concentration of aromatic structures similar to those of coal-tar pitches. Therefore, its main components should be large pericondensed molecules bearing hydroaromatic and naphthenic rings (Fig. 4). The low values of A240 seem to be due to the presence of considerable alkyl substitution.

Table 3. Percentage of Har and O-substitution

Pitch	Har	O-subs
CTP1	65.4	65.3
CTP2	68.3	72.8
LP	18.7	40.9
A240	29.4	42.7

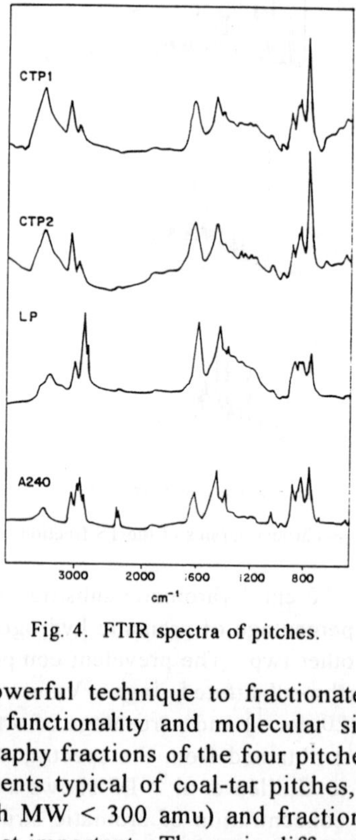

Fig. 4. FTIR spectra of pitches.

Extrography is a powerful technique to fractionate pitches into classes of compounds of different functionality and molecular size[10]. Fig 5 shows the distribution of the extrography fractions of the four pitches. CTP1 and CTP2 give a distribution of components typical of coal-tar pitches, with fraction II (neutral aromatic compounds with MW < 300 amu) and fraction IV (phenolic and basic components) being the most important. The main difference between both coal-tar pitches is that CTP2, the impregnating pitch, is quantitatively richer in those fractions[11]. Therefore, CTP2 may be described as a light and medium reactive pitch which generates a coke with an intermediate microstructure, whereas CTP1 gives a higher yield of denser and better structured cokes, although its high QI content makes predictions difficult.

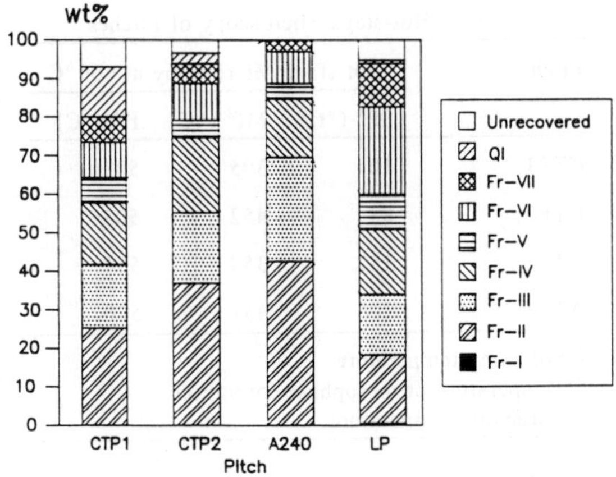

Fig. 5. Distribution of the extrography fractions, unrecovered material and quinoline-insoluble contents of pitches.

Most of the components of A240 petroleum pitch separate out into fractions II, III and IV (86 wt.%). This is a consequence of the high content of aliphatic structures, which causes a high solubility of the compounds. This is corroborated by the higher average molecular weights of these fractions compared with those of coal tar pitches[12]. In spite of this, A240 can be included in the group of light pitches, mainly because of its low TI content and the absence of high-molecular-weight unrecovered compounds. The lowest amount in fraction II and the highest amount in fractions VI and VII of LP pitch reveal a large amount of high molecular weight compounds, in agreement with its high softening point and the low ortho-substitution index, already mentioned. The naphthenic character of these compounds predicts a high carbon yield and a well-structured coke from LP.

Hot-stage microscopy (HSM) results are shown in Table 4. These results describe the temperature of pitch softening, and initiation of mesophase formation (temperature at which a rapid and continuous formation of mesophase is observed) and hardening. The formation of mesophase in LP takes place at the lowest temperature, and the hardening at the highest temperature. The predominance of high molecular compounds with high content in aliphatic hydrogen, brings about the removal of free radicals formed during pyrolysis by combining with the hydrogen, so that the net effect of temperature on the system is a dehydrogenation rather than a molecular growth. In light pitches, such as CTP2 and A240, mesophase formation has to be preceded by the generation of mesogenic molecules by polymerization. When the pitch system has a low hydrogen-donor ability (CTP1 and CTP2), polymerization occurs rapidly, giving rise to a short interval of mesophase evolution.

Table 4. Hot-stage Microscopy of Pitches

Pitch	Hot-stage Microscopy at 600°C		
	SP[1] (°C)	Mf[2] (°C)	H[3] (°C)
CTP1	104	395	503
CTP2	83	452	503
LP	182	358	554
A240	106	451	520

[1] Softening temperature
[2] Temperature of mesophase formation
[3] Hardening temperature

As for the suitability of these pitches as matrix precursors, A240 is very fluid at temperatures at which pressing of the prepreg takes place. This can produce the exudation of pitch from the composite[13].

2.2.2. *Characterization of Matrix Microstructure*

The study of matrix microstructure was carried out on the polished surface of a cross-section of the composites prepared from each of the four pitches. Resultant composites were labeled as CC1 from CTP1, CC2 from CTP2, CC-LP from LP and CC-A240 from A240. Fig. 6 shows optical micrographs of representative areas of cross-sections of composites and a SEM micrograph of CC1 composite is shown in Fig. 7. CC1 presents a development of mesophase hindered not only by the presence of fibers, but also by the high QI content present in parent pitch. This produces areas of mosaic optical texture (Figs. 6a and 7, position B), alternated with domains (Figs. 6a and 7, position A), orientated around the fiber. Fig. 6a shows also a very good matrix/fiber bonding, debonding cracks only representing 0.4 vol.% of the composite. CC2 presents a dominant domain texture (Fig. 6b) orientated around the fiber, with only a small amount of mosaics. Intramatrix cracks were also observed in this composite (Fig. 6b, position C), and a good but lower matrix/fiber bonding than in CC1, debonding cracks representing 2.6 vol.% of the composite. CC-LP shows a domain optical texture and shrinkage cracks within the matrix (Fig. 6c, position D) and also located at the fiber/matrix interface (Fig. 6c, position E). These cracks were generated during carbonization of the composite at 1200°C, as confirmed by the optical microscope observations of the green composite obtained at 750°C (Fig. 8), where cracks are almost absent. As for CC-A240, optical microscope observations confirm that pitch seeps out during the pressing of the prepreg, which was already anticipated from the chemical characterization and hot-stage observations of the pitch. The matrix content of CC-

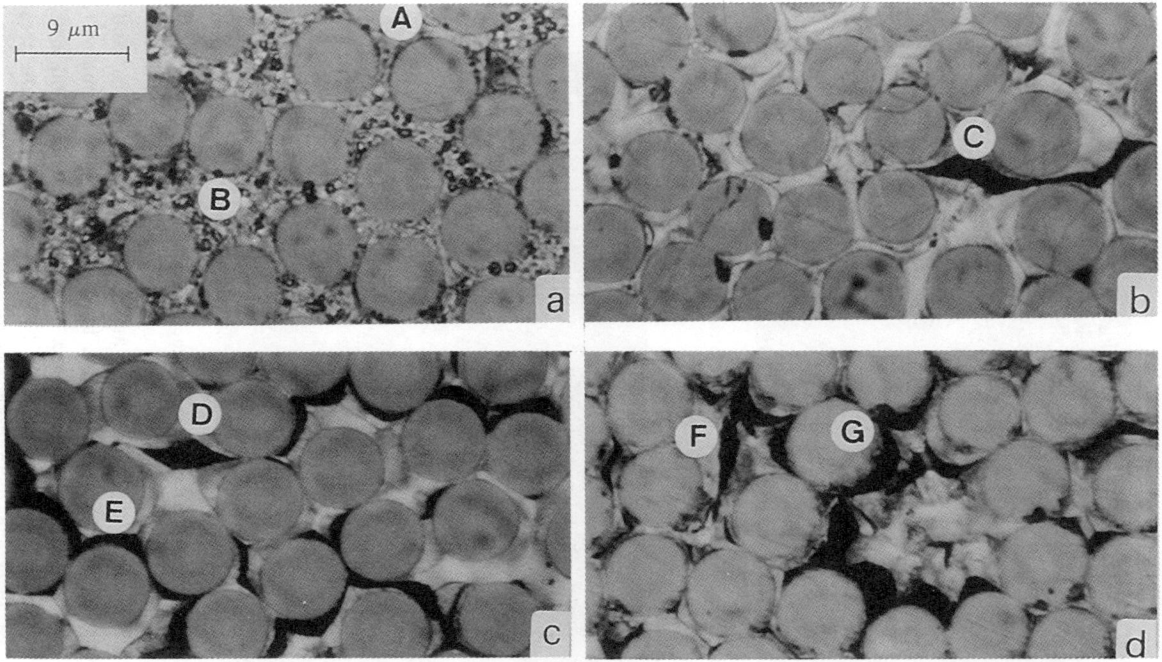

Fig. 6. Optical micrographs of composites, (a) CC1, (b) CC2, (c) CC-LP and (d) CC-A240.

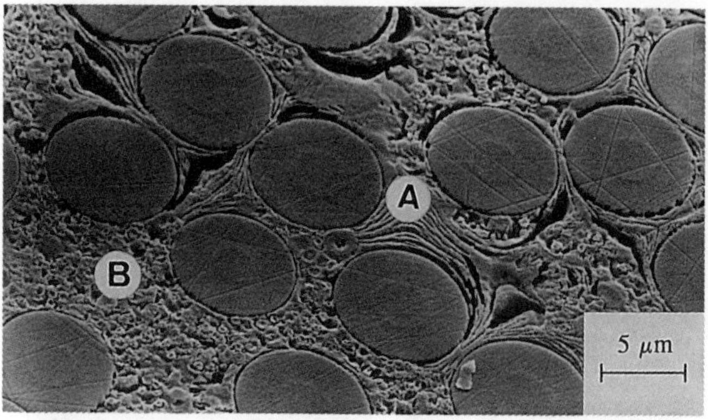

Fig. 7. SEM micrograph of composite CC1, after etching with chromic acid: (A) domains and (B) mosaics.

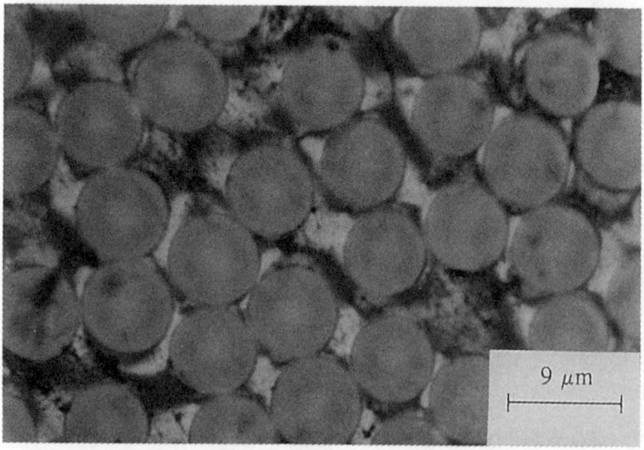

Fig. 8. Optical micrograph of the green composite CC-LP.

A240 is 14 vol.%, much lower than CC1 and CC2 (about 28 vol.%) and CC-LP, which is 23 vol.% (Table 5). The matrix texture shows flow domains (Fig. 6d) parallel to the fiber surface, with fiber/matrix interface cracks (Fig. 6d, position G), and intramatrix cracks (Fig. 6d, position F). CC-A240 composite presents the highest value of interface debonding cracks, 6 vol.%, as determined by light microscopy (Table 5). The high fiber content (72.0 vol.%) could have favored the formation of intramatrix cracks[14].

Table 5. Properties of Composites

Composite	FC[1]	d[2]	TP[3]	ILSS[4]	FS[5]
CC1	61	1.59	10.6	22.9	331
CC2	54	1.52	16.0	-	236
CC-LP	58	1.50	18.8	17.2	386
CC-A240	72	1.46	13.8	13.5	-

[1] Fiber content (vol.%)
[2] Density (g cm^{-3})
[3] Total porosity (vol.%)
[4] Interlaminar shear strength (MPa)
[5] Flexural strength (MPa)

2.2.3. *Porosity of Composites*

The presence of voids (pores and cracks) in C/C composites is an important factor controlling their properties. Voids represent an access path for oxidizing agents with subsequent degradation of composite properties, as well as centers of initiation of fracture in composites. Voids can be generated by an incomplete penetration by liquid impregnants during the fabrication of the composite and by volatiles released during carbonization or solvent removal. In addition, cracks are generated by shrinkage of the matrix during carbonization or by stresses on cooling. Porosities of composites determined by light microscopy are given in Table 5. CC1 shows the lowest porosity (10.6 vol.%). Porosities of CC-LP (18.8 vol.%) and CC2 (16.0 vol.%) are mainly derived from shrinkage which opens up lamellar cracks and increases the tendency for the matrix to separate from the fiber surfaces (Fig. 6d); 2.6 vol.% of the matrix appears separated from the fiber. This effect, more pronounced in the case of CC-A240, could be due to its lower matrix content.

2.2.4. *Density of Composites*

Table 5 shows the bulk density of the four composites. CC1 presents the highest value, 1.59 g cm^{-3}, followed by CC2 (1.52 g cm^{-3}) and CC-LP (1.50 g cm^{-3}), CC-A240 having the lowest (1.46 g cm^{-3}). The lower density of CC-LP compared to CC1 and CC2, despite the higher carbon yield of the parent pitch, is due to the extensive generation of cracks during carbonization at 1200°C. The composite obtained from A240 exhibits the lowest density, as could be expected from its carbon yield and the high aliphatic composition, as shown by FTIR analysis.

2.2.5. Mechanical Properties of Composites

Table 5 summarizes the results of interlaminar shear strength (ILSS) and flexural strength (FS) tests and the fiber content of the composites obtained from the four pitches without any impregnation treatment. CC1, CC2 and CC-LP have similar values of carbon fiber content, which allow a more realistic comparison from the point of view of the influence of the matrix microstructure on the mechanical behavior of the composite. Fig. 9 shows the load-displacement curves of the interlaminar shear tests, for CC1, CC-LP and CC-A240. CC1 exhibits a more brittle behavior, whereas CC-A240 and CC-LP have a pseudo-plastic behavior. The higher interlaminar strength of CC1 (22.9 MPa) is related to the absence of appreciable cracking between fiber and matrix and the presence of mosaics in the matrix as observed from light microscopy examinations. The lower values exhibited by CC-LP (17.2 MPa) and CC-A240 (13.4 MPa) seem to be related to the shrinkage interphase cracks produced during carbonization of the composites (Fig. 6).

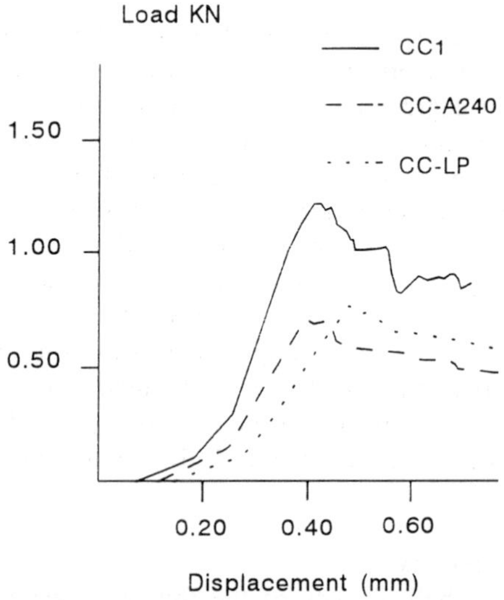

Fig. 9. ILSS curves of CC1, CC-LP and CC-A240.

Given that ILSS is a good measure of the interfacial fiber/matrix bonding, it can be expected that composites with a higher decoupling as observed by optical microscopy, have lower interlaminar shear strengths. This is the case of CC-LP and CC-A240, which show 2.6 vol.% and 6 vol.% of debonding cracks. This debonding, produced between 750 and 1200°C, is related to the strength of the fiber/matrix bonding, and the extent of the bulk shrinkage of the matrix during carbonization. With increasing pyrolysis temperature, the macromolecules increase their aromaticity by the removal of aliphatic structures and heteroatoms, producing more planar molecules and, therefore, decreasing the interlaminar distance between the graphitic structures. As a consequence, shrinkage is produced.

Interfacial bonding increases with aromaticity of the pitch and this improves the wettability of the pitch. In contrast, the shrinkage of the matrix increases with the aliphatic content of the pitch, and the texture of the matrix also influences formation of intramatrix cracks. Thus, those pitches which produce cokes with the larger size of optical texture (LP and A240) are the most prone to develop intramatrix cracks. The results obtained in this study agree with these predictions. The low amounts of debonding of CC1 could be caused by the high content of QI of the parent pitch.

Fig. 10 shows flexural test (FS) curves of composites. As seen in Table 4, CC-LP has the highest value of FS, which can be related to its pseudoplastic behavior derived from the accommodation of the applied stresses by the cracks, as observed in Fig. 8.

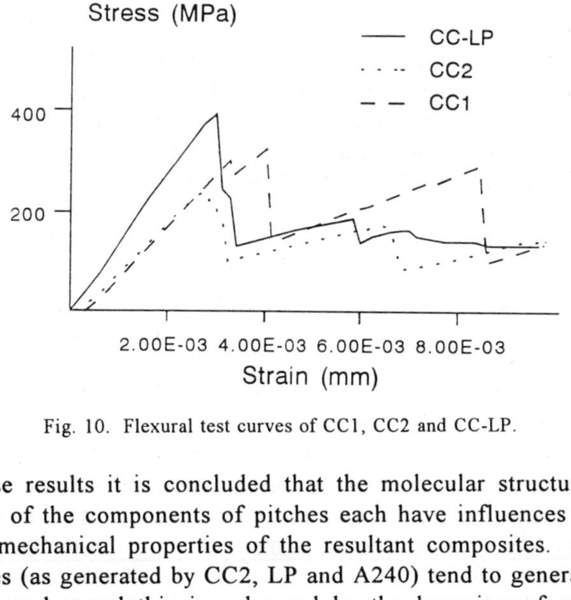

Fig. 10. Flexural test curves of CC1, CC2 and CC-LP.

From these results it is concluded that the molecular structure, nature and molecular size of the components of pitches each have influences on the microstructure and mechanical properties of the resultant composites. Matrices with domain textures (as generated by CC2, LP and A240) tend to generate intramatrix and interface cracks, and this is enhanced by the lowering of matrix content. Primary QI, present in parent pitches, influences mesophase formation and, consequently, the optical texture of the resultant matrix, as seen in CC1 where mosaics dominate. Composites obtained from this pitch show an optimum fiber-matrix union, as expected from its industrial utilization.

As well as the optical texture of the matrix, the weight loss on pyrolysis is mainly responsible for the lowering of density and increased porosity in composites. Cracks produced during carbonization by shrinkage depend on pitch aromaticity and seem to be responsible for the decrease in ILSS.

3. Composites from Treated Pitches

The above results lead to a better understanding of the properties of pitch which have a real influence on the mechanical behavior and properties of resultant composites. As commented above, the second objective of this research concerns the study of the optimization of commercial pitches for their use as matrix precursors, by air-blowing and thermal treatment, as well as by carbon black additions.

3.1. Air-blowing Treatment

3.1.1. Experimental

For this study, two industrial coal-tar pitches were used, CTPA (a binder pitch) and CTPB (an impregnating pitch) whose analysis and specifications are in Table 6. The oxidation treatment was similar to that described by Barr and Lewis[2]. Pitches were air-blown in a 2-liter Pyrex reactor fitted with a stirrer. The reactor was heated with an electric mantle and the temperature regulated with a Honeywell controller with a thermocouple immersed into the pitch.

Table 6. Properties of Pitches

Pitch	Treatment	Elemental Analysis (daf, wt%)					C/H	SP	CY	TI	QI
		C	H	N	S	O*		(°C)	(wt.%)	(wt.%)	(wt.%)
CTP	--	93.7	4.1	1.1	1.1†	-	1.9	72	48.4	35.7	13.0
CTP	250°C/Air	93.8	3.5	1.1	0.3	1.3	2.2	151	70.8	59.8	39.3
CTP	275°C/Air	94.0	3.5	1.1	0.4	1.0	2.2	171	72.1	62.6	49.2
CTP	300°C/Air	93.2	3.4	1.0	0.5	1.9	2.3	223	79.4	70.0	53.5
CTPB	--	92.0	4.5	1.1	0.5	1.9	1.7	54	35.2	20.8	3.0
CTPB	250°C/Air	92.7	4.0	1.1	0.4	1.8	1.9	148	62.4	52.7	39.2
CTPB	275°C/Air	93.1	4.1	1.2	0.4	1.2	1.9	164	64.4	53.8	42.9
CTPB	300°C/Air	92.7	4.0	1.1	0.6	1.6	1.9	219	67.9	60.5	45.9

*, by difference
C/H, atomic ratio C/H
CY, carbon yield (1000°C)
QI, quinoline content
†, sulfur and oxygen
SP, softening point TMA
TI, toluene content

1000 g of pitch, < 6 mm particle diameter, was preheated to 175°C for 30 min in nitrogen, in order to soften the pitch to an adequate viscosity for the stirring operation. Heating was continued to 250°C at 3°C min^{-1} under an air flow of 80 l h^{-1}, and maintained until the viscosity of the pitch did not allow further stirring. CTPA and CTPB required 14 and 18 h, respectively. The pitch was then quench-cooled to room temperature. The same conditions were used at 275 and 300°C. Resultant pitches were ground and sieved to < 0.4 mm particle size.

The characterization of pitches included, as well as the main properties (elemental analysis, softening point, carbon yield, solubility in toluene and quinoline), the fractionation by extrography, XPS, FTIR and X-ray diffraction, as previously described[15]. Mesophase development of pitches was monitored by optical microscopy of pitch pyrolysis products obtained in a horizontal tube furnace. For this, about 4 g of each of the 8 pitches were placed in a ceramic pot, which was flushed with nitrogen from 25 to 450°C at 5°C min^{-1} and maintained at the maximum temperature for 30 min. The sample was then removed from the furnace and rapidly cooled under nitrogen.

Cokes were obtained by heating each pitch in a horizontal furnace, from 25 to 1000°C at 1°C min^{-1}, with 30 min soak time. Characterization of cokes was performed by optical microscopy (optical texture and porosity), helium densimetry, mercury porosimetry and a microstrength test, the later similar to that of Ragan and Marsh[16]. Experimental details have been previously described[15].

Composites were prepared as described in Section 2.1.2., from the original and treated pitches. An optimization of the impregnation procedure was achieved by improving the alignment of the fiber and avoiding the seepage of pitch. Moreover, some modifications in the hot-pressing equipment allowed a gradual application of pressure, with subsequent improvement of materials quality. It was not possible to prepare a composite from CTPA3 pitch, because of its low solubility in THF. Characterization of the resultant materials was made by optical microscopy in terms of their porosity and the optical texture of the matrix. Interlaminar shear strength and flexural strength tests were made on the composites according to ASTM standards (Section 2.1.4.). Fracture surfaces of composites were observed by SEM.

3.1.2. Results

3.1.2.1. Properties and Chemical Composition of Pitches

The results from characterization of the parent pitches CTPA and CTPB by extrography show that CTPB contains higher amounts of light neutral compounds (fraction II), and of phenolic and nitrogen basic compounds (fraction IV) than CTPA. Unrecovered material of CTPA is as high as 20 wt.%, and is mainly composed of primary QI. Taking into account that CTPB is an impregnating pitch and considering the analytical data in Table 6, which show a higher content of oxygen for CTPB, fraction IV of this pitch is richer in phenolic and other oxygen containing compounds.

Fig. 11 shows TG and DTG curves of CTPA and CTPB. The weight loss of CTPB is higher than that of CTPA at all temperatures, these differences becoming more important at temperatures above 500°C. The maximum weight loss rate of CTPB at 395°C is due to its higher content of light compounds. Besides the principal maximum weight loss rate at 420°C, the DTG curve of CTPA shows a significant weight loss at 480 and 535°C.

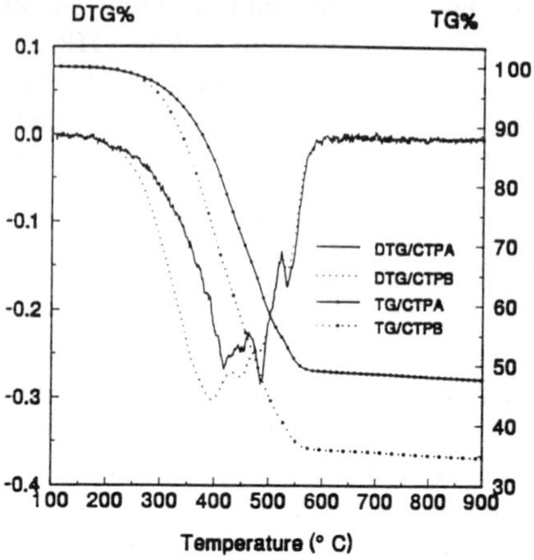

Fig. 11. TG and DTG curves of CTPA and CTPB pitches.

Air-blowing has no significant influence on pitch yields, which were similar for all pitches. Air-blowing produced a significant increase in the SP of CTPA and CTPB. The similar SP of both series of air-blown pitches is a consequence of the experimental conditions used (different times). Coking capacity of pitches, as measured by the carbon yield (CY) obtained on pyrolysis at 900°C, increases in the same way as the SP, varying at the initial stage of the air-blowing treatment from 48.4 to 79.4 wt.%, and from 35.2 to 67.9 wt.% for CTPA and CTPB, respectively (Table 6). As the temperature increased, CY variations became smaller, showing in the range of 250–300°C only an increase of about 8 and 6 wt.% for CTPA and CTPB series, respectively.

Solubility tests (Table 6) show an increasing content in TI and QI for air-blown pitches. Parent pitches have a toluene-insoluble content smaller than the lowest-temperature air-blown pitches by a factor of 2.0 and 2.5 for CTPA and CTPB, respectively. Variations in QI content are even higher, specially for the impregnating pitch, where the increase is by a factor of 13. From this it can be inferred that chemical reactions caused by air-blowing transform a considerable amount of TS and TI-QS into QI. However, despite the strong increase of QI content, these air-blown pitches are totally isotropic, as observed by optical microscopy.

The oxygen content of both series shows a slight variation (Table 6). CTPA and CTPB seem to have different behavior during treatment, as deduced from the data obtained. While CTPA exhibits a continuous increase in oxygen content with increasing air-blowing temperature, CTPB shows an irregular variation. The two pitches show a minimum oxygen content after air-blowing at 275°C. XPS was used to determine C/O ratios for the CTPB series. Results vary in the same way as data from elemental analysis, showing a maximum of C/O ratio for the air-blown pitch at 275°C (Fig. 12). This variation could be indicative of a new pathway of oxidation-polymerization, which starts at temperatures near 275°C[17]. The strong increase of the SP between 275 and 300°C supports this hypothesis.

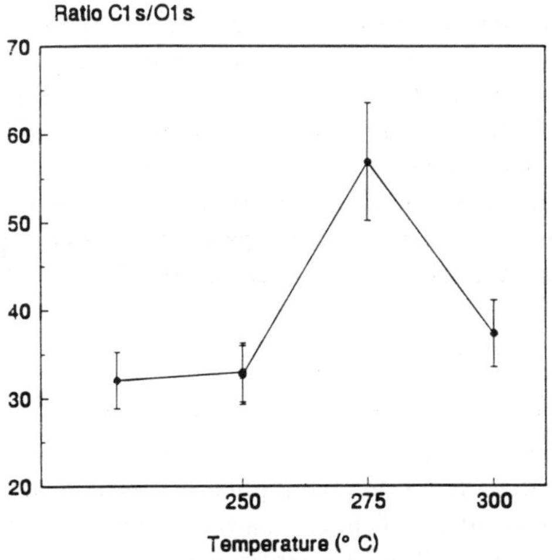

Fig. 12. XPS Analysis of CTPB series.

FTIR spectra of air-blown pitches are similar to those of parent pitches already shown in Fig. 4, presenting a decrease only in the aliphatic stretching absorbances. Ortho-substitution indices and aromaticity were calculated as described in Section 2.4. The results are shown in Table 7. With air-blowing, there is a decrease in the ortho-substitution index and an increase in the aromaticity index. Values of the ortho-substitution index, for both series, corroborate that the condensation degree of components in CTPA is higher than in CTPB, and the trend of air-blown pitches is the same. As for the aromaticity index, it is higher for CTPB, and the changes observed with air-blowing follow a clearly increasing trend for the two series. The

Table 7. Har and Ortho-substitution Index of Pitches

Pitch	Treatment	O-subs	Har
CTPA	--	0.65	0.66
CTPA1	250°C/Air	0.56	0.75
CTPA2	275°C/Air	0.54	0.81
CTPA3	300°C/Air	0.43	0.76
CTPB	--	0.73	0.68
CTPB1	250°C/Air	0.63	0.70
CTPB2	275°C/Air	0.60	0.81
CTPB3	300°C/Air	0.57	0.77

increases in aromaticity index and C/H atomic ratio (Table 6) suggest that polymerization involves significant loss of aliphatic hydrogen.

In summary, variations in chemical and physical properties of pitches with air-blowing suggest the concurrence of two main mechanisms in air-blowing polymerizations. One leads to the formation of cross-linked olygomers[2,3] and the other to the formation of large planar macromolecules as a result of extensive ring condensation[18], this second mechanism leading to better structured aromatic units, which accelerate mesophase formation. The relative participation of these reactions depends on the chemical composition of the parent pitch and the temperature of air-blowing[15], the first mechanism prevailing for the CTPB pitch, and the second mechanism for the CTPA pitch.

3.1.2.2. Coke Optical Texture and Properties

Although structure and properties of pitch-based carbon matrices are influenced by the presence of fiber in composites, and also by the different conditions used for composite preparation involving pitch pyrolysis, useful information about the structure and properties of pitch matrix carbon can be obtained from a preliminary study of cokes. Previous experience[6] shows that the dominant optical texture of matrix carbon is the same as that of the coke, except for some preferred orientation of domain-like textures around the fibers and for the intramatrix location of mosaics, when both structures are present. Similarly, it is assumed that strength and reactivity of matrix carbon in composites will follow the same trends as the single cokes.

CTPA cokes show a dominant optical texture of domains (Fig. 13, position A) and agglomerates of primary QI and anisotropic material, which cannot be considered as mosaics because of the high contents of QI (Fig. 13, position B). Air-blowing of CTPA produced an increase in carbon yield (Table 6) along the series, and causes a decrease in the relative proportions of QI in the cokes (Table 8) and a more homogeneous distribution. As a result, there is a smaller contribution of the

Table 8. Primary QI Content of Pitches and Cokes

Sample	CY (wt.%)	QI content (wt.%)	
		Pitch	Coke
CTPA	48.4	13.0	26.9
CTPA1	70.8	14.8	20.9
CTPA2	72.1	14.8	20.5
CTPA3	79.4	14.8	18.6
CTPB	35.2	3.0	8.5
CTPB1	62.4	3.5	5.6
CTPB2	64.4	3.5	5.4
CTPB3	67.9	3.5	5.1

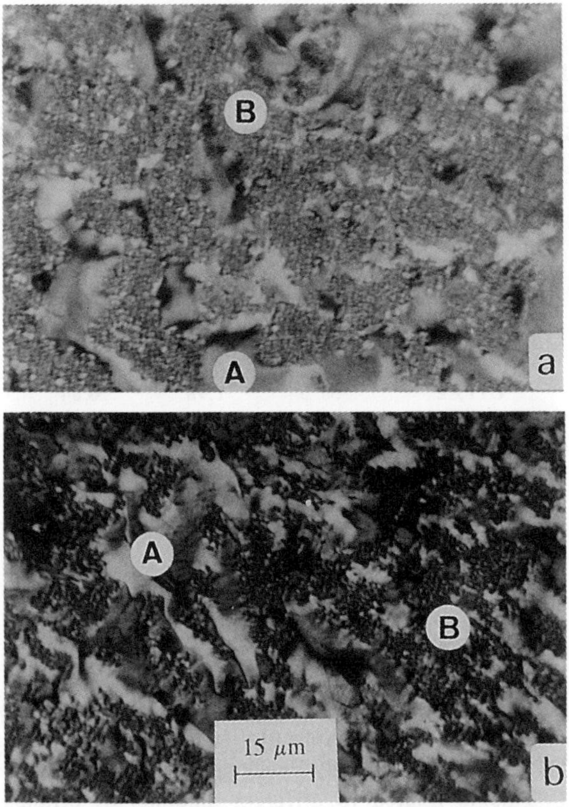

Fig. 13. Optical micrographs of cokes, HTT 1000°C, from (a) CTPA and (b) CTPA2: (A) domains and (B) agglomerates.

QI agglomerates to their optical texture. In addition, the structural changes of pitch produced by air-blowing contribute to increases in amounts of domains along the series. Air-blowing of CTPA leads to the formation of better-structured poly-condensed aromatic units, which accelerate mesophase formation, and generate a coke of improved microcrystalline structure.

Pitches of the CTPB series differ completely from those of CTPA. First, the concentration of QI is lower (Table 8), and has a smaller influence on the optical texture of cokes. Variations must be related to physical/chemical changes derived from the air-blowing treatment, itself. An increase in the contribution of mosaics and a decrease in the domains (Fig. 14) were observed when increasing the severity of the air-blowing. A decrease is also observed in the size of the domains of cokes from air-blown pitches. The structures of small size present in the coke from the treated pitches are similar to QI agglomerates observed in CTPA cokes (Fig. 14, position B). Air-blowing of pitches produced a substantial decrease of coke porosity, as determined by light microscopy (Table 9). Additionally, air-blowing causes a significant increase in the bulk density of cokes at the initial stage of the treatment (250°C), which increases slightly with the severity of the air-blowing treatment (275 and 300°C), as is shown in Table 9.

Table 9. Density of Carbonaceous Matrices

Sample	P (vol.%)	d_t (g cm^{-3})	d_b (g cm^{-3})
CTPA	92.8	1.80	0.13
CTPA1	55.4	1.96	0.87
CTPA2	57.6	1.98	0.84
CTPA3	50.0	1.96	0.98
CTPB	97.2	1.92	0.05
CTPB1	56.5	2.00	0.87
CTPB2	52.7	2.02	0.95
CTPB3	46.8	1.99	1.06

P, porosity
d_t, true density
d_b, bulk density

Variations of true density of cokes with pitch treatment are smaller and both series show a regular increase (Table 9). It is interesting to note that the true densities of CTPA cokes are slightly lower than those observed for the respective CTPB cokes. This is related to different amounts of primary QI in the two samples. The strong bonding between QI particles and the anisotropic coke could favor the formation of pores of small size, as seen in the porosity data of Table 9.

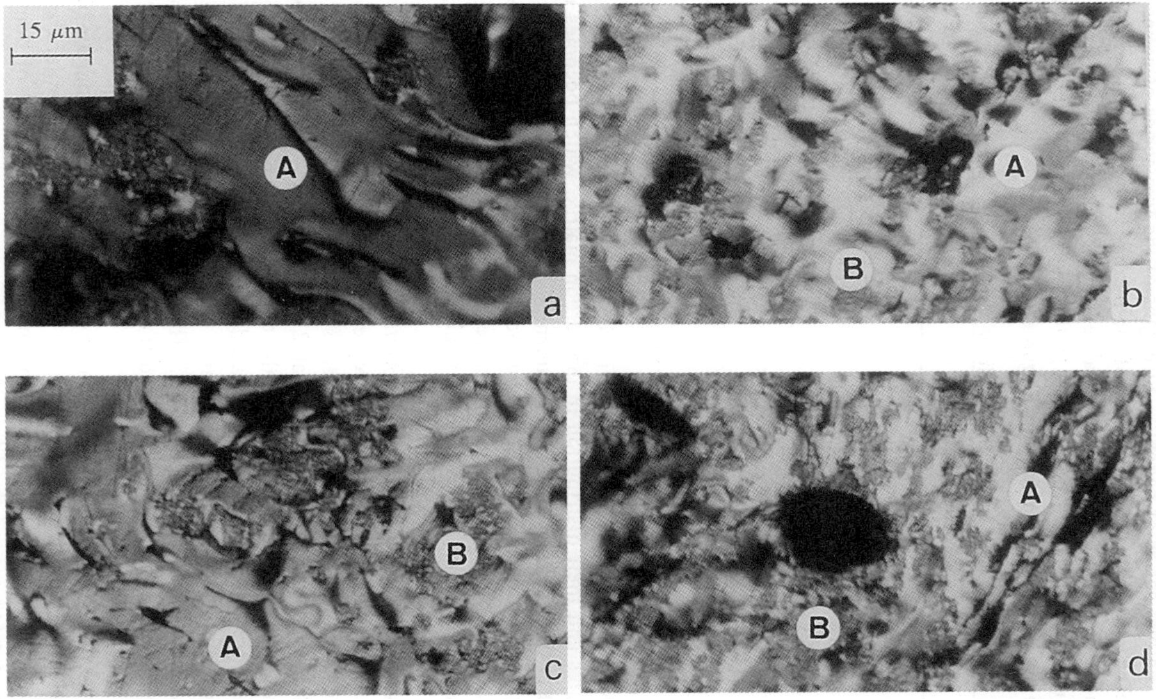

Fig. 14. Optical micrographs of cokes, HTT 1000°C, from (a) CTPB, (b) CTPB1, (c) CTPB2 and (d) CTPB3: (A) domains and (B) agglomerates.

For coke strength, Table 10 shows values of the r_1, r_2 and r_3 indices and the r_1/r_3 ratio for the two series of cokes[16]. Strengths of the two series of cokes from air-blown pitches increase with the severity of the treatment, being higher for cokes of CTPA series. The decrease of the optical texture of cokes with the severity of the air-blowing of CTPB pitch contributes to the improvement of the strength of the cokes, this being in agreement with Ragan et al.[19]

Table 10. Microstrength Indices of Cokes (r = %).

Sample	r_1	r_2	r_3	r_1/r_3
CTPA	3.26	43.79	52.95	0.06
CTPA1	54.50	31.51	13.99	3.90
CTPA2	67.52	17.62	14.86	4.54
CTPA3	71.38	18.31	10.31	6.92
CTPB	0.0	40.23	59.77	0.0
CTPB1	31.10	47.00	21.90	1.42
CTPB2	53.59	30.97	15.44	3.47
CTPB3	61.94	21.27	16.79	3.69

The increase of strength for CTPA series could be related to the presence of QI agglomerates and their distribution in the carbon matrix, which becomes more homogeneous at the final stages of the treatment.

A strong bonding between QI/coke and the development of an optical texture more similar to that of mosaics at the interface, could explain the higher strength of CTPA cokes despite their lower bulk and real density, and their higher porosity.

3.1.2.3. Influence of Pitch Air-blowing on Matrix Microstructure

Fig. 15 shows optical micrographs of representative areas of cross-sections of the composites obtained from the CTPB series and some of CTPA. With CTPA air-blown pitches, mainly those obtained at 275 and 300°C, the preparation of composites was extremely difficult due to the extensive polymerization caused by the air-blowing treatment. Prepregging faced the problem of solids deposition which were not completely soluble in THF and, consequently, homogenous wetting of the fiber with the pitch was not possible. This is the reason why results shown in this section will be mainly concerned to the composites obtained from CTPB series.

In any case, the variation on the optical texture of composites from air-blown pitches CTPA1 and CTPA2, compared with the parent, followed a similar trend to that observed in the cokes. Air-blowing promoted the development of domains, which were orientated around the fiber. No significant variations were observed with the severity of pitch treatment.

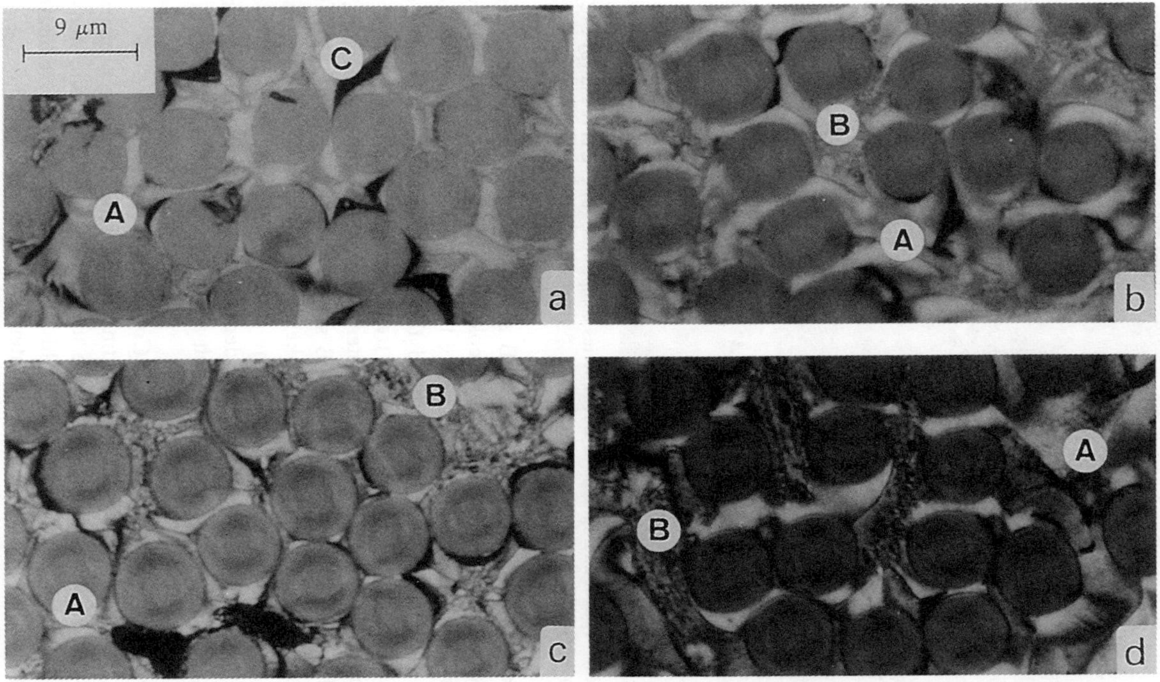

Figure 15. Optical micrographs of composites from CTPB series, (a) C-CTPB, (b) C-CTPB1, (c) C-CTPB2 and (d) C-CTPB3: (A) domains, (B) agglomerates, (C) shrinkage cracks.

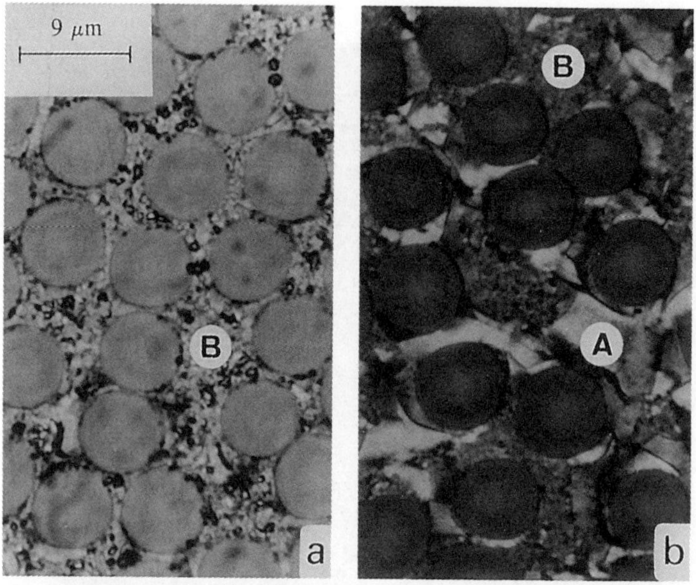

Fig. 15 *(continued)*. Optical micrographs of CTPA series composites, (a) C-CTPA and (b) C-CTPA2: (A) domains, (B) agglomerates.

In the case of composites from CTPB series, again matrix microstructural changes correspond to those of individual cokes. With pitch air-blowing there is a larger contribution of mosaics, intramatrix located, these being accompanied by the presence of domains orientated around the fiber. The amounts of mosaics slightly increased with the severity of pitch treatment.

The values of porosity (vol.%), as determined by optical microscopy, were very similar for all the composites of CTPB series (between 13 and 18%), which means that those changes observed on the cokes are minimized in the composite by the presence of the fiber and possibly by the operational conditions. However, it is interesting to indicate the substantial decrease in the occurrence of cracks (intramatrix and interface located) with pitch air-blowing treatment.

3.1.2.4. Influence of Pitch Air-Blowing on Composite Properties

From the effects of the air-blowing treatment on pitches, and the properties of the resultant cokes, an improvement in the mechanical properties and density of composites from these treated pitches should be expected. Table 11 contains the data on mechanical properties and fiber content of the composites.

Table 11. Properties of Composites

Composite	FC[1]	d[2]	ILSS[3]	FS[4]
C-CTPA	55	1.66	27	820
C-CTPA1	43	1.70	34	624
C-CTPA2	36	1.73	37	618
C-CTPB	54	1.53	19	486
C-CTPB1	57	1.59	36	644
C-CTPB2	56	1.58	21	609
C-CTPB3	42	1.57	32	432

[1] Fiber content (vol.%)
[2] Density (g cm^{-3})
[3] Interlaminar shear strength (MPa)
[4] Flexural strength (MPa)

In spite of the improvement of the density and strength observed for the series of cokes obtained from air-blowing of CTPA, flexural strength of the composites substantially decrease, from 820 MPa (untreated pitch) to 624 MPa (250°C/air) and 618 MPa (275°C/air). This decrease can be explained by the presence of higher viscosity systems which produce composites with an increase of matrix contribution. Moreover, the enhancement of the contribution of domains to the optical texture, induced by the treatment, brings about the generation of shrinkage cracks which can also contribute to the lowering of the flexural strength values. Finally, the wettability of pitch with respect to the fiber considerably decreases with CTPA air-blowing treatment, making difficult the preparation of the prepreg for some samples.

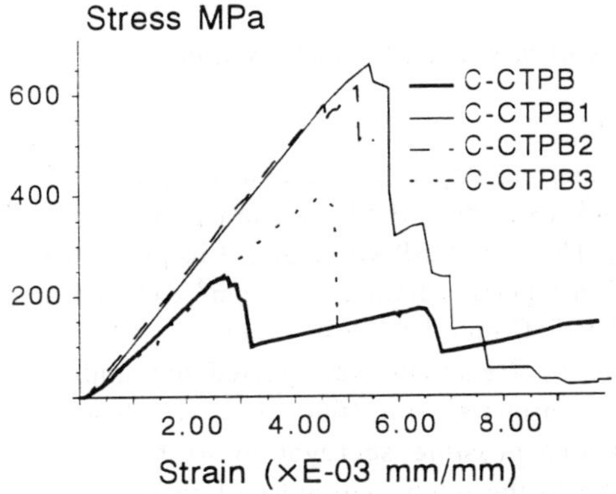

Fig. 16. FS curves of CTPB series composites.

With the CTPB series, the composite from 250°C/air pitch, exhibits a considerably higher value of flexural strength (644 MPa) than that obtained from the untreated pitch (486 MPa). This increase is related to the absence of shrinkage cracks, interface located, which were present in the untreated pitch composite and the increased amount of intramatrix mosaics brought about by the oxidation treatment (Fig. 16). However, the flexural strength value, according to the results obtained on the cokes, fell to 609 MPa and 432 MPa with 275°C/air and 300°C/air treatment, respectively. Although the three composites have very similar values of porosity, between 13 and 18%, and also similar matrix microstructure, being even slightly higher in amounts of mosaics with more severe treatment, the low value of 300°C/air composite can be explained by the loss of wettability of the pitch and the high volumetric contraction of the matrix. Both are due to the high viscosity of the system on pyrolysis, caused by the extensive polymerization in the air-blowing treatment. In the same way, the resultant low fiber content (42 vol.%) compared to the rest, which was about 57 vol.%, can also contribute to the decrease in the flexural strength.

The bulk density of composites increased from 1.66 g cm^{-3} for the untreated composite to 1.73 g cm^{-3} in CTPA series and from 1.53 g cm^{-3} for the untreated pitch composite to 1.59 g cm^{-3} for those from the treated ones, in the case of CTPB series. The beneficial effect of air-blowing on fiber/matrix bonding is confirmed by the increase in the interlaminar shear strength value in both series, from 27 MPa for the parent pitch composite to 37 MPa for the composites from treated pitches in the case of CTPA series, and from 19 MPa for the untreated pitch composite to values in the range of 32 to 36 MPa for those from the air-blown pitches in CTPB series. Fig. 17 shows ILSS curves for the composites from CTPB series. Fig. 18 is an SEM micrograph of a fracture surface of CTPB3 composite, showing an excellent adhesion between the carbon matrix and the fibers.

3.2. Carbon Black Addition and Thermal Treatment

3.2.1. Experimental

300 g of a commercial impregnating coal-tar pitch (CTP2) was blended with 9 g of a carbon black (Ketchem Black, 30 nm particle size, 800 m^2 g^{-1} of BET surface area, 360 ml/100 g of DBP value and 9 in pH). After mechanical mixing, samples were placed in a glass reactor of 500 cm^3 capacity and then heated to 400, 425, 430 and 450°C at 5°C min^{-1} for 5 h, except for 430°C which had a residence time of 10 h. Thermal treatment was carried out under nitrogen under reflux conditions, using a vigorous stirring (600 rpm), which was maintained during the cooling down to avoid possible segregation of the carbon black. The same treatment was applied to the pitch, without carbon black, to find out the effect of

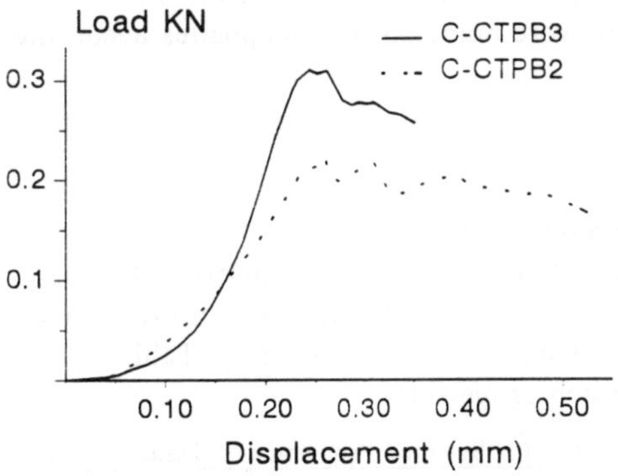

Fig. 17. ILSS curves of CTPB series composites.

Fig. 18. SEM micrograph of a fracture surface of C-CTPB3.

the thermal treatment itself. The result was a series of nine pitches composed of the untreated pitch, four thermally treated pitches and four thermally treated with carbon black. Resultant pitches were characterized by selective solvent extraction in Soxhlet (with hexane, benzene and pyridine), elemental analysis, softening point and carbon yield, as well as FTIR and ^{1}H-NMR of the benzene-soluble fraction.

Cokes from each pitch were obtained by heating in a horizontal furnace from 25 to 100°C at 1°C min^{-1} with 30 min soak time, and characterized by optical microscopy. Porosity of cokes was determined from the polished samples by a point-counting procedure on 500 points, statistically selected, using × 60 magnification.

Composites were prepared from each pitch as described in previous sections. The characterization of the resultant materials was made by optical microscopy in terms of porosity and optical texture of the matrix. Interlaminar shear strength and flexural strength tests were made on the composites according to ASTM standards (Section 2.1.4.).

3.2.2. *Results*

3.2.2.1. *Pitch Characterization*

Differences in pitch composition were observed for the different temperatures and times of treatment, due to their influence on the degree of polymerization of pitch and on the extents of volatile release. Table 12 shows how these two variables influence pitch properties.

Table 12. Selected Properties of Treated Pitches

Pitch Treatment	SP[1]	CY[2]	CY_c[3]	Elemental Analysis (wt.%)				HS	HI-BS	BS	BI-PS	PI
				C	H	N	S+O[4]					
-----	50	35.2	35.2	92.35	4.53	0.98	2.14	31.8	44.2	76.0	13.6	10.4
400 °C-5h												
Only pitch	70	43.7	42.4	93.05	4.44	1.01	1.51	30.5	39.7	70.2	9.3	20.5
3 wt.% CB	58	46.1	44.0	92.99	4.48	1.00	1.53	45.1	25.4	70.5	9.6	20.0
425 °C-5h												
Only pitch	80	53.3	48.5	93.33	4.33	1.06	1.29	23.4	37.5	60.9	8.2	30.9
3 wt.% CB	67	54.3	51.2	93.31	4.27	0.95	1.47	35.0	26.8	61.8	8.4	29.8
430 °C-10h												
Only pitch	92	66.3	57.0	93.96	3.91	0.92	1.22	22.8	24.0	46.8	6.6	46.6
3 wt.%	100	71.1	64.4	94.05	3.97	1.02	0.95	27.2	18.0	45.2	6.1	48.8
450 °C-5h												
Only pitch	108	67.4	56.6	94.09	3.87	0.92	1.12	---	---	43.0	6.1	50.8
3 wt.% CB	167	72.3	64.8	93.98	3.77	0.95	1.29	---	---	35.0	5.6	59.4

1 softening point by TMA (°C)
2 carbon yield (wt.%) at 1000°C, 1°C min^{-1}
3 carbon yield (wt.%) at 1000°C, 1°C min^{-1}, refered to original pitch and corrected from carbon black data
4 by difference

Kanno et al.[5] showed that carbon black additions of 5 wt.% to a mesophase pitch delayed to higher temperatures the initiation of weight loss and improved pitch carbon yield, this being evidence of the occurrence of pitch-carbon black interactions. In the present study, the trend followed by the variation of pitch yield with temperature, in the presence of CB, suggested that interactions between pitch and CB are more significant at the lower temperatures, which means that they could be of more physical nature than chemical, and probably involve adsorption of light components of pitch by the CB particles. However, the carbon yield, referred to parent pitch (CYc), corrected according to the weight loss caused by the treatment and the amount of CB added, linearly increases with temperature, the increase being higher when CB is added, especially above 425°C. This strongly suggests that CB acts as a catalyst for pitch polymerization at temperatures above 425°C. Concerning the variation of pitch softening point with treatment, as determined by TMA, it is interesting to note the lower value observed with CB at temperatures below 430°C, compared to that of pitches obtained by thermal treatment without CB (Table 12). With the increase of temperature, between 400 and 425°C, the increase of the softening point in the presence of CB is relatively small, about 5%. However, above this temperature it strongly increases, from 67 at 425°C, to about 167 at 450°C. From the variation of the content of the fractions obtained by solvent extraction (Table 12) it seems that CB has two opposite effects: 1) at temperatures above 425°C, specially 450°C, CB promotes a dehydrogenative polymerization which mainly depends on the temperature; 2) below 425°C, CB promotes the solubility in hexane of pitch components which initially were not soluble.

FTIR of BS fractions confirms the increase of aliphatic hydrogen (H_{al}), specially in that obtained with CB at 400°C. Fig. 19 shows the FTIR spectra of the BS fraction of parent pitch and the same treated at 400°C for 5 h, with and without CB. The increase of H_{al} in the fraction with CB is mainly due to -CH_2- groups as confirmed by the increase in size of the bands at 2800–3000 cm^{-1}, that

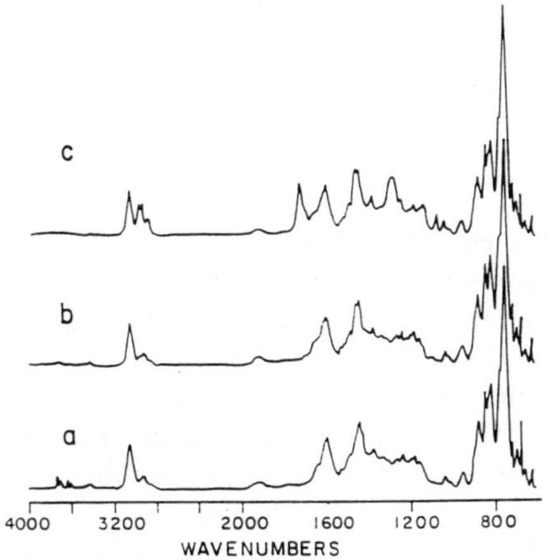

Fig. 19. FTIR spectra of the BS fractions of (a) original pitch, (b) 400°C-5h pitch and (c) CB 400°C-5h pitch.

of 1380 cm^{-1} being almost constant. The presence of new bands at 1725 cm^{-1} (C=O stretching vibrations), 1280 cm^{-1} (C-O stretching), 1120 cm^{-1} ($O-CH_2$ stretching) and 1080 cm^{-1} (arylesters, 6-membered rings ketones or 6-rings lactones), indicate progressive oxidation. CB with treatments below 425°C increases the H_β and H_γ contents of the BS fractions, as ^{1}H-NMR analysis shows (Table 13).

Table 13. Hydrogen Distribution (%) in the BS Fraction, Obtained by ^{1}H-NMR

Treatment	BS[1] (wt.%)	Har[2]	Hα	Hβ	Hγ	Har/Hal [3]
----	76.0	84.9	13.3	1.6	0.3	5.6
400 ℃-5h						
only pitch	70.2	85.0	12.6	1.9	0.5	5.7
3 wt% CB	70.5	76.1	11.4	9.2	3.2	3.2
425 ℃-5h						
only pitch	60.9	86.3	10.0	3.1	0.6	6.3
3 wt.% CB	61.8	82.7	10.8	4.9	1.6	4.8
430 ℃-10h						
only pitch	46.8	89.9	8.0	2.1	0.0	8.9
3 wt% CB	45.2	88.9	9.7	1.1	0.3	8.0
450 ℃-5h						
only pitch	---	90.4	8.4	1.2	0.0	9.4
3 wt.% CB	---	91.8	7.5	0.7	0.0	11.2

[1] values corrected from carbon black data
[2] aromatic hydrogen
[3] aliphatic hydrogen

Thus, it is confirmed that CB has two effects depending on temperature. Below 425°C, CB promotes a hydrogenation of the more reactive aromatic structures, i.e., hydroaromatics, lactones and quinones. At temperatures above 425°C, mainly at 450°C, CB contributes to the thermal polymerization of pitch molecules (tables 12 and 13). This is not surprising if it is considered that at temperatures above 400°C, the rate of formation of free radicals strongly increases with temperature and consequently, the polymerization is always dominant. This agrees with the variation of SP with temperature. The hydrogenated compounds produced at 400°C and 425°C (to a lesser extent) in the presence of 3 wt.% of CB, lower the SP of pitches (Fig. 20), without any significant effect on the BS content. In spite of the SP decrease, the carbon yield slightly increases. Because of that, these pitches show higher values of ΔCY/ΔSP than those obtained without CB (Fig. 21). At temperatures above 430°C, CB produced a sharp increase in the SP with

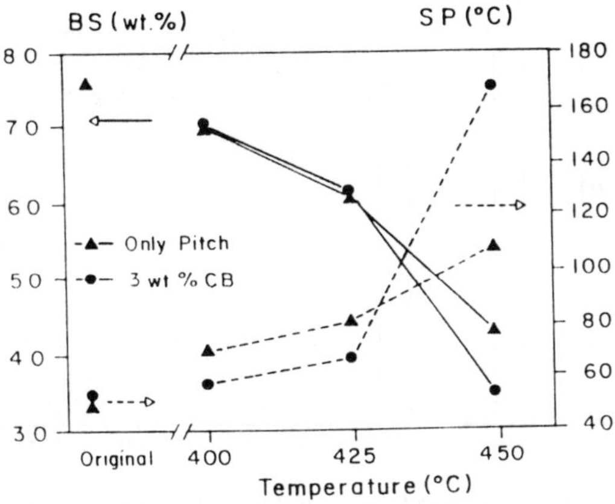

Fig. 20. Variation of benzene solubles content and softening point of pitches with treatment.

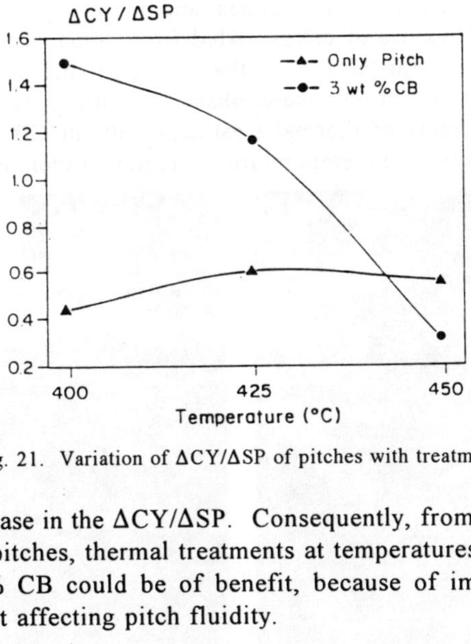

Fig. 21. Variation of ΔCY/ΔSP of pitches with treatment.

the subsequent decrease in the ΔCY/ΔSP. Consequently, from the point of view of the use of coal-tar pitches, thermal treatments at temperatures below 425°C in the presence of 3 wt.% CB could be of benefit, because of improvements of pitch carbon yield without affecting pitch fluidity.

3.2.2.2. *Effects of Treatment on Coke Structure*

Fig. 22 shows the variation of coke porosity, as determined by optical microscopy. The effect of the treatment on coke structure was a major decrease of porosity with CB addition at 400°C, from 98 vol.% to 53 vol.%, then to increase slightly as a consequence of the generation of closed porosity due to the high viscosity of the system. However, with thermal treatment without CB, the decrease

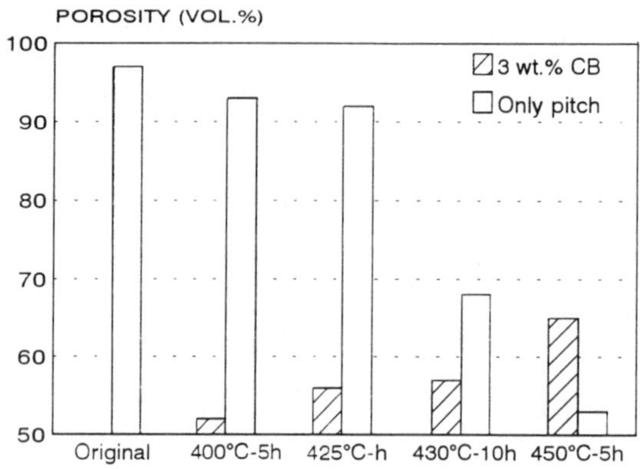

Fig. 22. Porosity of cokes from thermal treated and CB modified pitches.

of porosity was initially less pronounced and more significant at 450°C. As expected, the optical texture of cokes varied from domains for the coke from the original pitch to mosaics for those from the treated pitches with CB particles (Fig. 23). No significant variations were observed with different temperatures of treatment. As for effects of thermal treatment without CB, a slight increase in domain size was observed, as temperature of treatment increased.

Fig. 23. Optical micrographs of cokes from (a) original pitch and (b) CB 400°C-5h treated pitch.

3.2.2.3. Effects on Composite Microstructure

Fig. 24 shows optical micrographs of cross-sections of composites from treated pitches, with and without CB. Variations of optical texture of composites from thermally treated pitches are similar to the cokes, with an increase of domain size with temperature of treatment, these domains being orientated around the fiber.

Composites from CB treated pitches show a variation in optical texture similar to those of cokes. As seen in Fig. 24, a decrease in size of the optical texture is produced, from flow domains (untreated pitch) to mosaics with some domains for the treated pitch composites, mosaic content increasing with treatment. Porosity (vol.%), as determined by optical microscopy, decreases in initial stages of CB treatment (from 13% for the composite from original pitch to 11% for the 400°C-5h CB composite), then to increase in the next stages of treatment (18% for the 430°C-10h CB composite). This trend is the same as that observed for the cokes.

3.2.2.4. Influence on Composite Properties

Table 14 shows the values of the bulk density, fiber content and the results of the mechanical testing of the composites.

Table 14. Properties of Composites

Treatment	ILSS[1]	FS[2]	d[3]	FC[4]
----	14	486	1.53	57
400°C-5h				
only pitch	19	674	1.54	57
3 wt.% CB	20	736	1.52	53
425°C-5h				
only pitch	25	783	---	---
3 wt.% CB	26	676	1.59	50
430°C-10h				
only pitch	28	583	---	44
3 wt.% CB	33	484	---	41

[1] interlaminar shear strength (MPa)
[2] flexural strength (MPa)
[3] bulk density (g cm^{-3})
[4] fiber content (vol.%)
---, not evaluated

The preparation of a composite from the pitch treated at 450°C, with and without carbon black, was not possible due to the high viscosity of the pitch. The interlaminar shear strength of composites from treated pitches was considerably improved when compared with that of composites from untreated pitch, and increases with the severity of the treatment. This is because of better fiber/matrix

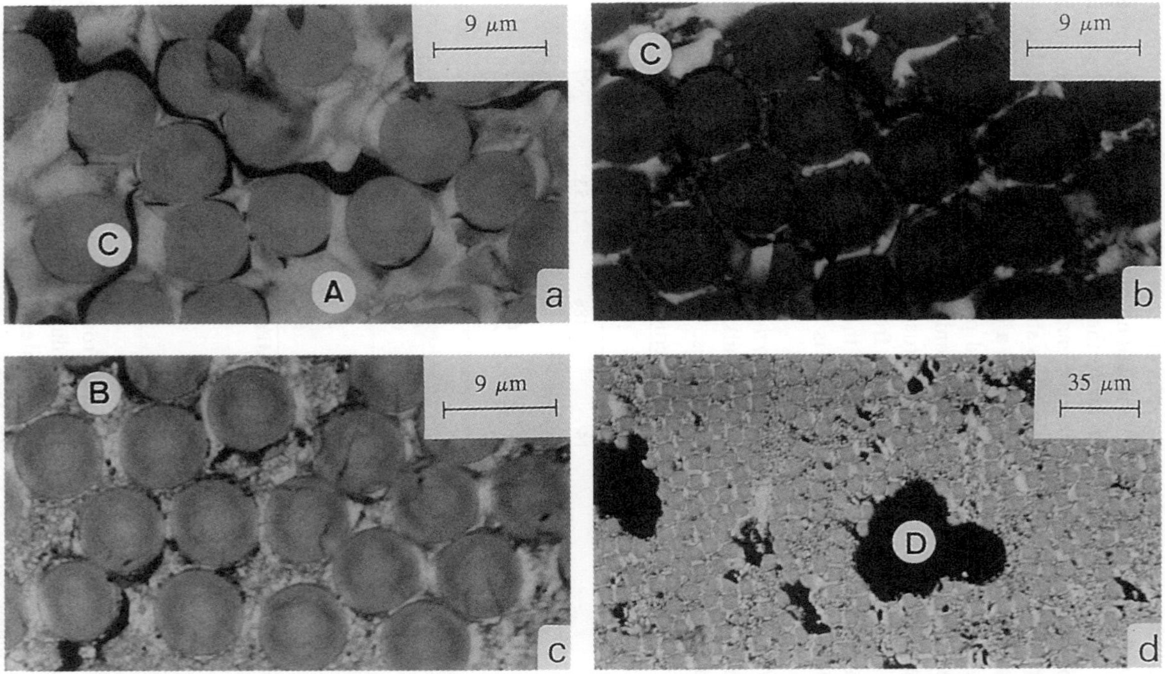

Fig. 24. Optical micrographs of cross-sections of (a) composite from untreated pitch, (b) composite from 430°C-10h thermally treated pitch, (c) composite from CB 400°C-5 h pitch and (d) composite from CB 430°C-10 h pitch: (A) domains, (B) mosaics, (C) interface cracks and (D) porosity.

bonding. With the CB treatment, the interface cracks present in the untreated pitch composite almost disappear (Fig. 24). The decrease in size of the optical texture produced by this treatment, from flow domains (untreated pitch) to mosaics with some domains also contributes to this effect.

The composite from 400°C-5h pitch has a considerably higher value of flexural strength (674 MPa) than the composite from the untreated pitch (486 MPa), then to decrease with further treatment (583 MPa for the 430°C-10h composite). This is due to the lower fiber volume fraction of the latter composite along with the enhancement of volumetric contraction of the matrix with treatment, caused by polymerization of molecules. Flexural strength of composites increases further with carbon black addition at 400 and 425°C, being higher for 400°C. At 430°C flexural strength is similar to that of the untreated pitch composite. The increase at the initial stage of treatment is related to lower porosity of the composite and the decrease in the optical texture of the matrix produced by the carbon black (fiber content is very similar in the two composites). The substantial decrease in flexural strength from 400 and 425 to 430°C is associated to loss of wettability of pitch, development of porosity (Fig. 24) and an increase in matrix content (Table 14), all these because of generation of a high viscosity system, highly polymerized. This is in agreement with predictions made from the variations of $\Delta CY/\Delta SP$ with pitch treatment.

4. Acknowledgments

This research was supported by CICYT (Project MAT94-0715) and FICYT.

5. References

1. E. Fitzer, *Carbon* **25** (1987) 163.
2. S. B. Barr and I. C. Lewis, *Carbon* **16** (1978) 439.
3. M. Hein, *Erdöl und Kohle. Ergas-Petrochemie-Hydrocarbon Technology* **43**, (1990) 354.
4. K. Yanagisawa and T. Suzuki, *Fuel* **72** (1993) 25.
5. K. Kanno, K. E. Yoon, J. J. Fernández, I. Mochida, F. Fortin and Y. Korai, *Carbon* **32** (1994) 801.
6. A. Figueiras, J. J. Fernández, M. Granda, J. Bermejo, E. Casal and R. Menéndez, *Journal of Microscopy* **177** (1995) 218.
7. M. A. Forrest and H. Marsh, in *Coal and coal products: analytical characterization techniques,* vol. 205, (American Chemical Society, Washington, DC, 1982).
8. M. A. Forrest and H. Marsh, *J. Mat. Sci.* **18** (1983) 973.

9. M. D. Guillén, M. J. Iglesias, A. Domínguez and C. G. Blanco, *Energy & Fuels* **6** (1992) 518.
10. M. Granda, J. Bermejo, S. R. Moinelo and R. Menéndez, *Fuel* **69** (1990) 702.
11. M. Granda, R. Menéndez, S. R. Moinelo, J. Bermejo and C. E. Snape, *Fuel* **72** (1993) 19.
12. R. Menéndez, M. Granda, J. Bermejo and H. Marsh, *Fuel* **73** (1994) 25.
13. B. Rand. *Fuel* **66** (1987) 1491.
14. E. Fitzer and B. Terwiesch, *Carbon* **10** (1972) 383.
15. J. J. Fernández, A. Figueiras, M. Granda, J. Bermejo and R. Menéndez, *Carbon* **33** (1995) 295.
16. S. Ragan and H. Marsh, *Fuel* **60** (1981) 522.
17. S. Otani, *Carbon* **5** (1967) 219.
18. S. M. Zeng, T. Maeda, K. Tokumitsu, J. Mondori and I. Mochida, *Carbon* **31** (1993) 413.
19. S. Ragan, A. Grint and H. Marsh, *Fuel* **60** (1981) 646.

THE SURFACE TENSION OF BINDERS AND THEIR MEASUREMENT METHODS

M. KALOČ and E. LINZER
VŠB-Technical University of Ostrava
708 33 Ostrava-Poruba, Czech Republic

D. KOŠŤÁL
DEZA a.s. Valašské Meziříčí - Czech Republic

ABSTRACT

This paper presents the comparison of two methods for measuring the surface tension of pitches: the lying drop and the so-called maximal pressure in the gas bubble methods. The method of lying drop is known. It is very simple; however, results can be influenced by the way the technique is carried out. The gas bubble experimental apparatus has simple construction and guarantees the reproducibility and the accuracy of the results. By this method, it is also possible to determine the density of the liquid binder. The calibration of the apparatus was performed using glycerol as a standard. This provided adequate precision for these purposes. One of the most important factors in this method is the kinetics of the formation of the gas bubble. Four binding pitches of interest to the carbon industry were used as samples. The temperature interval was 190–260°C. The experimental results show that the surface tension depends on the group composition of the pitch, especially QI. The penetration ability was determined using the penetrating drop techniques as applied to a coke powder briquette. The effects of temperature changes were studied on the photos of the system "briquette-pitch." The depth of pitch penetration and other relations were studied in the polished section of this system. These results show that the penetration is also independent of QI content. Comparison of the results obtained using the gas bubble technique and the lying drop method show good agreement.

1. Introduction

The experimental methods used when testing basic surface tension can be categorized into:

- methods of capillary elevation and depression
- the drop weight method
- detaching methods
- the hanging or lying drop method
- the "maximal pressure in the gas pocket" method.

Many factors must be considered when choosing experimental methods for the determination of surface tension of the liquid phase. Most important are the character of the liquid phase under examination and its changes with temperature, measurement temperatures, and supposed values of surface tension.

For the group of liquids and melts of the type of tar or pitch, three fundamental groups of methods are worth considering: detaching methods, the lying drop method, and the maximal pressure on the gas pocket method.

2. Detaching Methods

These methods are based upon the measurement of the force needed for detaching a corpuscle from the surface of the liquid-melt under examination. There are many variants of this method, differing in the corpuscle type used, e.g., vertical plate, hollow roller, horizontally suspended ring. The immersion depth of the corpuscles at the equilibrium of forces (when using a vertical plate or a hollow roller) and the force needed to detach the corpuscle from the liquid level are alternately measured.

These methods are considered to be very accurate, but the measurements require considerable care. For example, the correct geometrical location of a measuring corpuscle towards the liquid level must be known, and a constant liquid level must be maintained, i.e., without shock, releasing the gaseous phase, etc. Another important basic requirement is good wettability of the corpuscle ($\Theta \ll 90°$) as well as the inertness of the corpuscle material to the liquid-melt studied (especially in the case of melts of metals and slags).

The calculation of surface tension uses the following equation:

$$W = W_{plate} + 2 \cdot \sigma \cdot (x + y) \tag{1}$$

where W_{plate} is the mass of the plate, W is the detaching force, x is the width of the plate, and y is the immersion depth.

3. The Lying Drop Method

The principle consists in the determination of chosen geometrical parameters of a drop of the observed liquid-melt placed on a solid plate. The plate used must not react with the liquid phase and must be mechanically rigid at the temperature of the experiment.

The method enables the simultaneous assessment of the density of the liquid phase (ρ) and of a wetting angle (θ). The implementation of the method differs primarily in the manner of recording the shape of the drop and processing of the image. Depending upon the design of the measuring apparatus, X-ray radiation or photographic recordings have been used. The accuracy of measuring depends on the enlargement.

The parameters evaluated are shown in Fig. 1.

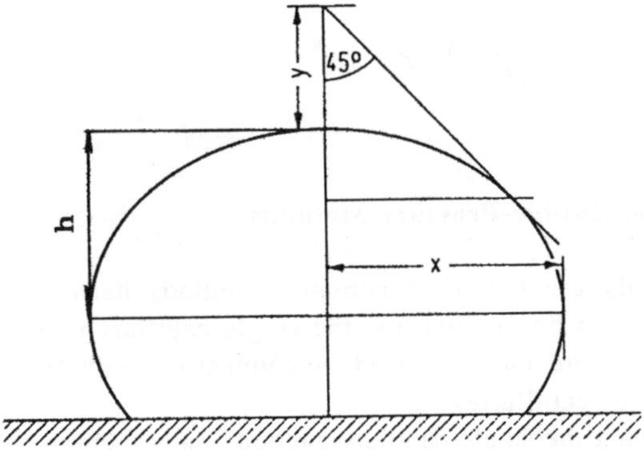

Fig. 1. Method of evaluating parameters.

At the Department of Physical Chemistry and Theory of Technology Processes of the Faculty of Metallurgy and Materials Engineering at VŠB-Technical University, a CCD camera is used and the images are digitally processed (Fig. 2). The details are given elsewhere[1].

Fig. 2. Example of digital processing.

According to Fig. 1, the surface tension may be calculated by the Dorsey equation[2]:

$$\sigma = g \cdot \Delta\rho \cdot x^2 \cdot \left[\frac{0.052}{f} - 0.12268 + \frac{0.0481}{f} \right] \qquad (2)$$

where $f = (x/y) - 0.41421$.

or by the equation:

$$\sigma = \frac{4 \cdot x^2}{H} \cdot \Delta\rho \cdot g \qquad (3)$$

where $1/H = f(x/2h)$.

4. The Maximum Bubble-Pressure Methods

These methods are the most common methods because of their technical accessibility. They can be divided into the single-capillary method and the double-capillary method. The latter is used in connection with the same or differing immersion depths of capillaries.

When employing the single-capillary method, many methodical problems occur. The accuracy of measuring surface tension is limited by the accuracy of measuring the immersion depth of the capillary. It is necessary to introduce the correction of meniscus curvature of melt in the crucible as a result of wettability of its wall.

The double-capillary method with the same immersion depth of capillaries of various diameters was, for the first time, used by Sudgen in 1922[3]. The starting relation to calculate surface tension is then:

$$P_{\max} = P_{hyd} + P_{\sigma} \qquad (4)$$

where $P_{hyd} = \rho \cdot g \cdot h$, i.e., it depends upon the liquid density and immersion depth of the capillary; and P_{σ} is given by the capillary constant and the surface curvature of a rising gas bubble. Because surface tension does not depend upon the immersion depth of liquid phase from the gradient of line dependence, $P_{max} = f(h_l)$.

In addition, according to the demands on accuracy when measuring at higher temperatures, corrections, including the changes in geometrical parameters of the crucible and capillaries on connection with expansion of used materials, can be allowed for in the evaluation.

5. References

1. E. Linzer and B. Horák, *Aplikace metody ležící kapky s použitím počítačoví analýzy obrazu*, in Mezinárodní konference VŠB-TU Ostrava, sekce č.9, září 1995, s.26.
2. N. E. Dorsey, *J. Wash. Acad. Sci.* **18** (1928) 505.
3. S. Sudgen, *J. Chem. Soc.* **121** (1922) 1482.

USABILITY OF COAL TAR FOR THE PREPARATION OF INDUSTRIAL CARBON

M. KALOČ, D. KOŠŤÁL, and M. JANÍK
Faculty of Metallurgy and Material Engineering
Technical University of Mining and Metallurgy
tř. 17. listopadu, 708 33 Ostrava - Poruba, Czech Republic

ABSTRACT

Coal tar presents a very significant raw material for the production of various kinds of pitch, which may be used as the binder for different industrial carbons. According to the subsequent processing and the use of the final component, various requirements are put on its quality which affects some technical and technological factors.

This contribution evaluates the properties of a number of Czech coal tars. The tars have been ascertained to contain a great amount of the substances insoluble in toluene (TI) and quinoline (QI). Their presence affects the processing needed to form coke.

The properties of the final pitch depend on the origin, the properties and the behavior of the compounds insoluble in quinoline. These are always contained in the coal tar and they remain in the pitch after its distillation. These QI are divided into the primary and secondary parts; both parts have a specific influence on the final quality parameters.

1. General

High-temperature coal tar originating in the amount of 2.5 to 4.5% during the chamber carbonization of suitable mixtures of black coals can be included among the important resources for the production of industrial carbonaceous materials. The yield of tar depends upon, among other matters, the content of volatile constituents in coal (see Table 1).

Table 1. Yield of Crude Coal Tar and Crude Benzol (according to content of volatile matter)

V^{daf} (%)	Tar (%)	Benzol (%)
20 – 22	2.5 – 2.7	0.8 – 0.9
23 – 25	2.9 – 3.2	0.9 – 1.0
26 – 27	3.5 – 4.0	1.1 – 1.2
30 – 32	4.2 – 4.6	1.3 – 1.5

A basic factor affecting the amount and composition of tar within the coking process is the temperature of carbonization. In Fig. 1a, the course of changes in yields of products of carbonization is given with regard to the temperature. The changes shown are the summary of all the carbonization courses. It follows from Fig. 1b that, in addition to the drop in the total amount of tar, together with increasing temperature, the change in tar composition occurs, which is characterized by its group composition. The higher coking temperature provides the structural transformations of primary products in the coking chamber with more heat energy. It is especially the case of condensation to more complex compounds, or to infusible and insoluble polyaromatic complexes, including almost elementary carbon in the form of soot particles and pyrocarbon simultaneously with the formation of low-molecular non-aromatic compounds and hydrogen.

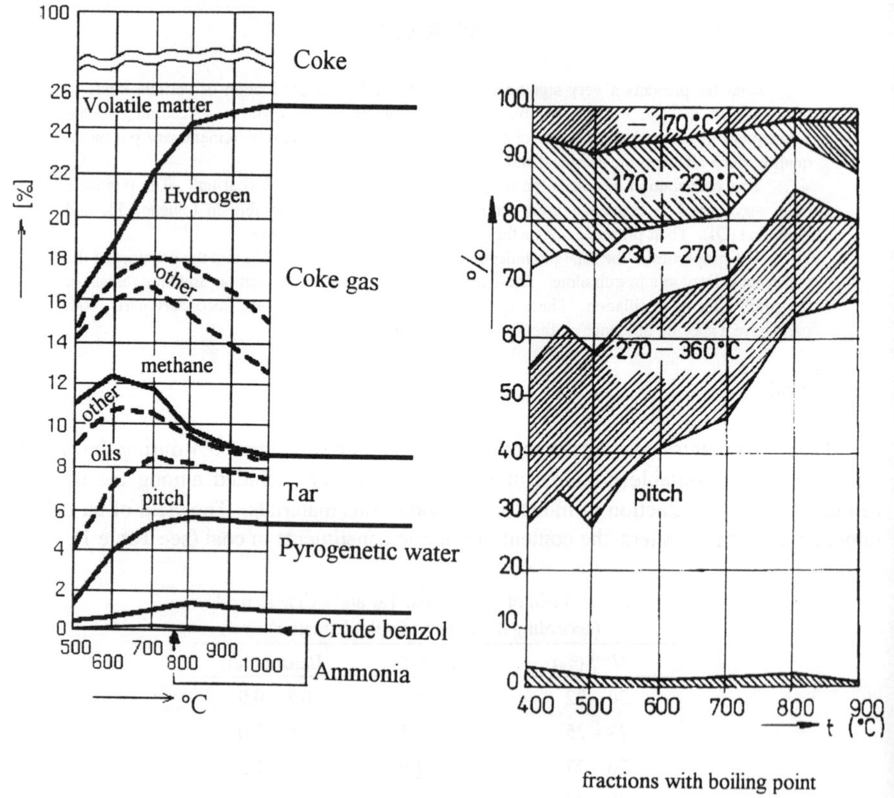

a) b)

Fig. 1. (a) Yield of products by coking, (b) Changes of tar composition versus coking temperature.

The quality of tar is influenced also by the type and arrangement of the coke oven, including its momentary technical state (e.g., tightness of masonry, etc.).

A decreased tightness of masonry within the coke oven has probably the worst effect upon the quality of tar, because air or waste gases sucked into the chambers enable the formation of the matters that usually rank with the TI group, and especially the QI group. Some components of combustion products (e.g., NO_x) even catalyze polymerization and polycondensation of aromatic and unsaturated compounds.

To form an idea of usability of tars from our coking plants for preparation of binding pitches, our production of the last year was evaluated according to the following criteria: TI, QI, and pitch content. Values found are presented in Table 2.

Table 2. Main Characteristics of Czech Coal Tars 1986 and 1994

Plant	Year	TI (%)	QI (%)	Pitch (%)
Svoboda	1986	15.6	10.7	64.6
	1994	7.8	5.5	62.8
Šverma	1986	16.4	12.1	62.9
	1994	6.7	4.1	62.3
Armáda	1986	14.2	9.0	65.0
	1994	23.8	19.8	77.3
Třinec	1986	7.2	3.7	61.2
	1994	11.0	8.4	66.2
Vítkovice	1986	8.5	4.2	62.3
	1994	7.1	4.0	61.9
Nová huť	1986	6.9	3.6	58.6
	1994	4.9	2.6	62.0

What plays an important role is the content of matters insoluble in chinoline that, according to the manner and phase of their origin, can be divided into the primary QI and the secondary QI[1]. This is schematically illustrated in Fig. 2.

These QIs influence another application of pitch because: (a) influence the sedimentary separation of water from the tar-effectivity of the technology; (b) influence a binding function of the pitch (see ref. 2); and (c) influence the forming of anisotropic mesophase, which is, in some cases, a necessary precondition for getting the final product of required quality.

For the purpose of rationally, technically and economically correct utilization of coal tar, or coal tar pitch, it is necessary to pay great attention to this circumstance, and the following may be considered as highly profitable:

- correct control of the coking process
- maintenance of the technical state of masonry at the required level
- selective storage and processing of tars depending upon their input quality.

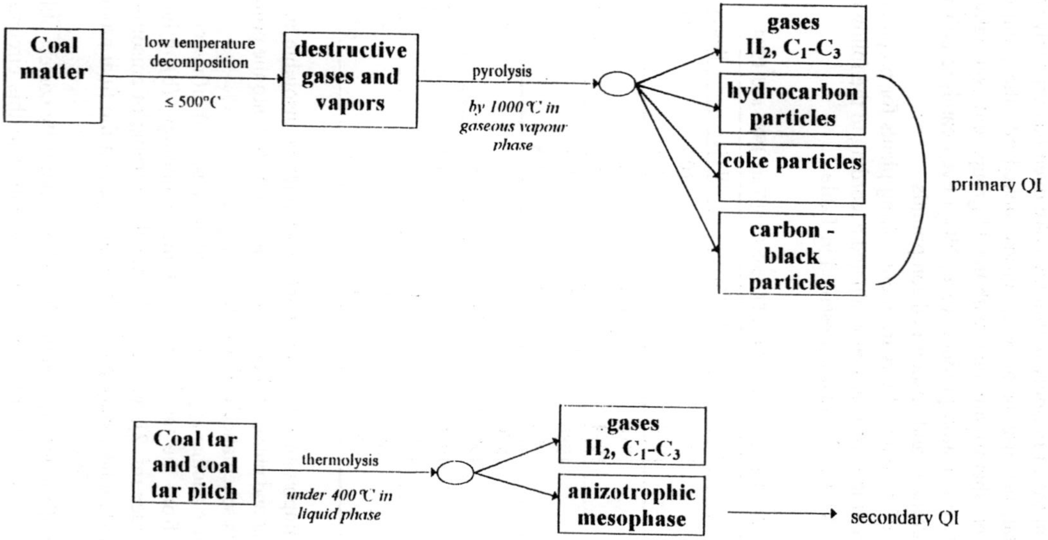

Fig. 2. Room temperature resistivity of various pitch-based carbon fibers versus in-plane coherence length.

2. References

1. M. Janík and M. Kaloč, Hutnické listy **40** (1985) 566.
2. M. Kaloč, D. Košťál, and M. Janík M, Some Properties of Coal Tar Binders, in *Abstracts of Carbon and Carbonaceous Composite Materials: Structure-Property Relationships*, (1995).

Interface Effects

SURFACE ENERGETICS, INTERFACE AND MECHANICAL PROPERTIES OF CARBON/CARBON COMPOSITES

O. P. BAHL and T. L. DHAMI
Carbon Technology Unit
National Physical Laboratory, Dr. K. S. Krishnan Road
New Delhi - 110 012, India

ABSTRACT

The effect of surface energetics on the pyrolysis behavior, mechanical properties and fracture behavior of carbon/carbon composites developed using different reinforcements possessing varying amounts of surface energetics and phenolic resin and coal tar pitch as matrix precursors has been studied. Cross-sectional shrinkage is found to be directly proportional to the amount of surface energetics present on carbon fibers. Flexural strength, in general, is maximum (1700 MPa) for HT fiber based composites at polymer stage. On carbonization, flexural strength decreases; the decrease is proportional to the amount of surface energetics present on carbon fibers. Fracture mode changes from pure shear (no surface treatment) to tensile cum shear as the carbon fiber surface energetics increase and eventually transforms to pure tensile in the case of maximum surface energetics.

On graphitization the strength values of carbon/carbon composites increase; the maximum increase is for HM fiber based composites (800 MPa). Carbon/carbon composites developed using Panex fibers (oxidized PAN fibers) and coal tar pitch as matrix show comparable mechanical properties (without densification) when heat treated to 2700°C to carbon/carbon composites using high strength (T-300 Toray) carbon fibers. On the other hand, in resin based carbon/carbon composites, fiber/matrix interactions are so strong that they lead to formation of hollow carbon fibers. On HTT 2700°C the composite became so fragile that it became almost difficult to handle.

1. Introduction

In the recent past we have witnessed the development of highly sophisticated steels and super alloys which have contributed enormously to the overall development of various technologies affecting every walk of life. Carbon has been known to mankind a long time, initially used as a material for writing ink. With development of carbon industry, it is being used in different forms such as carbon black, activated carbon, electrodes, brushes, cinema arc carbons, etc.

Carbon holds a special place in the periodic table. It is one of the lightest refractory materials and holds its mechanical properties up to 2000°C. It is a unique material in the sense that it can exhibit a range of properties, as soft as graphite form and as hard as diamond. This is possible as it can possess different structures. Fig. 1 shows different properties exhibited by this material.

Carbon material possesses different properties, but, because of its brittle nature, it is difficult to convert it into different complicated shapes in its monolithic form. During the 1960s carbon filaments were invented which possessed superior mechanical properties, thus

CARBON CAN BE		
• EXTREMELY STRONG HARD AND STIFF	:	ALSO SOFT AND DUCTILE
• HIGHLY POROUS	:	ALSO IMPERVIOUS TO LIQUIDS AND GASES
• POSSESSING LARGE SURFACE AREA	:	ALSO WITHOUT ACTIVE AREA
• EXTREMELY BRITTLE	:	ALSO PSEUDO-PLASTIC WITH HIGH FRACTURE TOUGHNESS

Fig. 1. Different properties of carbon.

overcoming the deficiencies experienced by the monolithic form of carbon. Since then, different varieties of carbon fibers, ranging from high strength, intermediate modulus to high modulus, have been developed and are being used in different varieties of composites such as carbon fiber reinforced plastics and carbon fiber reinforced carbon (carbon/carbon composites).

Carbon/carbon composites consist of fibrous carbon embedded in a carbonaceous matrix[1,2]. The matrix carbon can be generated by different routes, such as chemical vapor deposition[3], thermosetting resins[4] or thermoplastic pitches[5]. These are being used in a number of strategic applications such as defense, space, nuclear energy and high-tech engineering. Carbon/carbon composites are amongst the top ranking ceramic composites considered for high temperature applications because their mechanical and thermal properties are far superior to the conventional materials.

The performance of fiber reinforced composites depends on the mechanical properties of the reinforcing fibers, matrix system and fiber/matrix interactions. As far as polymer composites are concerned, the first two factors play dominant roles. In the case of carbon/carbon composites, the reinforcement and the matrix system are brittle in nature; thus, the fiber/matrix interactions play a predominant role in deciding the ultimate performance of the composite. The function of the fiber/matrix interface is to transfer the stress from one fiber to another and to decide the fracture behavior which governs the mechanical properties of the carbon/carbon composites.

The fiber/matrix interactions in both polymer composites and carbon/carbon composites depend on the surface energetics of the carbon fibers, the precursor used for generating the matrix carbon and the processing parameters, particularly the final heat treatment temperature in case of carbon/carbon composites.

Numerous studies[6-8] have been conducted on the development of carbon/carbon composites using different matrix precursors. The main emphasis of the studies has been the effect of processing parameters on the microstructure of the matrix carbon and its relation to the mechanical and thermal properties of the carbon/carbon composites.

The results are presented detailing the effect of surface energetics on the overall performance of carbon/carbon composites. Different types of reinforcements commercially available [high strength, high modulus (surface treated and non-surface treated) and intermediate modulus] and Panex fibers (manufactured in the lab) and two types of matrix precursors (phenolic resin and coal tar pitch) were used for the development of carbon/carbon composites.

2. Experimental

2.1. Reinforcements

The following carbon fibers were used as reinforcements :

1. HTA, high strength carbon fibers
2. HM-45, high modulus carbon fibers
3. IM-500, intermediate modulus carbon fibers
4. X-340, high modulus carbon fibers without any commercial surface treatment
5. PR-I, PR-II and PR-III, oxidized PAN fibers (developed in lab) with different isothermal times in a oven at 250°C.

2.2. Matrices

1. Resole type phenol formaldehyde (phenolic) resin developed by VSSC, Trivandrum, India, and marketed by IVP Limited, India.
2. HGD preforming pitch 3/85 and HGD impregnating pitch 18/85 (product of M/s. HGD, France)

2.3. Fabrication and Characterization of Composites

Unidirectional carbon fiber reinforced polymer composites were prepared using prepregging and compression molding technique. These were carbonized up to 1000°C at 10°C/h under an inert atmosphere. These carbonized composites were porous in nature and had very low density. These were then densified using multiple impregnation/carbonization technique. Porous carbon/carbon composites were impregnated at a pressure of 25 bars with impregnating coal tar pitch having a coke yield of about 45%. Composites were graphitized up to 2700°C in an inert atmosphere.

The composites were characterized for density, pyrolysis behavior, mechanical properties, microstructure and fracture behavior.

3. Results and Discussion

3.1. Carbon/carbon Composites Developed with Commercially Available Carbon Fibers

3.1.1. Surface energetics of carbon fibers

The surface groups present on various carbon fibers have been evaluated by thermogravimetric analysis; details are given in Fig. 2. The amount of carboxyl groups which decompose up to the temperature range of 600–650°C is maximum (4 to 5%) on high strength fibers, as is expected. IM-500 carbon fibers have about 3.5–4.0% and HM-45 carbon fibers possess a negligible amount (0.1 to 0.2%) of carboxyl groups. The amount of carboxyl groups present on X-340 carbons fibers is almost negligible.

The surface groups present in the form of carbonyl and hydroxy (decompose in the temperature range of 900–1000°C) are almost the same in HTA and IM-500 carbon fibers (9–10.5%). HM-45 carbon fibers possess about 5–6% of these surface groups whereas X-340 carbon fibers possess least amount of carbonyl groups (4–5 %).

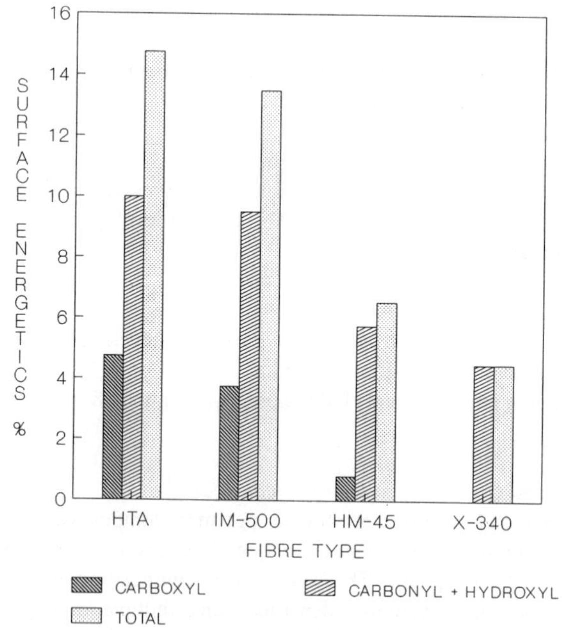

Fig. 2. Surface energetics of various carbon fibers.

3.1.2. Pyrolysis behavior of resin and coal tar pitch based polymer composites

It is well known that the pyrolysis behavior of polymer composites[4,9,10] which is dependent on the coke yield of the resin used, fiber/matrix interactions and processing parameters, affects to a large extent the microstructure of matrix carbon and mechanical properties of carbon/carbon composites[2,9,11]. The pyrolysis behavior of polymer composites made with different carbon fibers has been studied. During carbonization of polymer composites, which is generally done at very low heating rates (10°C/h), the thermosetting resin decomposes with rearrangement of molecular chains resulting in shrinkage, which changes the macrostructure of the matrix[12,13]. The shrinkage is more in the direction perpendicular to the fibers leading to formation of cracks and voids compared to shrinkage in the direction parallel to the fibers. Fig. 3 shows the cross-sectional shrinkage observed during pyrolysis of polymer composites (resin based). The cross-sectional shrinkage is maximum (18–20%) in composites made with high strength carbon fibers whereas composites made with intermediate modulus carbon fibers exhibit somewhat less cross-sectional shrinkage (13–15%). Composites made with high modulus carbon fibers exhibit 7–8% cross-sectional shrinkage and X-340 carbon fibers exhibit the least cross-sectional shrinkage (5%). The cross-sectional shrinkage observed in composites made with different carbon fibers is on the expected lines. Carbon fibers having maximum surface groups show maximum shrinkage, indicating very strong fiber/matrix bonding whereas carbon fibers having a minimum amount of surface groups show minimum shrinkage, indicating weak fiber/matrix bonding.

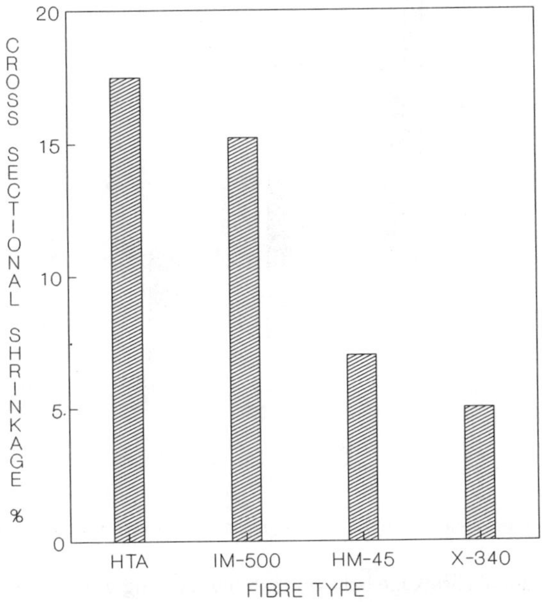

Fig. 3. Cross-sectional shrinkage of polymer composites (resin) during pyrolysis.

Since coal tar pitch is thermoplastic in nature, it was not possible to measure the cross-sectional shrinkage during the carbonization stage.

3.1.3. Cross-sectional shrinkage during graphitization

During graphitization of composites up to 2700°C, reorientation of graphitic planes takes place resulting in appreciable shrinkage. Fig. 4 shows cross-sectional shrinkage exhibited by resin based carbon/carbon composites and coal tar pitch based carbon/carbon composites. The pattern of cross-sectional shrinkage exhibited by resin based carbon/carbon composites is the same as that of polymer composites during carbonization. High strength carbon fibers based composites show maximum shrinkage (20–22%) whereas high modulus carbon fibers based composites show shrinkage of the order of 12–14%. Minimum shrinkage is observed in composites fabricated with non-surface treated carbon fibers, i.e, X-340 (5–7%).

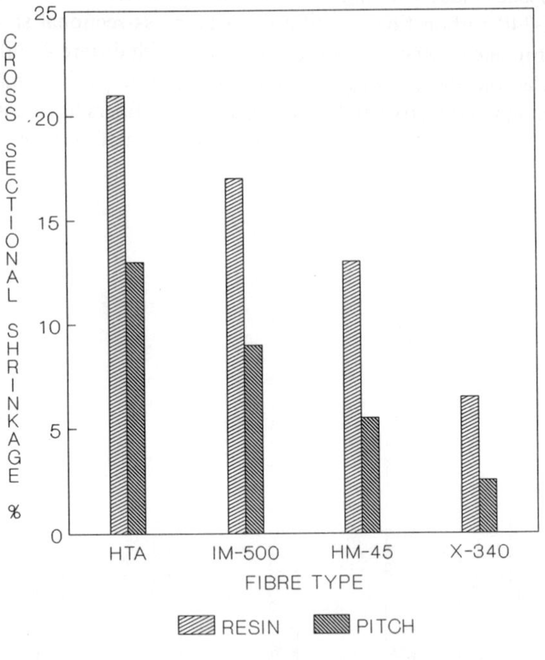

Fig. 4. Cross-sectional shrinkage during graphitization of carbon/carbon composites.

The cross-sectional shrinkage behavior exhibited during graphitization of coal tar pitch based composites is the same as observed with resin based composites although the amount of cross-sectional shrinkage exhibited is somewhat less. This may be due to the inherent difference in macrostructure of the carbon matrix derived from thermosetting resins and thermoplastic pitch.

It appears that the fiber/matrix[2,12,14] bond formed during the polymer stage of composites ultimately decides the cross-sectional shrinkage exhibited by carbon/carbon composites during the carbonization and graphitization stages. Fiber/matrix bonding seems to be directly proportional to the amount of surface groups present on the carbon fiber surface.

3.1.4. Mechanical properties and fracture behavior of resin based carbon/carbon composites

Fig. 5 shows flexural strength of resin based carbon/carbon composites made with different carbon fibers (HTA, IM-500, HM-45 and X-340) at different stages of fabrication. As was observed during carbonization of polymer composites, an appreciable weight loss and cross-sectional shrinkage take place, leading to formation of cracks and voids. All the above observed steps lead to reduction in flexural strength of the composites during carbonization. The fall in the flexural strength is proportional to the cross-sectional shrinkage exhibited by the polymer composites during carbonization. Composites exhibiting maximum shrinkage show maximum decrease in flexural strength whereas those exhibiting minimum shrinkage show minimum decrease in flexural strength.

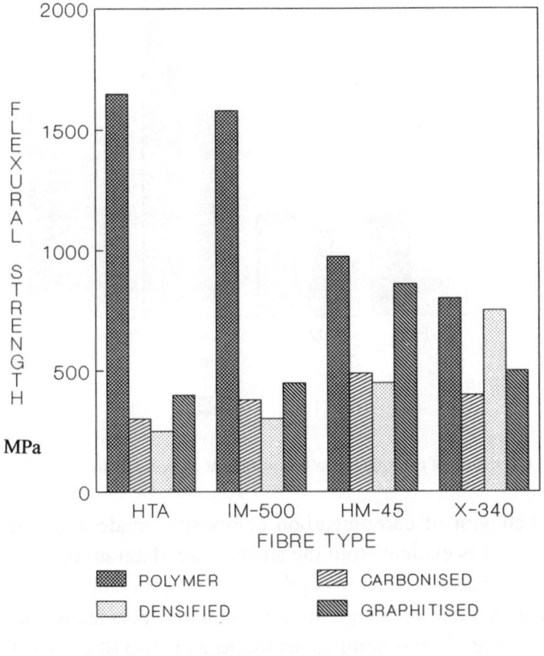

Fig. 5. Flexural strength of carbon/carbon composites at polymer, carbonized, densified and graphitized stages (resin based).

Fig. 6 shows the fiber strength translation of resin based composites made with different carbon fibers at both carbonized and graphitized stages. Maximum translation is observed for HM-45 fiber based composites whereas X-340 fiber based composites show about 35% translation. HT and IM fiber based composites show < 20% translation. The behavior is the same at the graphitized stage although the magnitude of translation changes to about 80% for

HM fiber based composites. X-340 fiber based composites show about 42–43% translation. This clearly shows that for good performing carbon/carbon composites, a little amount of surface treatment, which forms the right degree of bonding with matrix, is a must.

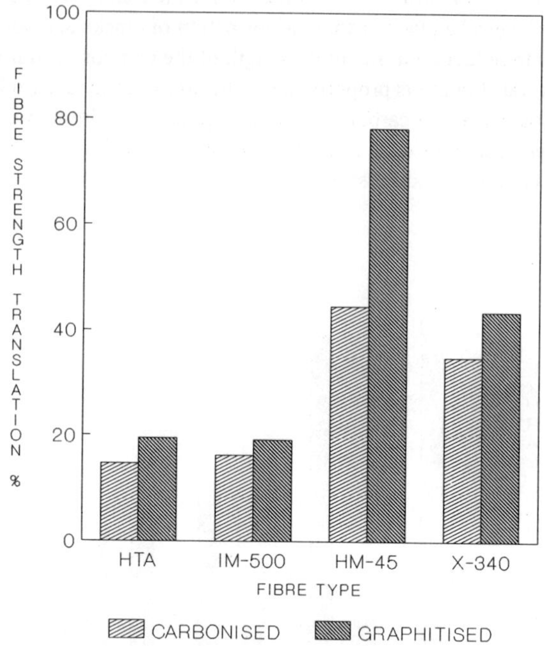

Fig. 6. Fiber strength translation of carbon/carbon composites at carbonized and graphitized stages (resin based).

The fracture behavior of carbon/carbon composites made with four different carbon fibers is quite distinct as is evident from the stress/strain diagram of carbonized composites (Fig. 7a). Composites made with high strength carbon fibers exhibit pure tensile failure (Fig. 7a, curve c); composites made with high modulus carbon fibers show shear cum tensile type of failure (Fig. 7a, curve a); and composites made with X-340 carbon fibers show almost shear failure (Fig. 7a, curve d). Composites made with intermediate modulus carbon fibers show tensile cum shear in steps (Fig. 7a, curve b). The fracture faces of these composites under SEM are shown in Fig. 8. High strength carbon fiber based composites show almost brittle or catastrophic failure with no fiber pull out (Fig. 8) whereas high modulus and X-340 carbon fiber based composites show shear cum tensile failure with lots of fiber pull out (Fig. 9). Intermediate modulus carbon fibers based composites exhibit a comparatively high amount of fiber pull out (Fig. 8) compared to high strength carbon fiber based composites.

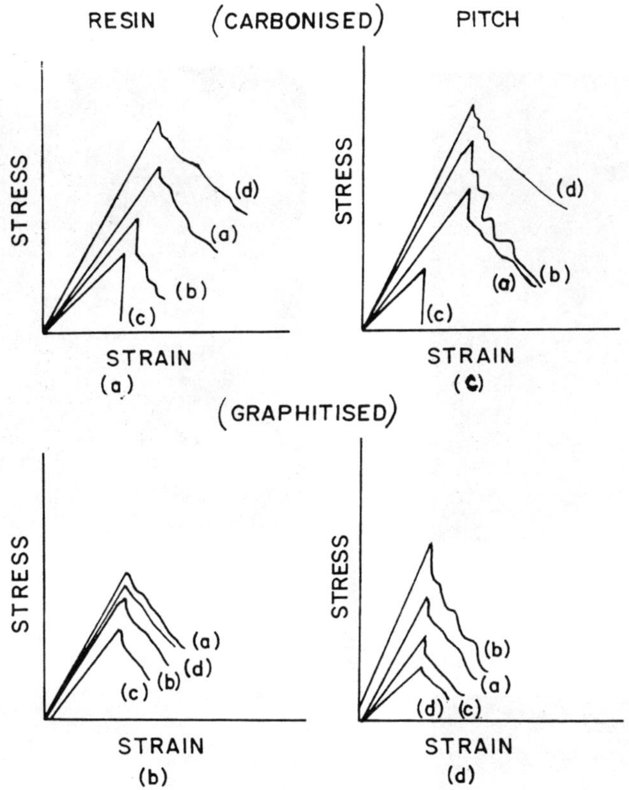

Fig. 7. Stress-strain curves of carbonized and graphitized composites made with (a) HM, (b) IM, (c) HT and (d) X-340 fibers and resin and pitch as matrix.

The differences in the fracture behavior of carbon/carbon composites (carbonized stage) made with different carbon fibers is due to differences in the fiber/matrix bonding and in the macrostructure, as both fibers and matrix at this stage are brittle in nature. Carbon/carbon composites made with high strength carbon fibers show very strong fiber/matrix bonding (Fig. 8) which will lead to propagation of cracks immediately once a crack is initiated, thus showing brittle failure. These cracks are initiated due to shrinkage of matrix resin during carbonization. In the case of carbon/carbon composites made with high modulus (HM-45) and X-340 carbon fibers, the fiber/matrix bonding is not too strong; hence, cracks deviate at the fiber/matrix interphase due to the presence of annular gaps at the fiber/matrix interphase (Fig. 9). This leads to the failure of these composites in tensile cum shear or shear mode (Fig. 7a, curves a and d).

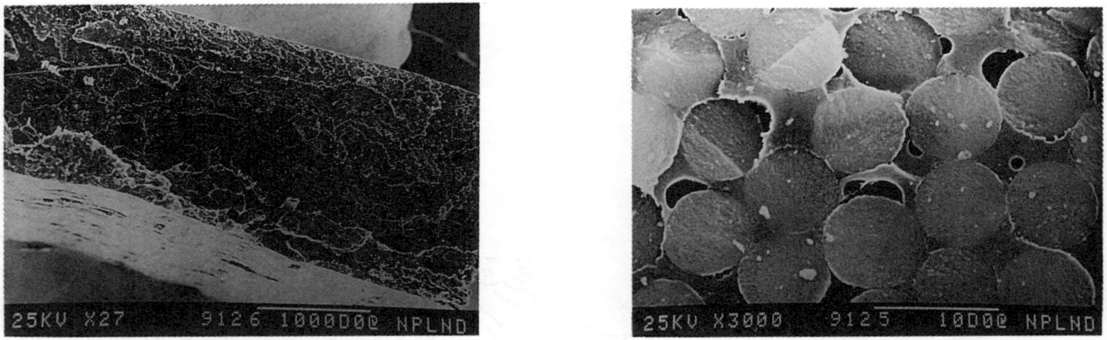

HTA (Carbonised)

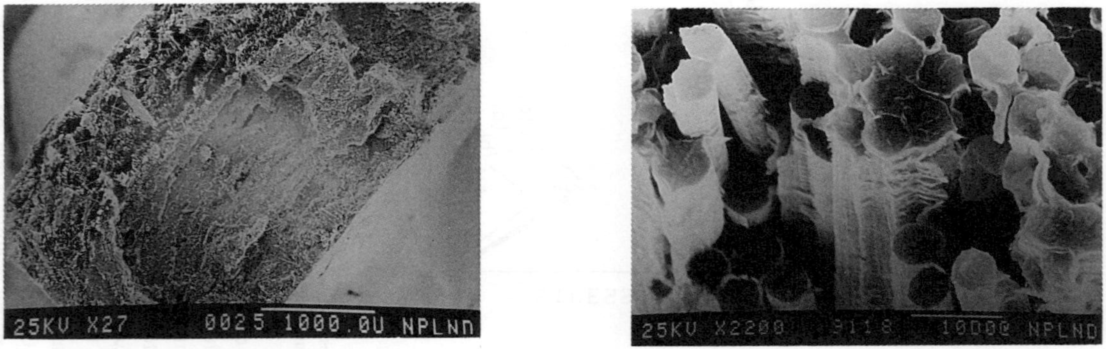

IM (Carbonised)

Fig. 8. Fracture faces of resin based carbonized composites.

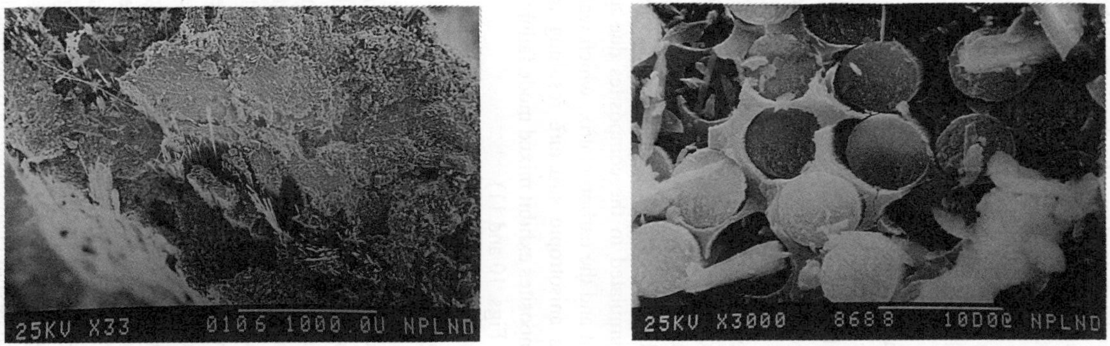

HM-45 (Carbonised)

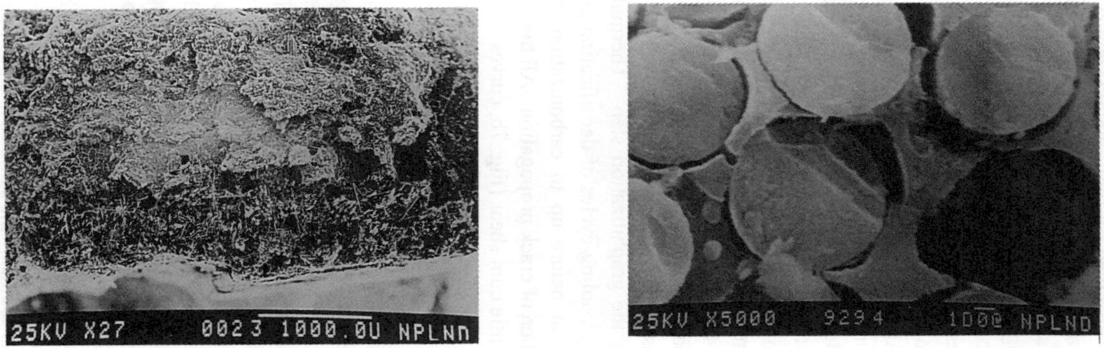

X-340 (Carbonised)

Fig. 9. Fracture faces of resin based carbonized composites.

On densification of porous composites, the flexural strength values decrease. The flexural strength decreases more for high strength and intermediate modulus based composites than for high modulus and X-340 fiber based composites. The decrease in the flexural strength is attributed to strong fiber/matrix bonding and mismatch of the thermal expansion of fibers and matrix leading to accumulation of thermal stresses during the heating/cooling cycles of densification. In the case of HM-45 and X-340 fiber based composites, annular gaps at the fiber/matrix interface are able to counteract the thermal stresses generated during the heating/cooling cycles; hence, the decrease in the value of flexural strength is not less.

As observed in Fig. 5, the flexural strength of carbon/carbon composites made with different carbon fibers increases on graphitization. The increase is almost double with high modulus carbon fiber based composites compared to its carbonized values. In the case of high strength and intermediate modulus carbon fiber based composites, the increase in the value of flexural strength is quite appreciable compared to its carbonized value. This is also evident in Fig. 6 where fiber strength translation increases from (carbonized to graphitized) about 40–45% to 80% in the case of HM fiber based composites; in the case of HT and IM fiber based composites, the increase is very negligible, i.e., from 15–17% to about 20%. The increase in the value of flexural strength is due to modification of fiber/matrix interactions during the graphitization step. Thermal stresses accumulated in the composites due to heating/cooling cycles of densification get annealed out and the carbon matrix, which was brittle in nature up to carbonization stage, develops anisotropic structure resulting in deflection of crack propagation. All the graphitized composites exhibit mixed mode failure, i.e., brittle cum shear (Fig. 7b, curves a, b, c and d; and Figs. 10 and 11).

3.1.5. Mechanical properties and fracture behavior of coal tar pitch based carbon/carbon composites

Fig. 12 shows the value of flexural strength of carbon/carbon composites made with different carbon fibers at different stages of development (carbonized, densified and graphitized). The flexural strength values are less than 100 MPa for high strength carbon fiber based composites compared to 200 MPa for intermediate modulus carbon fiber based composites. The values in the case of high modulus carbon fiber based composites are about 175–180 MPa; X-340 fiber based composites show a value of about 190–200 MPa. The value of flexural strength is minimum for high strength carbon fiber based composites since cracks and voids will be formed during the carbonization stage as a result of strong fiber/matrix bonding. The composites fail in brittle mode as shown in stress-strain curve (Fig. 7c, curve c). An SEM photograph of fractured face with almost no fiber pull out is shown in Fig. 13. A high magnification SEM photograph (Fig. 13) shows strong bonding between the fiber and matrix, leading to crack propagation without any hindrance. This behavior is exactly the same as has been observed for high strength carbon fibers resin based carbon/carbon composites.

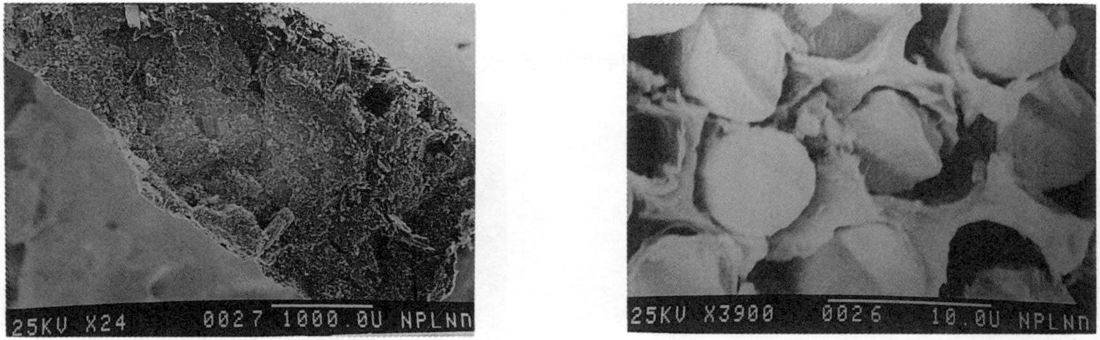

HTA (Graphitised)

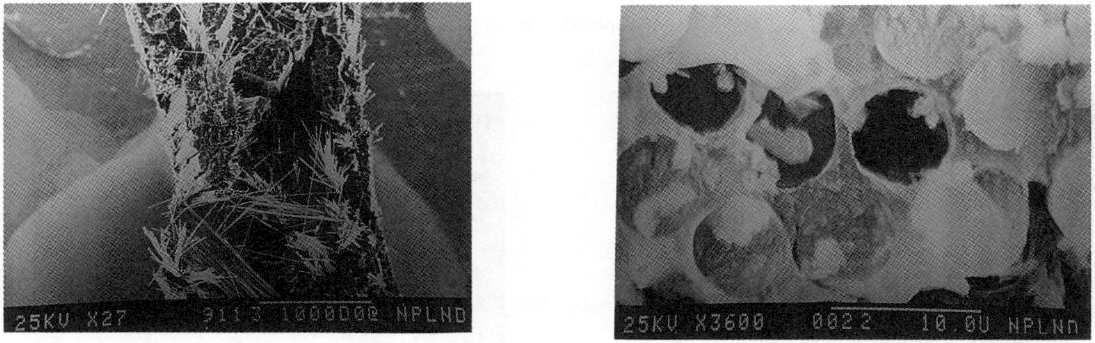

IM (Graphitised)

Fig. 10. Fracture faces of resin based graphitized composites.

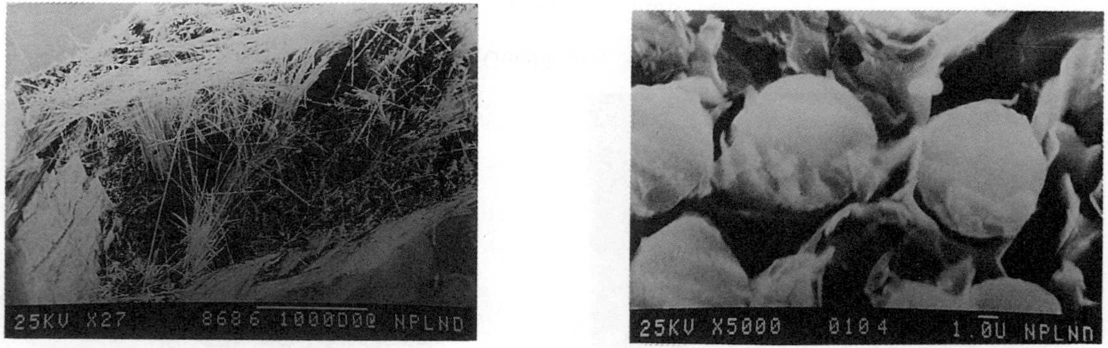

HM-45 (Graphitised)

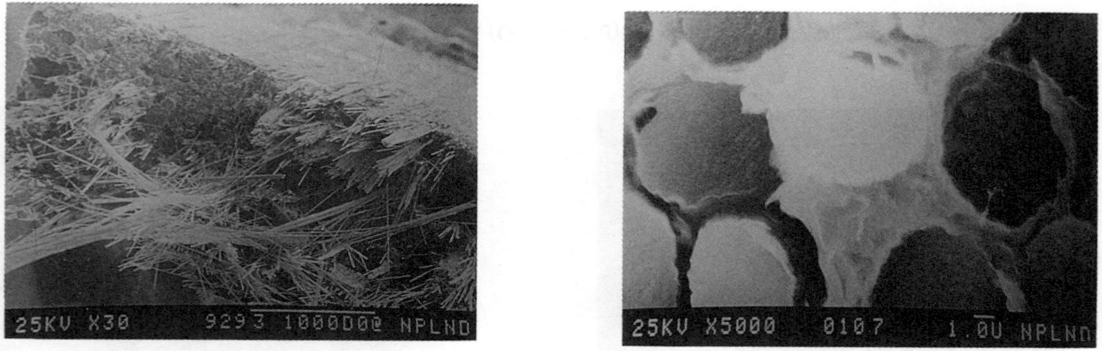

X-340 (Graphitised)

Fig. 11. Fracture faces of resin based graphitized composites.

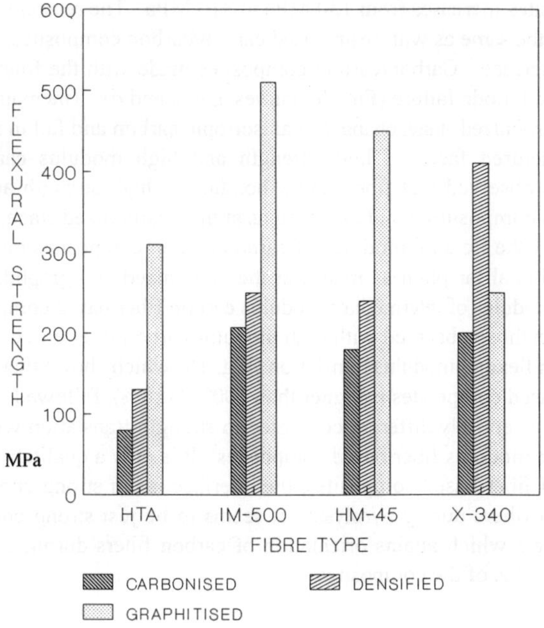

Fig. 12. Flexural strength of carbon/carbon composites at carbonized, densified and graphitized stages (coal tar pitch based).

The behavior of high modulus and X-340 carbon fiber based carbon/carbon composites at carbonized stage is exactly the same as observed for high modulus carbon fiber resin based carbon/carbon composites at carbonized stage. Its fracture behavior (Fig.7c, curves a and d) is similar to the one observed for resin based composites. Fig. 13 shows an SEM micrograph of the high modulus carbon fiber based composites. The figure shows strong fiber/matrix bonding with annular gaps at the fiber/matrix interface. The bonding is not as strong as with high strength carbon fiber based composites. This annular gap will deviate the crack and resist its passing straight through the fibers. This is the reason why high modulus carbon fiber based composites show high flexural strength.

The fracture behavior of intermediate modulus based carbon/carbon composites is same as with high modulus carbon fibers based composites except that it is tensile cum shear with steps (Fig. 7c, curve b).

On graphitization of these composites, their flexural strength increases (except in the case of X-340 fiber composites) and becomes almost double when compared to that of densified composites. The flexural strength (510 MPa) of intermediate modulus carbon fiber based composites becomes even higher than that of high modulus carbon fiber based composites after graphitization (450 MPa). The flexural strength of high strength carbon fiber based composites increases from 130 MPa to 310 MPa. The reason for this increase in flexural strength is the same as with resin based carbon/carbon composites, i.e., modification of fiber/matrix interface. Carbon/carbon composites made with the four fibers on graphitization show mixed mode failure (Fig. 7d, curves a, b, c and d). The matrix carbon, which was brittle at the carbonized stage, changes to anisotropic carbon and fail in shear mode. Fig. 14 shows the fractured faces of high strength and high modulus carbon fiber based composites. It is observed that fiber/matrix bonding in high strength and high modulus carbon fiber based composites is not as strong as at their carbonized stage.

Fig. 15 shows the flexural modulus of carbon/carbon composites made with different carbon fibers and coal tar pitch as matrix at the carbonized and graphitized stages. The values of flexural modulus of intermediate modulus carbon fiber based composites (176 GPa) are even higher than those observed with high modulus carbon fiber based composites. This is better reflected in flexural modulus translation (Fig. 16) which shows that translation in the case of IM fiber based composites is higher than 100% (118%), followed by HT fiber based composites which is entirely different compared to strength translation which is maximum in the case of high modulus fiber based composites. It is only a qualitative comparison. It seems that for IM fiber based composites, the interface is not strong enough to allow any damage to carbon fibers during pyrolysis. It seems to be just strong enough to keep the matrix around fibers which attains orientation of carbon fibers during graphitization thus enhancing the modulus of the composites.

3.1.6. *Interlaminar shear strength of carbon/carbon composites made with resin and coal tar pitch as matrix*

Fig. 17 shows the ILSS of resin and coal tar pitch based carbon/carbon composites at carbonized and graphitized stage. In the case of carbon/carbon composites, fiber/matrix interactions influence to a large extent the ultimate mechanical and fracture behavior of composites. ILSS is a measure of the fiber/matrix bonding in composites.

For resin based composites, the value of ILSS varies from 20 to 32 MPa at the carbonized stage. The value of ILSS is directly proportional to the presence of surface groups on different carbon fibers. A higher ILSS value indicates strong bonding formed between surface groups on the surface of carbon fibers and the functional groups present in the resin. On graphitization the value of ILSS increases compared to that of carbonized composites although the increase is much higher for IM-500 carbon fiber based composites compared with HT and HM-45 carbon fiber based composites.

The behavior of ILSS of coal tar pitch based carbon/carbon (Fig. 18) composites is the same as with resin based carbon/carbon composites except that the values of ILSS at the carbonized stage are much lower (5 to 12 MPa) and on graphitization the ILSS values become more than double compared to the values at the carbonization stage. The lower values of ILSS at the carbonization stage are due to the presence of fewer functional groups in coal tar pitch compared to that in thermosetting resin and hence less strong bonding between fiber and matrix.

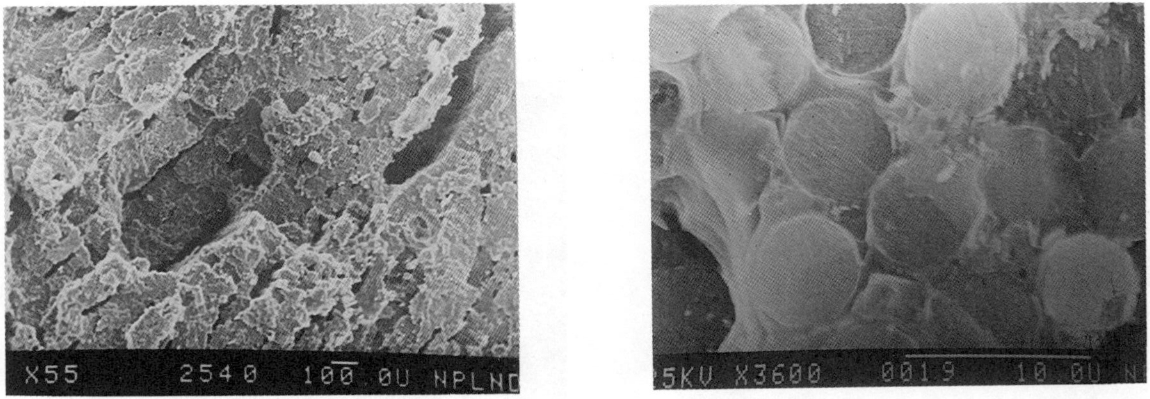

HTA (Carbonised)

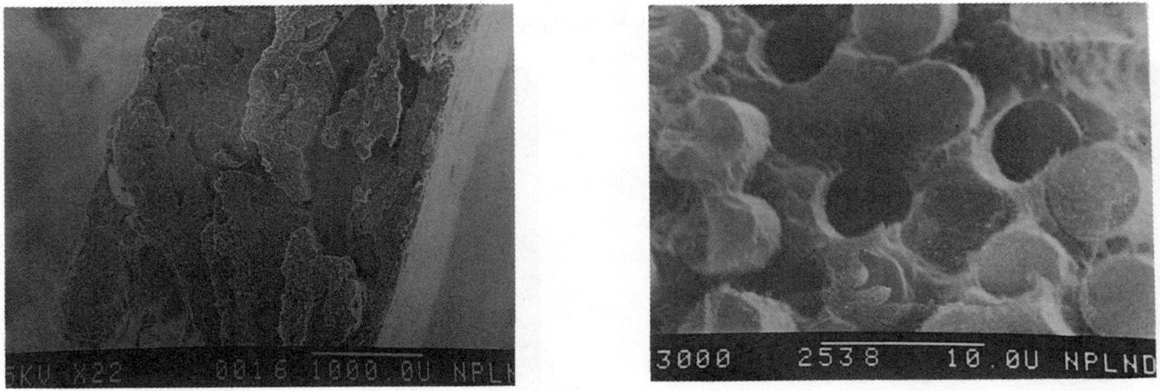

HM 45 (Carbonised)

Fig. 13. Fracture faces of coal tar pitch based carbonized composites.

HTA (Graphitised)

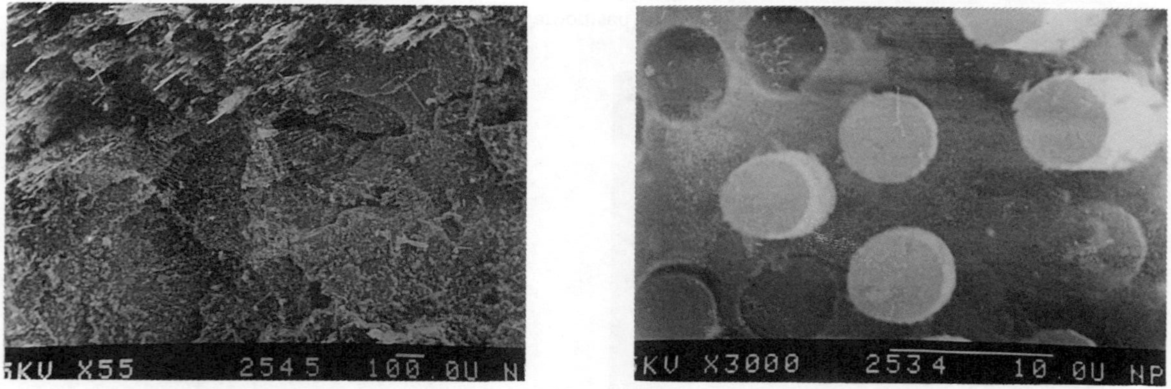

HM 45 (Graphitised)

Fig. 14. Fracture faces of coal tar pitch based graphitized composites.

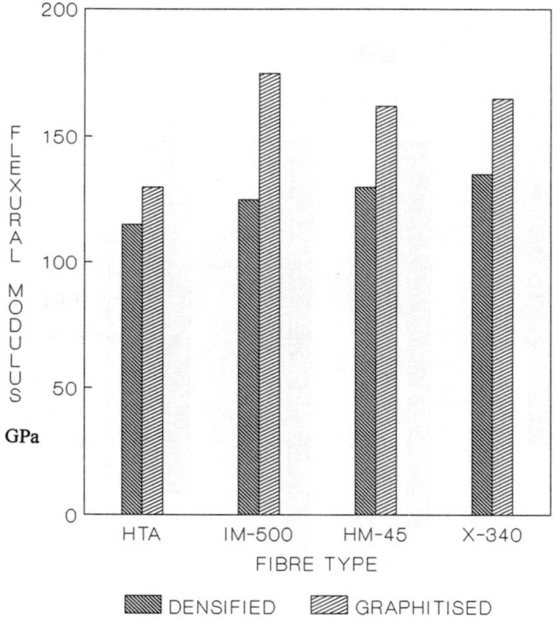

Fig. 15. Flexural modulus of carbon/carbon composites made with different carbon fibers and coal tar pitch.

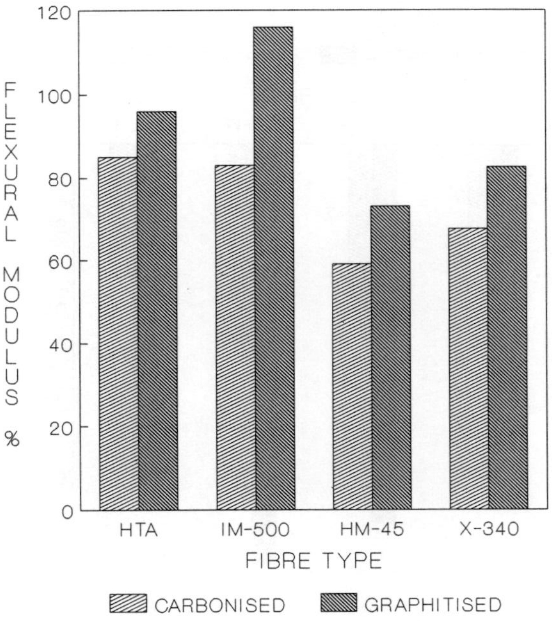

Fig. 16. Fibers modulus transition in carbon/carbon composites made with different carbon fibers and coal tar pitch as matrix.

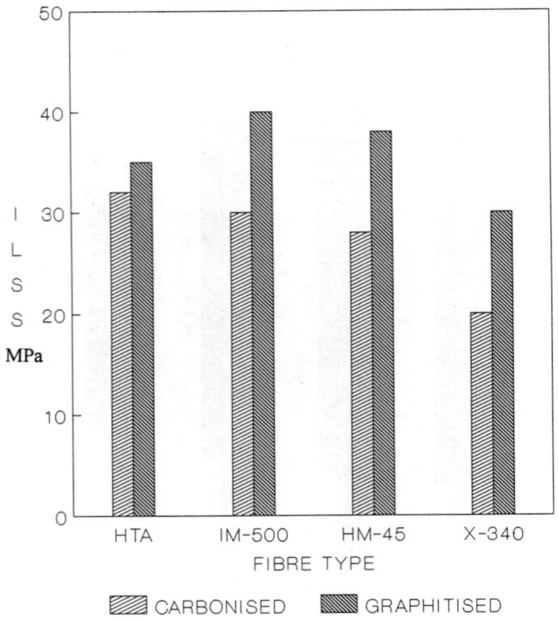

Fig. 17. ILSS of carbon/carbon composites made with different carbon fibers and resin as matrix.

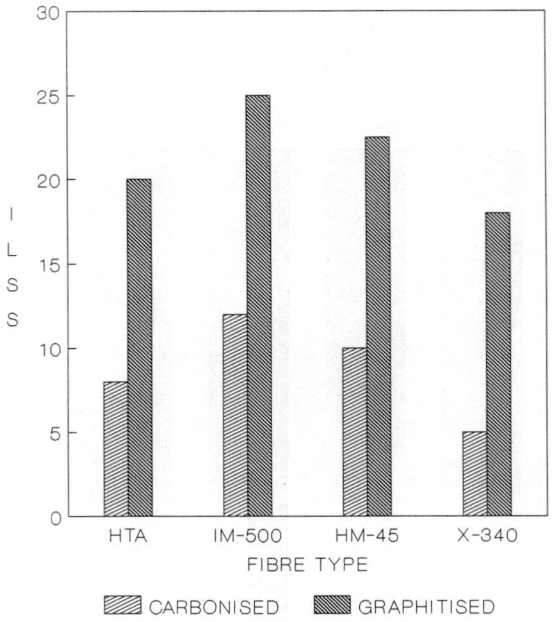

Fig. 18. ILSS of carbon/carbon composites made with different carbon fibers and coal tar pitch.

4. Panex (Oxidized PAN) Fiber Based Carbon/carbon Composites

4.1. Surface Energetics of Panex Fibers

As the Panex fibers have been prepared at 250°C for different intervals of time varying from zero to 120 min, they possess different surface energetics which are given in Table 1.

Table 1. Surface Groups on Different Panex Fibers

Fiber type	Phenolic (mEq/100 gm)	Lactonic (mEq/100 gm)	Carboxylic (mEq/100 gm)	Total (mEq/100 gm)
PR-I	116.0	32.0	10.0	158.0
PR-II	209.0	131.0	67.0	407.0
PR-III	316.0	130.0	89.0	536.0

4.1.1. Composites made with phenolic resin and coal tar pitch as matrix precursors

The polymer composites made with phenolic resin and coal tar pitch were carbonized in an inert atmosphere up to 1000°C. During the pyrolysis of polymer composites, significant stresses would be generated in the composites due to excessive evolution of hydrocarbons from the composites (matrix as well as from the Panex fibers) thus leading to large volume shrinkage and weight loss of the composites. Cross sectional shrinkage and weight loss of the polymer composites during pyrolysis are given in Table 2.

Table 2. Dimensional Changes and Weight Loss during carbonization to 1000°C of Panex Fiber Reinforced Polymer Composites

Fiber type	ΔL (%)	ΔB (%)	ΔT (%)	ΔW (%)
PR-I	-7.5	-18.0	-18.0	-40.0
PR-II	-8.0	-20.0	-22.0	-35.5
PR-III	-9.2	-24.0	-23.0	-32.5

Δ = change in L, length; B, width; T, thickness; and W, weight

As expected, composites made with PR-III fibers exhibited the highest shrinkage of 23–24% across the fiber axis, while PR-I fiber based composites exhibit the lowest shrinkage, 18%. However, weight loss of composites made with PR-I fibers is highest (40.0%) while PR-III fiber based composites exhibited only 32% weight loss. In general no shrinkage is observed along the longitudinal direction/fiber axis for carbon fiber based composites. However, in the present composites with Panex fibers, substantial shrinkage (7–9 %) is also observed in the longitudinal direction. Due to shrinkage in all directions, better compaction is expected in such composites. The bulk density of skeleton carbon-carbon composites is found to be between 1.30–1.4 gm/cc, which is unusually high. Density of composites prepared with high strength (HT) carbon fibers is found to be usually between 1.1 and 1.15 gm/cc[15]. Likewise carbon yield of the composites (HTT 1000°C), is also unusually high

(55 %). It is hypothesized that stronger interphase between Panex fibers and coal tar pitch affects the resin polymerization which results in higher yield. This is further confirmed by the results compiled in Table 3.

Table 3. Effect of Fiber/Matrix Interaction on Yield of Composites

Material	Yield at 1000°C (%)
PR-I fiber	50.5
Coal tar pitch	38.0
C/C composites, Vf = 50 ± 2%	50.0
Theoretical yield of composites	43.0

4.1.2. Mechanical properties of carbonized (HTT 1000°C) and graphitized composites

The mechanical/physical properties of pyrolyzed composites are shown in Table 4.

Table 4. Mechanical Properties of Panex Fiber Reinforced Carbonized Composites

	Resin based composites				Pitch based composites			
Composites made with	D (g/cc)	Vf (%)	F.S. (MPa)	F.M. (GPa)	D (g/cc)	Vf (%)	F.S. (MPa)	F.M. (GPa)
PR-I	1.35	48 ± 2	30.0	51.0	1.51	50 ± 2	55	31
PR-II	1.25	48 ± 2	16.5	44.0	1.55	50 ± 2	40	44
PR-III	1.20	48 ± 2	11.0	31.0	1.60	50 ± 2	35	51

D = density, Vf = fiber volume, F.S. = flexural strength, F.M. = flexural modulus

As seen from the table, flexural strength of resin based composites comes to 11–30 MPa, while for coal tar pitch based composites it is 35–55 MPa. Apparently, strong fiber/matrix bonding in polymer composites persists throughout the pyrolysis stage since the fibers and matrix both are brittle in nature and, once a crack is initiated, it propagates straight through the interphase, breaking the fibers (Fig. 19). This is further confirmed from the fractured face of the composites (Fig. 19) where fiber and matrix seem to be merging at some places. This leads the composites to fail at low load. The decrease in flexural strength is expected because during pyrolysis, lots of volume shrinkage and weight loss of the composites takes place, leading to generation of stress concentration centers which become the crack initiators. Flexural modulus of phenolic resin based composites shows the same behavior as flexural strength whereas flexural modulus of pitch based composites increases through PR-I to PR-III fiber based composites (Table 4). This may be due to alignment of the matrix along the fiber axis as pitch is thermoplastic in nature and passes through the liquid phase during pyrolysis and secondly the matrix carbon obtained from coal tar pitch is known to be anisotropic in nature, even at 1000°C[16]. Strong fiber/matrix bonding and brittle failure of phenolic resin based composites leads to still lower mechanical strength; thus, these composites are not suitable for structural applications at this stage of heat treatment (1000°C).

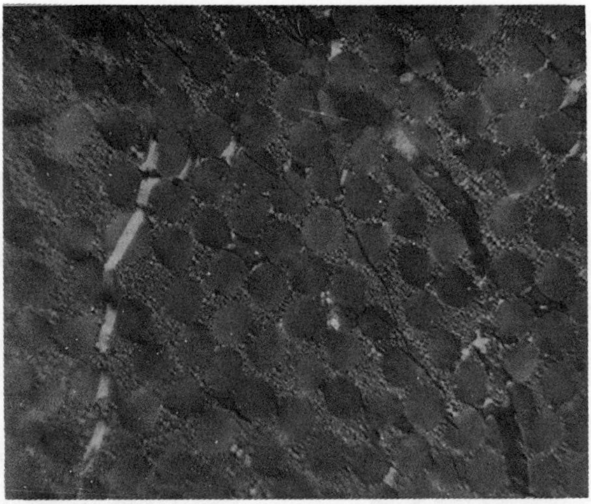

X 800

Fig. 19. Microstructure of the coal tar pitch based carbon/carbon composites HTT 1000°C.

The fiber/matrix interactions are so strong in resin based composites that during pyrolysis of the composites the resin matrix does not allow the fiber to shrink away from it and it pulls the fiber towards it. The result is that, during pyrolysis, the Panex fiber becomes hollow from the center as shown in Fig. 20. This hollowness is thought to be due to simultaneous pyrolysis of fiber and matrix, and as the Panex fiber is not well stabilized and the fiber/matrix bonding is too strong, the matrix resin pulls the fiber towards it. This is shown schematically in Fig. 21. Similar hollowness of fibers has also been reported by Markovic et al.[17,18].

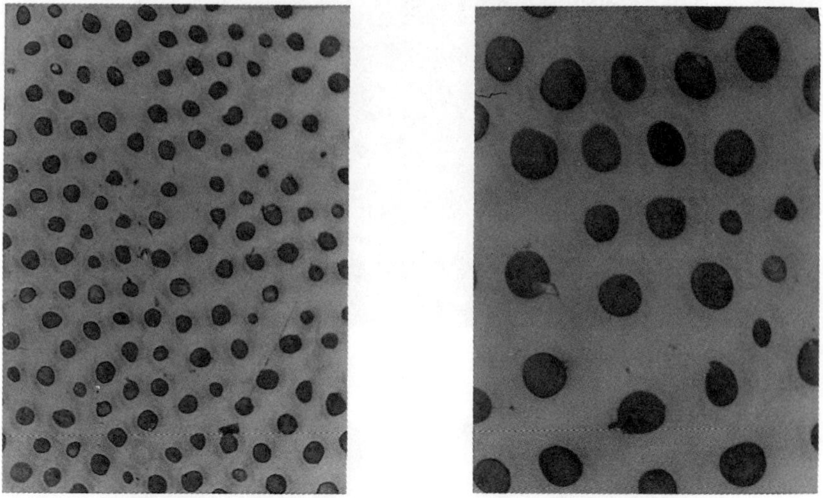

X 800 X 1600

Fig. 20. Optical micrographs of resin based carbon/carbon composites (HTT 1000°C) showing hollow carbon fibers.

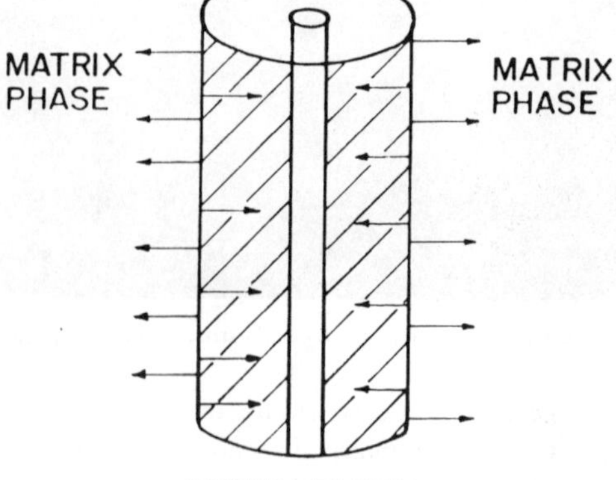

Fig. 21. Schematic representation of formation of holes in core of carbon fiber during pyrolysis.

Mechanical properties of graphitized composites are compiled in Table 5.

Table 5. Mechanical Properties of Graphitized (HTT 2700°C) Composites

	Resin based composites				Pitch based composites			
Composites made with	D (g/cc)	Vf (%)	F.S. (MPa)	F.M. (GPa)	D (g/cc)	Vf (%)	F.S. (MPa)	F.M. (GPa)
PR-I	1.48	55 ± 2	45.0	80.0	1.51	55 ± 2	243	100
PR-II	1.55	60 ± 2	50.0	90.0	1.53	55 ± 2	190	110
PR-III	1.60	60 ± 2	53.0	95.0	1.56	55 ± 2	226	120

D = density, Vf = fiber volume, F.S. = flexural strength, F.M. = flexural modulus

Flexural strength of resin based composites is of the order of 45–53, MPa which is not at all attractive. However, coal tar pitch based composites show a remarkable increase in flexural strength and modulus on graphitization. The flexural strength of the composites increases 4 to 6 fold compared to their values at the carbonized stage; the flexural modulus values show a 2 to 3 fold increase. The values of mechanical properties (flexural strength and flexural modulus) increase in graphitized composites because fiber/matrix interactions get modified upon heat treating carbon-carbon composites to 2700°C. This is reflected in the mode of fracture of composites which has changed from tensile type in carbonized composites to tensile cum shear in graphitized composites (HTT 2700°C). The fractured face of a graphitized composite is shown in Fig. 22, which shows pull out of the fibers.

Fig. 22. SEM of fracture face of coal tar pitch based carbon/carbon composites (HTT 2700°C).

The absolute values of flexural strength are in the range of 190 to 240 MPa whereas those of flexural modulus are in the range of 100–120 GPa. These values of flexural strength and flexural modulus are in the range of values generally seen for carbon/carbon composites using commercial carbon fibers (HT carbon fibers). Composites made with high strength carbon fibers (HT) and phenolic resin exhibit flexural strength (after graphitization) of around 300 MPa[(19)] whereas Panex fibers reinforced composites the value is only 45–53 MPa. This strength value is just about 20% of what high strength carbon fiber based composites exhibit; the reason for the low values being very strong fiber/matrix interactions which persist up to graphitization stage[4,20] and lead to brittle failure at low load. Therefore, carbon-carbon composites made with Panex fibers and phenolic resin as matrix precursor cannot be considered as viable candidates for structural applications.

4.1.3. Microstructure of the composites

Fig. 23 shows optical micrographs of Panex fiber reinforced pitch based carbonized composites (HTT 1000°C), which appear to be well bonded as they do not reveal noticeable gaps between fibers and matrix. The microstructure of the matrix carbon in between the carbon fiber filaments appears to be columnar in nature and is optically active. This is more evident in the magnified view in Fig. 23. On heat treating the composites to 2700°C the degree of anisotropy enhances further.

The microstructure of resin based carbonized composites is shown in Fig. 20. In this case the matrix carbon and carbon fibers have mingled with each other so that there is no demarcation of fibers and matrix. There is coalescence into a continuum of fibers and matrix so that the components are not readily recognizable. This confirms the earlier findings of Markovic et al.[17,18] who have reported the coalescence of fiber and matrix in resin based composites due to reactions taking place between the surface of the fiber and resin matrix.

An SEM micrograph of pitch based composites (carbonized stage) is shown in Fig. 19. One finds a lot of porosity present around the fibers. A large number of cracks are also seen in the matrix, but the interphase between the fiber and matrix is continuous. However, on heat-treatment of these composites to 2700°C, the microstructure has changed (Fig. 24). The fibers show the radial type of microstructure with well defined ridges or steps which is common to see on the fractured surfaces of PAN-based fibers heated to such a high temperature. Some ordering in the crystallites of the matrix is also seen. There is a small annular gap at the fiber/matrix interphase between some of the carbon fibers. Composites (HTT 2700°C) made with high strength carbon fibers (HT), are different than Panex fiber based composites heat treated to the same temperature. It appears as if the strong fiber/matrix interaction is responsible for the changes in the microstructure of fiber during HTT to 2700°C. This change can be due to high HTT taking place under stress due to strong fiber/matrix bonding developed at the initial stage of formation of composites and which persisted during carbonization and graphitization (HTT 2700°C) of composites[4,20].

X 800

X 1600

Fig. 23. Optical micrographs of pitch based carbon/carbon composites (HTT 1000°C).

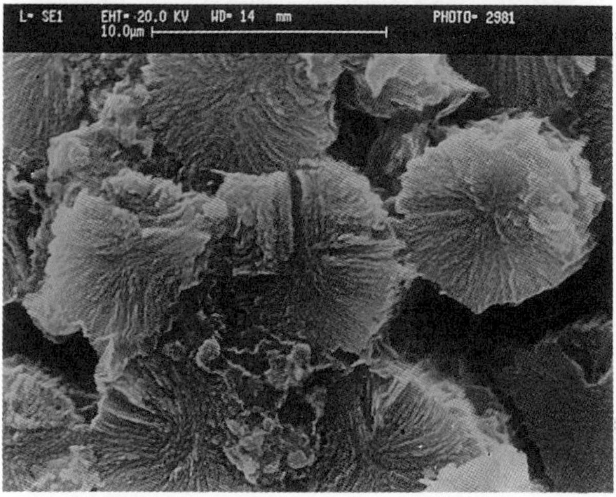

Fig. 24. Fractured face of coal tar pitch based carbon/carbon composites (HTT 2700 °C).

4.1.4. Effect of co-carbonization/interphase on diameter of fibers

Fig. 25 shows the effect of co-carbonization and interphase on carbon fiber diameter. Stabilized PAN when carbonized generally yields carbon fibers with diameter of 7.5 um and 6 um at 1000°C and 2700°C respectively. However during co-carbonization with matrix, carbon fibers exhibit an approximate diameter of 10 um and 9 um respectively, approximately 40% higher than normal. Panex fibers were not allowed to shrink towards its center due to strong fiber/matrix interactions in coal tar pitch based composites. In resin based composites, fiber/matrix interactions are so strong that they lead to hollow fibers, as seen in Fig. 20.

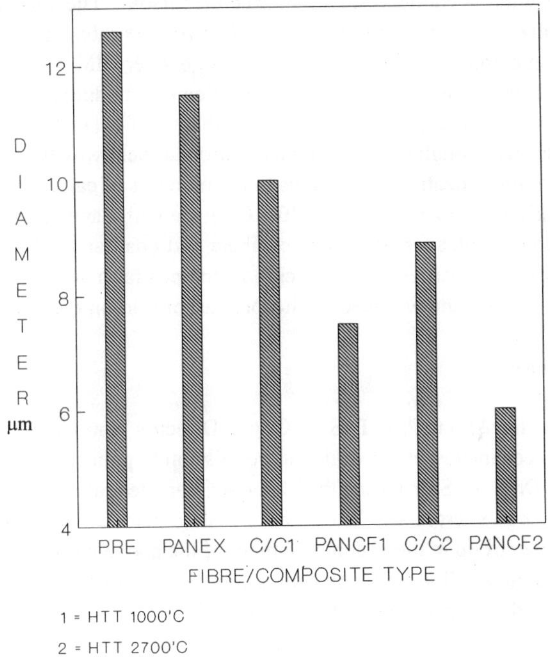

Fig. 25. Effect of co-carbonization/interface on fiber diameter.

5. Conclusions

The surface energetics of different carbon fibers affect to a large extent the fiber/matrix interactions and hence the mechanical properties of carbon/carbon composites. In resin based carbon/carbon composites, the shrinkage behavior of polymer composites up to carbonization stage is directly proportional to the amount of surface groups present on carbon fibers. The mechanical properties of polymer composites (1000–1800 MPa) are also found to be proportional to the amount of surface groups present on the surface of carbon fibers.

The mechanical properties decrease on carbonization which is consistent with the cross-sectional shrinkage they experience. On densification and graphitization the mechanical properties increase (300–800 MPa) due to modification of fiber/matrix interactions and change of fracture mode; the increase being maximum in case of high modulus carbon fiber based carbon/carbon composites which is also evident from fiber strength translation value of about 75%.

Flexural strength of carbonized composites made with coal tar pitch as matrix precursor is found to increase on densification, which is contrary to what has been observed in composites made with phenolic resin as matrix precursor. The mechanical properties increase on graphitization due to modification of fiber/matrix interactions. IM-500 carbon fiber based carbon/carbon composites have an edge over HM-45 carbon fiber based carbon/carbon composites in flexural strength and flexural modulus values. The value of modulus translation has been found to be more than 100% (120%) which shows that interphase is just strong enough to keep the matrix around them which attains orientation of carbon fibers during graphitization which adds to the modulus of carbon/carbon composites.

For carbon/carbon composites (HTT 2700°C) made with Panex fibers, it is possible to make carbon/carbon composites using Panex fibers and coal tar pitch as matrix showing comparable mechanical properties to carbon/carbon composites made with high strength (T-300) carbon fibers. This route will lead to cheaper carbon/carbon composites.

6. Acknowledgments

The authors are thankful to Prof. E. S. R. Gopal, Director, National Physical Laboratory, New Delhi, for his keen interest in the work and permission to publish the same. The authors are also thankful to Dr. S. K. Sharma and the Material Characterization Group, NPL, for their help in scanning electron microscopy. The authors wish to place on record the help extended by members of carbon group in the accomplishment of the above work particularly Dr. P. K. Jain and Mr. S. R. Dhakate. Special thanks are to Mr. V. K. Parashar and Mr. R. K. Seth for the help extended in the preparation of the document.

7. References

1. L. E. Mc Allister and W. L. Lachman, in *Handbook of Composites, vol. 4*, eds. A. Kelly and S. T. Mileikopub (Elsevier Science Publisher, 1983), p. 109.
2. E. Fitzer, *Carbon* **25**, (1987) 163.
3. W. V.Kotlensky, et al., *Sampe Quarterly* (April 1977) 1.
4. L. M. Manocha and O. P. Bahl, *Carbon* **26** (1988) 13.
5. E. Fitzer, W. Huettner and L. M.Manocha, *Carbon* **18** (1980) 291.
6. R. A. Meyer and S. R. Gyetvay, in *Proc. 16th Biennial Conf. Carbon* (Univ. of California, San Diego, 1983) p.505.
7. L. M. Manocha, *Carbon* **32** (1994) 213.

8. S. Kimura, E. Yasuda and Y. Tanabe, in *Proc. Inter. Conf. on Composites Materials (Tokyo ICCM-IV*, 1982) p. 1601.
9. L. M. Manocha, E. Yasuda, Y. Tanabe and S. Kimura, *Carbon* **26** (1988) 333.
10. E. Fitzer, K. H.Geigl and W .Huettner, *Carbon* **18** (1980) 265.
11. E. Fitzer, W. Huettner, L. M. Manocha and D. Walter, in *Proc. 15th London Int. Conf. on Carbon and Graphite* (1978), p.454.
12. E. Fitzer, K. H. Geigl, W. Huettner and R. Weiss, *Carbon* **18** (1980) 389.
13. W. G. Broadshow, et al., *Ceramic Bulletin* **57** (1978) 193.
14. L. M. Manocha, O. P. Bahl and Y. K. Singh, *Carbon* **27** (1989) 381.
15. O. P. Bahl, L. M. Manocha, G. Bhatia, T. L. Dhami and R. K. Aggarwal, *J. Sci. & Ind. Res.* **50** (1991) 533.
16. T. L. Dhami, O. P. Bahl and L. M. Manocha, *Carbon* **31** (1993) 751.
17. V. Markovic and S. Marinkovic, *Carbon* **26** (1988) 329.
18. V. Markovic et al., in *Proc. 15th Bienn. Conf. on Carbon* (1980) p.272.
19. J. Chlolpeck and S. Bizewicz, *Carbon,* **29** (1991) 127.
20. L. M. Manocha, O. P. Bahl and Y. K. Singh, *Carbon* **29** (1991) 351.

INTERFACIAL REACTIONS IN CERAMIC COMPOSITES REINFORCED WITH CARBON

V. BESSON, P. DUPEL, and A. HENDRY
Metallurgy and Engineering Materials Group,
University of Strathclyde, Scotland, UK

ABSTRACT

Sintering and interfacial reactions have been studied in sialon or zirconia matrix composites reinforced with different types of carbon fibers or particles. In sialon composites reinforced with short pitch fibers it has been shown that the sintering temperature needs to be increased to 1550°C in order to get a total $\alpha \rightarrow \beta'$ transformation in the matrix but that for a temperature superior to 1500°C an interfacial reaction occurs between matrix and reinforcement. The mechanism of this reaction is explained. In pressureless sintered zirconia matrix composites reinforced with mesocarbon microbeads, the final densities are poor due to mismatch between the thermal expansion coefficients and at temperature superior to 1450°C an interfacial reaction occurs between carbon and zirconia. A mechanism for this reaction is proposed. It is explained why if the same composite is hot pressed no reaction occurs at 1550°C. The matrix of such composites is analyzed in order to identify the different present phases and the results are discussed. Samples hot pressed and reinforced with pitch short fibers or VGCF are observed and compared. Advantages of VGCF over other types of fibers are explained.

1. Introduction

The development of ceramic matrix composites has for a long time been of great interest for industry and researchers in view of their interesting properties and numerous matrices and reinforcement have been studied at the University of Strathclyde[1–5]. Considering the fact that short pitch fibers were the first readily available high temperature ceramic fibers, and that their relatively low price and interesting performances made them attractive to the ceramic composite industry, the development of composites reinforced with these fibers became an important objective. These fibers offer the possibility to improve the very low fracture toughness of ceramics in general and of improving their Weibull modulus making them more reliable and industrially acceptable in engineering applications. A carbon reinforcement also offers the possibility to improve thermal and electrical conductivity of the ceramic and, using a high volume fraction of reinforcement, it is possible to make them electromachinable.

In 1993, the Ceramic Group at Strathclyde University became involved in an EEC Human Capital and Mobility Program on the production and applications of vapor grown carbon fibers and other ceramic fibers and filaments. Previous published knowledge[1–5] of the fabrication of carbon fiber composites was used to develop composites with different matrices reinforced with not only pitch fibers and mesocarbon microbeads but also with vapor grown carbon fibers. The results of these studies are presented in this paper.

2. Experimental Procedures

The composites were prepared using, for the sialon matrix, β'–sialon 201 supplied by Cooks on Group and containing unreacted α–Si_3N_4, 6 wt % Al_2O_3, 8 wt % Y_2O_3 (as a densifying aid) and 10 wt % AlN polytypes. During sintering, this material forms a β'–sialon of composition $Si_{5.25}$ $Al_{0.75}$ $O_{0.75}$ $N_{7.25}$ (β'; z = 0.75). For zirconia matrix composites, ZrO_2 powder with a mean particle size of 0.8 μm, stabilized with 5.4% yttria, supplied by Goodfellow Metals Ltd., was used.

As a reinforcement, mesocarbon microbeads with diameters of 6 and 20 μm and short carbon pitch fibers supplied by Osaka Gas and vapor grown carbon fibers (VGCF) supplied by Applied Sciences Inc. in the form of a unidirectional thin mat or a 'fur' like tow have been used.

In the case of sialon or zirconia reinforced with pitch short fibers and zirconia reinforced with microbeads, the matrix powder was wet-mixed with the reinforcement in propan-2-ol with alumina balls for 2 hours. The powders were then dried and dry-mixed for another hour. Proportions of 10 and 20 vol % of reinforcement were used. After mixing, the samples were uniaxially cold compacted in a steel die into a disk shape with a green density around 60% of the theoretical density.

Samples made with sialon were hot-pressed in a pseudo-isostatic hot-press. The samples were embedded in boron nitride powder inside a graphite die. They were then sintered under a pressure of 30 bars for 1 hour, or shorter times in certain cases in order to study the kinetics of the reaction. The sintering temperatures were 1500, 1550 and 1600°C.

Some of the samples made with zirconia were sintered in a similar way but under a pressure of 20 bars and at temperatures of 1450, 1500 and 1550°C. Other samples with a zirconia matrix were pressureless sintered in a horizontal induction furnace in an atmosphere of 50% CO - 50% N_2. In a previous study[6] it has been shown that using this atmosphere inhibited the formation of zirconium carbides at 1450°C. The compacted samples were placed in an alumina boat, introduced to the furnace and sintered for 1 hour.

The fabrication of the samples containing VGCF will be discussed further in this paper.

Crystalline phases were identified using either a Hagg-Guinier focusing camera (CuKα, 50kV, 20mA) or a diffractometer. The microstructure of the samples has been studied by SEM and by TEM (Phillips 400T) equipped with a windowless energy dispersive X-ray analysis system (EDAX).

3. Results

3.1. Sialon Matrix Composites

Sialon samples were hot pressed under a pressure of 30 bars for 1 hour. Table 1 gives the final densities as well as the weight loss of samples containing 10 vol % or 20 vol % of short carbon pitch fibers sintered at different temperatures.

Table 1. Density and Weight Loss of Sialon Matrix Composites
Reinforced with Short Carbon Pitch Fibers Sintered at Different Temperature

temperature °C	density (% of theoretical)		weight loss (% of initial weight)	
	10 vol % fibers	20 vol % fibers	10 vol % fibers	20 vol % fibers
1500	81	83	0.2	0.4
1550	92	94	1.3	1.7
1600	94	95	2.0	2.5

As the temperature increases, the final density and the weight loss also increase, while increasing the reinforcement content causes an increase in the weight loss. SEM observation of the samples shows that at 1500°C the fibers are intact, at 1550°C they start partially reacting with the matrix, and at 1600°C the structure of most of the fibers is extensively altered by this reaction. This evolution can be seen in Figs. 1a, 1b and 1c.

In addition to this reaction, X-ray results show that at 1500°C the a → β' transformation is far from being completed, α–Si_3N_4 being the main phase, whereas at 1600° C the main phase is β'–sialon. Finally, TEM observation of the samples confirms the presence of a large quantity of large α–grains in the matrix whereas, at 1600°C, the matrix is more dense and no α–grains are observed.

Fig. 1a. SEM photograph of a cross section of a pitch fiber in a sialon composite sintered at 1500°C for 1 hour.

Fig. 1b. SEM photograph of a cross section of a pitch fiber in a sialon composite sintered at 1550°C for 1 hour.

Fig. 1c. SEM photograph of a cross section of a pitch short fiber in a sialon composite sintered at 1600°C for 1 hour.

At 1500°C the interface between matrix and carbon fiber is clean and EDAX analysis of the carbon fiber gives a carbon peak and a small sialon peak. At 1600°C, two main structures are observed. Firstly, some clean interfaces where the carbon contains some silicon and the matrix contains a large amount of liquid phase containing Y, Si, Al and O. Small grains are embedded in this solidified liquid and few silicon carbide grains are observed dispersed in the matrix (Fig. 2a). But, if we observe a sample sintered at 1600°C for only 5 min, some large areas containing SiC grains are present in the matrix as in Fig. 2b. The other structures observed are very small SiC grains within the carbon fibers as in Fig. 3. In all the samples bubbles are often observed in the liquid phase at the interface.

3.2. Pressureless Sintering of Zirconia Matrix Composites

Samples with a zirconia matrix reinforced with 20 vol % of mesocarbon microbeads of 6 or 20 μm diameter were pressureless sintered at different temperatures. Table 2 gives the values of final densities (% of theoretical density) and weight loss of these samples.

The very poor final densities are explained when observing the samples on the optical microscope. The carbon particles are surrounded by small cracks, probably due to thermal expansion of the carbon at an early stage when the matrix is not sintered. Attempts to achieve a better green compaction using a cold isostatic press for green forming did not solve the problem, again indicating that the low final density is due to sintering effects.

Table 2. Weight Loss and Final Density of Zirconia Matrix Composites Reinforced with 6 and 20 μm Diameter Mesocarbon Microbeads Sintered at Different Temperatures

temperature °C	weight loss (% of initial weight)		final density (% of theoretical)	
	6 μm	20 μm	6 μm	20 μm
1450	1.5	1.5	83	83
1500	2.3	2.1	87	82
1550	6.3	3.7	93	82

The increasing weight loss as the sintering temperature increased seems to indicate that a reaction occurs consuming some material. This reaction is more extended for 6 μm particles than for 20 μm particles. X-ray results in Fig. 4 show that at 1550°C there is formation of zirconium carbide. They also show that the formation of these carbides goes with the disappearance of the monoclinic and tetragonal phases in the matrix. In the case of pure zirconia with no reinforcement, the phases are unchanged between 1450 and 1550°C.

Because of the presence of so many cracks around the particles, when the sample is broken the particles stay intact, the bonding between the particles and the matrix being very weak. Therefore, it is impossible to observe on the SEM the internal structure of the particles and see if it has been altered by the reaction with the matrix. Figs. 5a and 5b show the surface of the carbon particles at 1450°C and at 1550°C. The surface of the particle sintered at 1550°C is more porous than the surface of the particle sintered at 1450°C which could indicate an alteration of the structure of the carbon.

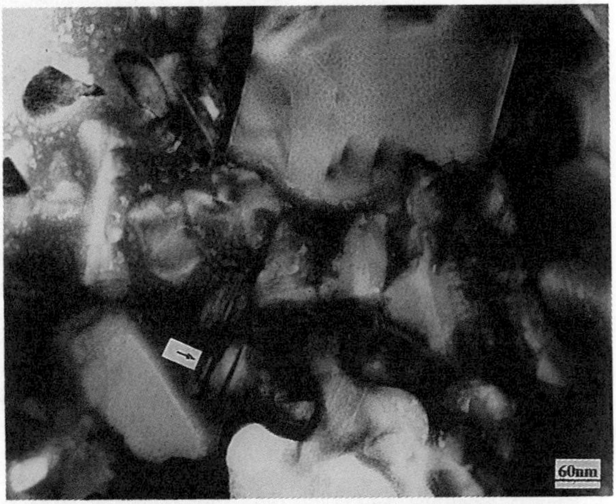

Fig. 2a. TEM bright field micrograph of the precipitation of SiC from the liquid phase in a pitch fiber-sialon composite sintered at 1600°C for 1 hour.

Fig. 2b. TEM bright field micrograph of the precipitation of SiC from the liquid phase in a pitch fiber-sialon composite sintered at 1600°C for 5 minutes.

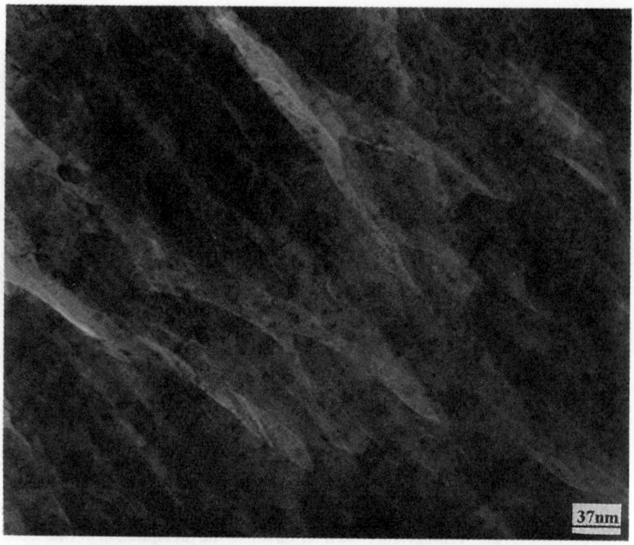

Fig. 3. TEM bright field micrograph of a reacted fiber in a pitch fiber-sialon composite sintered at 1600°C for 1 hour.

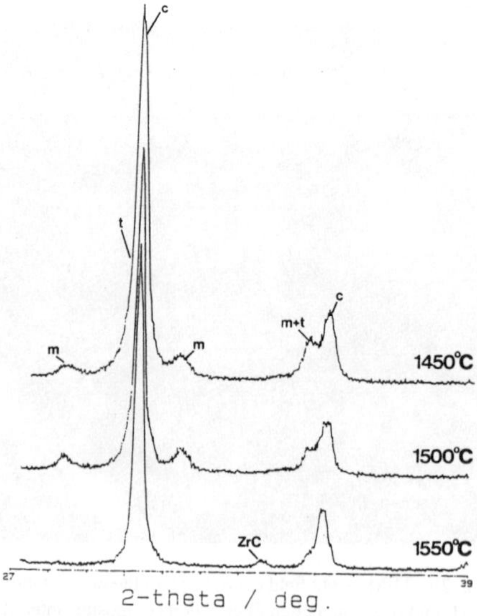

Fig. 4. X-ray diffraction results of a mesocarbon microbeads-zirconia composite pressureless sintered at different temperatures for 1 hour. m, t and c stand for, respectively, monoclinic, tetragonal and cubic zirconia.

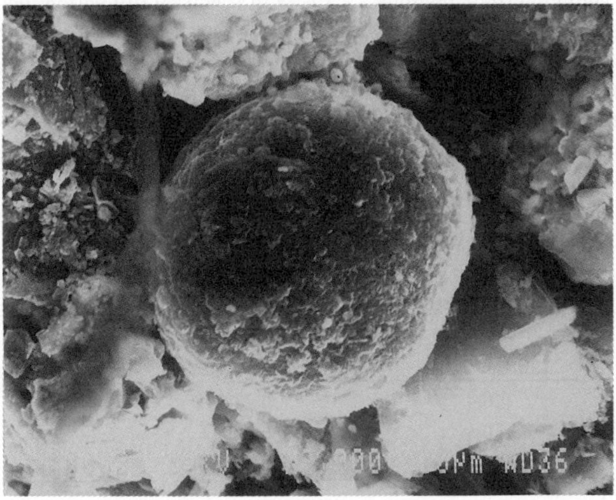

Fig. 5a. SEM photograph of the surface of a mesocarbon microbead in a zirconia matrix composite pressureless sintered at 1450°C for 1 hour.

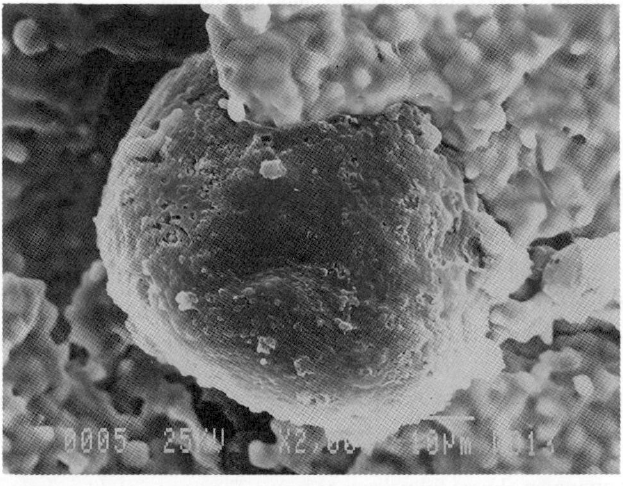

Fig. 5b. SEM photograph of the surface of a mesocarbon microbead in a zirconia matrix composite pressureless sintered at 1550°C for 1 hour.

3.3. Hot Pressing

Samples of zirconia matrix containing 20 vol % of 6 and 20 μm carbon particles and samples containing 10 vol % of short pitch fibers were hot pressed for 1 hour under a pressure of 20 bar. All of them were sintered to a final density of around 98% and observation of these samples on the optical microscope show that they are essentially fully dense. Table 3 gives the weight loss for these samples. There is no significant weight loss even at 1550°C. The X-ray results confirm that there is no reaction occurring leading to the formation of zirconium carbide.

Table 3. Weight Loss of Zirconia Matrix Composites
Reinforced with 20 Vol % of 6 and 20 μm Diameter Mesocarbon Microbeads and
10 Vol % of Short Carbon Pitch Fibers Hot Pressed at Different Temperature

temperature °C	weight loss (% of initial weight)		
	20 vol % 6μm	20 vol % 20μm	10 vol % fibers
1450	0.4	0.7	0.4
1500	0.4	0.5	0.4
1550		0.4	0.7

SEM observation of short fiber samples confirms that there is no interfacial reaction. The cross section of the fibers show the structure of the pitch fibers and that it is not affected by the matrix at 1450°C or at 1550°C as seen in Figs. 6a and 6b. The very good density of the samples allowed us to prepare samples for TEM observation. Fig. 7 shows the interface between the zirconia matrix and a carbon particulate. We can see that the interface is perfectly clean and that the carbon has not reacted with the zirconia.

3.4. VGCF

The vapor grown carbon fibers are present in a unidirectional mat or in the form of a fur like tow both made with fibers about 2 cm long and because of this the processing route used for short fibers could not be used. The following process was finally adopted.

The tow was firstly wet with a very thin zirconia slurry in order to flatten it and get it into an almost unidirectional mat like shape and left to dry to be more easily manageable. The following steps are common to both mat and tow presentations. Either were cut into squares which were then painted on one side with a thick paint made of zirconia and water. Once dry they were assembled by slightly wetting the carbon side with the fine zirconia slurry, covering it with the paint and assembling it with the next square while still wet. Once the desired number of layers was assembled pressure was uniformly applied to the sample by hand in order to help the zirconia wet the fibers. Once dry, the sample was hot pressed at a temperature of 1450°C under 20 bars for 1 hour.

Fig. 8 shows the fracture surface of one of these samples. We can see that the zirconia fully wet the fibers giving an almost fully dense sample. The tree trunk like structure of the

Fig. 6a. SEM photograph of a cross section of a pitch fiber in a zirconia matrix composite hot pressed at 1450°C for 1 hour.

Fig. 6b. SEM photograph of a cross section of a pitch fiber in a zirconia matrix composite hot pressed at 1550°C for 1 hour.

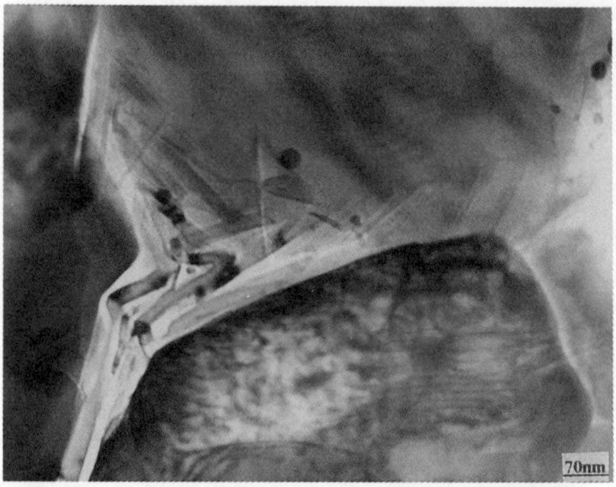

Fig. 7. TEM bright field micrograph of the interface between a carbon pitch fiber and the zirconia matrix in a sample hot pressed at 1450°C for 1 hour.

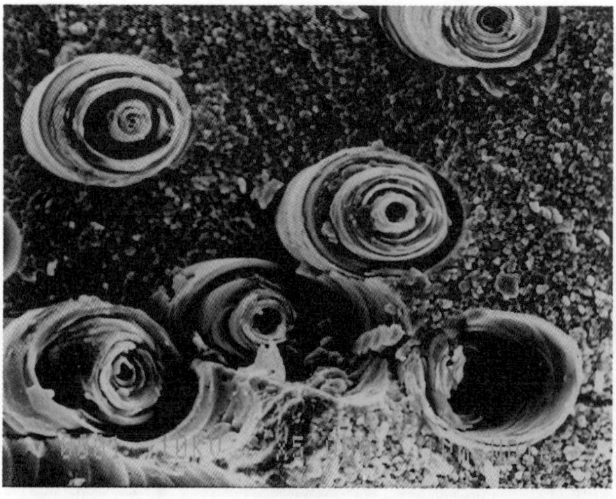

Fig. 8. SEM photograph of a fracture surface of a VGCF-zirconia composite hot pressed at 1450°C for 1 hour.

fibers is clearly visible with a layer of pyrolitic carbon on the outside and layers of partially graphitized carbon in the inside. It is noticeable that, when fractured, these fibers break in a very distinctive manner. Each of the carbon layers breaks at a different level creating a pull-out action within the fiber itself as shown in Fig. 9.

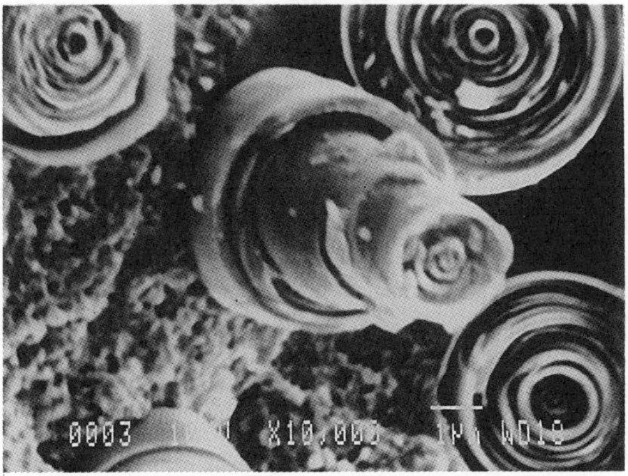

Fig. 9. SEM photograph of a fractured VGCF in a zirconia matrix composite hot pressed at 1450°C for 1 hour.

4. Discussion

4.1. Sialon Matrix Composites

As the temperature increases the attack on the fibers becomes more pronounced. It has two effects on the fibers. Firstly, there is surface degradation of the fiber due to the attack at the liquid phase-fiber interface, increasing the porosity in the fiber. Secondly, there is a reaction within the fiber itself, which converts the carbon into silicon carbide. There is also formation of silicon carbide grains in the matrix and, because of the relatively short time within which these reactions occur, some gaseous transport is very likely to be at the origin of these reactions at the surface and within the fibers.

In the hot press, the atmosphere is made of approximately 35% CO and 65% N_2 from decomposition of air on the graphite die. In a previous paper it has been shown that the reactions involved in this case are:

$$SiO_2 + C_{(s)} \leftrightarrow SiO_{(g)} + CO_{(g)}$$

where SiO_2 represents silica at reduced activity in the liquid phase. Carbon is consumed from the surface of the fibers increasing their porosity. Then, $SiO_{(g)}$ penetrates the fiber to react with the carbon:

$$SiO + C \leftrightarrow SiC + CO_{(g)}$$

producing silicon carbide inside the fibers.

At the same time, at the interface, silicon reacts with the carbon to form SiC:

$$SiO_2 + 3C \leftrightarrow SiC + 2CO$$

As sintering progresses, closed pores develop and the internal gas pressure increases inhibiting these reactions. This explains why the reactions are limited and not all the fibers are affected. As sintering goes on over periods up to 1 hour, silicon carbide grains redissolve as they are not stable at 1600°C with respect to carbon and the liquid phase.

In this case therefore, the stability of the interface and the extent of reaction is a function of liquid activity and gas pressure, and of the temperature of the processes. Control of the interfacial activity at a given temperature of sintering is therefore dictated by the purity of the sialon matrix and the level of concentration of the sintering aid employed. In this case yttrium oxide was used and this gives transient liquid phases at temperatures above 1350°C. The extent of reaction is a balance between the liquid reaction rate above this temperature and the rate of pore closure by sintering which governs the pore pressure of the gas (SiO) and the rate of mass transport.

4.2. Zirconia-Carbon Interface

In the case of pressureless sintered zirconia reinforced with mesocarbon microbeads, as the sintering temperature increases to 1550°C a reaction occurs. It is noticeable that at a same temperature the weight loss for a sample reinforced with 6 μm diameter particles is double that for a sample reinforced with 20 μm diameter particles. This indicates that the reaction occurs at the interface between carbon and matrix, the surface area for a same volume of the 6 μm reinforcement being much higher than the one of 20 μm.

The X-ray analysis of samples sintered at 1550°C indicates the presence of zirconium carbide. In a paper about the kinetics and mechanism of the oxidation of ZrC, Rao[7] proposes an oxidation mechanism including the formation of intermediate oxycarbides, the oxidation of which leads to the formation of free carbon. In this work ZrO_2 and carbon are sintered in a CO and N_2 atmosphere and it is possible that the same oxycarbides are formed and then transformed into zirconium carbide. This intermediate compound is suggested because the reaction appears to be entirely in the solid state and is thermodynamically and kinetically favorable above 1450°C. If the reaction is a solid-solid contact then direct contact of the zirconia matrix and carbon would form a layer of carbide with release of carbon monoxide gas through the pores, following the equation:

$$ZrO2 + 3C \rightarrow ZrC + 2CO$$

The carbide layer would then act as a barrier to further reaction which could only proceed by diffusion of carbon through the carbide layer to react with the matrix. The reduced carbon activity would then result in a layer of zirconium oxycarbide at the interface between the initial formation of ZrC and the zirconia matrix. Growth of the carbide layer would therefore be controlled by the diffusion of carbon and hence be slow. The present observations seem to show that only a thin layer of carbide is formed at the carbon surface.

The interface reaction in the case of zirconia/carbon is therefore of a quite different character to that in sialon/carbon; controlled by solid state processes in the former case and by formation of transient liquid phases and gas in the latter.

During hot pressing, no reaction occurs due to the fact that applying a pressure to the samples during sintering causes the pores to close very quickly, stopping any gas evolution. At high temperature, the inside pressure in the pores increases rapidly and the influence of any reaction producing gas on the formation of zirconium carbide is inhibited.

4.3. VGCF

The failure of VGCF has often been described as a "sword in sheath" failure[8,9]. A crack will propagate only through a certain number of layers and then propagate along the circumference of the fiber causing the outside layers to break before the inside ones. If tension is still applied to the fiber, it is supported only by the inside layers which are pulled out from the outside ones and will eventually break, possibly at a different level. It gives a very different failure behavior than PAN and pitch fibers that break in a brittle way[9,10].

The graph given in Fig. 10[8] shows a comparison of the different fracture behaviors for different fibers. It is noticeable that for VGCF, heat treatment at 2200°C improves the

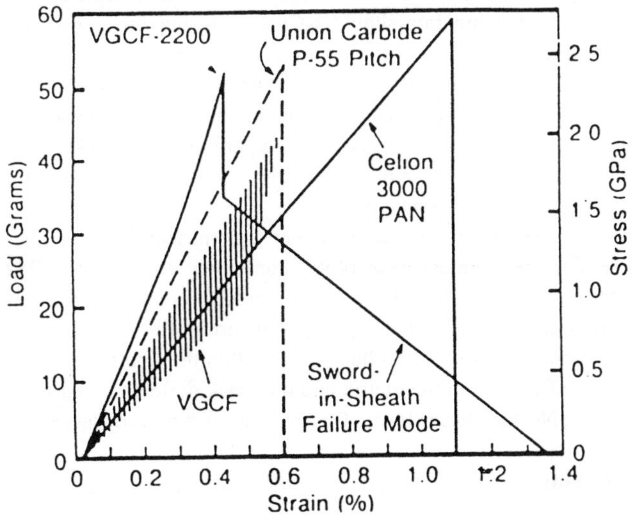

Fig. 10. Fracture behavior of VGCF compared with that of PAN and pitch based carbon fibers. VGCF 2200 represents VGCF heat treated at 2200°C[8]. Reprinted by permission of Kluwer Academic Publishers.

tensile properties of the fiber and modifies its fracture behavior. It is due to the graphitization of the carbon layers increasing the pull out phenomenon.

The fibers reinforce the matrix in three different ways. Firstly, there is a pull out of the fibers from the matrix which is also observed in the case of other types of carbon fibers. Secondly, there is a pull out action within the fiber itself due to the specific structure of VGCF and therefore not existing in the case of PAN and pitch fibers. Finally, reinforcement also occurs because of the sandwich conception of the samples' layers containing many fibers alternate with layers of pure matrix. When a crack propagating through the matrix reaches a layer of fibers it does not only break the fibers but it is also deflected because of the density of obstacles. This deflection reduces the energy of the crack.

5. Conclusion

Good quality carbon fiber reinforced sialon composites can be prepared although interfacial reaction occurs at temperatures superior to 1500°C. A sintering temperature above 1550°C is needed to complete the α to β' transformation in the matrix.

The interfacial reaction between carbon and the sintering liquid phase results after a short time in the formation of silicon carbide within the matrix which is redissolved after a longer sintering time. Formation and transportation of silicon monoxide gas within the fibers result in the conversion of the carbon into crystalline silicon carbide.

In order to stop the interfacial reaction occurring at the same time as the α to ' matrix transformation, a rapid heating rate should be used causing the pores to close quickly stopping any gas circulation.

During pressureless sintering of zirconia reinforced with mesocarbon microbeads the difference in thermal expansion coefficients of carbon and zirconia generates many cracks in the matrix leading to a poor final density. At temperature above 1450°C an interfacial reaction occurs between carbon and zirconia creating zirconium carbides within the matrix. This reaction is controlled by solid state processes.

During hot pressing of zirconia reinforced either with mesocarbon microbeads or carbon pitch fibers or VGCF, no reaction occurs at 1450°C or higher temperatures. The samples are fully dense and the structures of the different reinforcement are not altered.

The tree trunk like structure of the vapor grown carbon fibers increases their ability to reinforce the matrix by a pull out action happening within the fibers themselves in addition to the pull out of the fibers from the matrix and the crack deflection that are involved with other types of fibers. No significant level of interfacial reaction is observed.

6. Acknowledgements

The work described in this paper concerning VGCF is part of a European Union Human Capital and Mobility Programme on "Production and applications of vapor grown carbon fibers and other ceramic fibers and filament" and is coordinated by CENIM, Madrid, Spain.

7. References

1. Hanson et al., in *Materials for Advanced Technology Applications*, eds. M. Buggy & S. Hampshire (Trans. Tech. Pub., Switzerland, 1992).
2. Smith et al., in *Materials for Advanced Technology Applications*, eds. M. Buggy & S. Hampshire (Trans. Tech. Pub., Switzerland, 1992).
3. Anya, *Journal of the European Ceramic Society*, **12** (1993) 297.
4. Dupel and A. Hendry, *Journal of the European Ceramic Society*, **15** (1995) 801–809.
5. Smith et al, in *Proc. of Silicon Nitride '93*, eds. M. Hoffman, P.F. Becker & G. Petzow (Trans. Tech. Publications, Brookfield, VT, USA, 1993).
6. F. Foct, MsC, 1993 University of Strathclyde, Glasgow UK.
7. G. A. Rama Rao and V. Venugopal, *Journal of Alloys and Compounds*, **206** (1994) 237.
8. Tibbets, in *Proc. of the NATO Advanced Study Institute on Carbon Fibers and Filaments* (Alvor, Portugal, 1989).
9. A. Madronero, in *Proc. of Materials Development in Rail and Tire*, (AIMXXIV Biennial Congress, Genoa, Italy, 1992).
10. Daumit, *Carbon*, **27(5)** (1989) 759.

PROCESSING, STRUCTURE AND PROPERTIES OF CARBON AND CARBON/CERAMIC COMPOSITES

L. M. MANOCHA, J. R. PATEL, and S. B. SHAH
Institute of Materials Science, Sardar Patel University
Vallabh Vidyanagar-388 120, Gujarat, India

ABSTRACT

Carbon based solid materials have been conveniently fabricated using sinterable carbons prepared through mechanical-chemical treatment for appropriate time and compacted under high pressure at room temperature. Ceramic powders, SiC and B_4C have been mixed with sinterable carbon to give carbon/ceramic composites which exhibit 30–50% improvement in density and mechanical properties. The oxidation resistance is remarkably improved. Even carbon/carbon composites could be fabricated by using high strength carbon fibers and oxidized PAN fibers. With the present technique, oxidized PAN fiber reinforcement has been found to be more advantageous than high strength carbon fibers.

1. Introduction

Carbon materials possess a unique combination of mechanical, thermal and electrical properties not exhibited by any other single material. Therefore, these materials are preferred for various mechanical and electrical applications especially at high temperatures under non-oxidizing atmosphere. These products are generally fabricated through filler-binder routes wherein cokes are used as fillers and resins or pitches are used as binders. However, decomposition and molecular rearrangement of the binder phase during carbonization and subsequent high temperature heating result in the formation of pores and voids. These pores and voids can be partially filled in during subsequent impregnations, but the process is quite lengthy and costly. Moreover, secondary impregnation may not result in uniformly dense and strong products, necessary for mechanical and structural applications. Therefore alternative methods such as carbon mesobeads, self-sinterable carbon, etc., have been explored to fabricate dense carbon products without the need for reimpregnation[1,2]. Though technological viability has been demonstrated for some carbon forms, these methods have yet to be exploited for making carbon/carbon composites with fibrous reinforcements.

Another requirement in developing carbon products for mechanical and structural applications is to improve mechanical properties of carbon, especially hardness and bending strength, and thermal stability under oxidizing conditions without diminishing the electrical and thermal properties. This can be effectively achieved through incorporation of ceramics. The products are so called carbon/ceramic composites[3].

Incorporation of ceramics with carbon results in significant improvement in oxidation resistance, mechanical strength and hardness while the electrical resistivity and thermal conductivity are slightly reduced. Fig. 1 shows the relative improvements and changes in various properties of carbon as a result of ceramic addition.

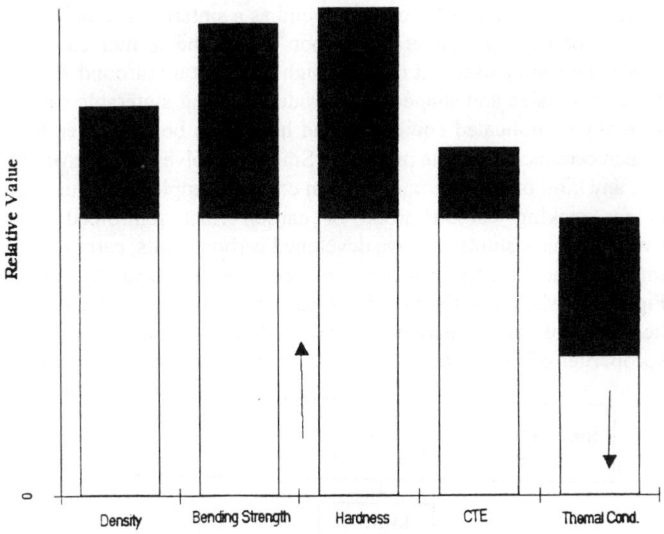

Fig. 1. Relative change in various properties of carbon as a result of ceramic addition.

Fig. 2 gives an overview of different methods employed to make carbon/ceramic composites. All of these methods have been used commercially for manufacturing products, but suffer from various drawbacks especially reproducibility and quality assurance. These processes are also inconvenient for fabricating large products and those with complicated shapes. Powder metallurgical routes can be employed to make carbon/ceramic composites.

TYPE	*REINFORCEMENT*	*PROCESSING*
FIBEROUS COMPOSITES	**Ceramic or Carbon**	*Vapour phase* (CVD) *Liquid phase* (Polymer pyrolysis)
PARTICULATE	**Ceramic**	*Liquid phase (Reinforce)* (Imperegnation + carbothermal reaction) *Liquid phase (matrix)* (mixing,forming,pyrolysis) *Solid phase (R & M)* (mixing, carbothermal) *Solid phase (R & M)* (mixing with sintering aid Pressure sintering) **Solid phase (R & M) (Pressureless sintering)**

Fig. 2. Processing of carbon/ceramic composites.

This can be done either by using a boron compound as a sintering aid and using a pressure sintering technique or by using sinterable carbon[4,5]. In the former case, commercially available calcined coke can be used but requires high temperature (around 2000°C) pressing which further limits the size and shape of the product. Using sinterable carbon, however, does not require any complicated equipment and hence can be conveniently employed to manufacture carbon/ceramic composite products. Since it involves room temperature mixing and compaction, any kind of ceramic combination could be employed. This process may be ideally suited for making carbon/carbon or carbon fiber reinforced carbon/ceramic composites as well. In our Institute we have developed carbon solids, carbon/SiC composites and fiber reinforced carbon and carbon/ceramic composites using the sinterable carbon technique. Fig. 3 is a sketch of the procedure used for making carbon and carbon based products. The effects of various parameters involved in processing these materials on the structure and properties of the products have been studied in detail and are discussed in this paper.

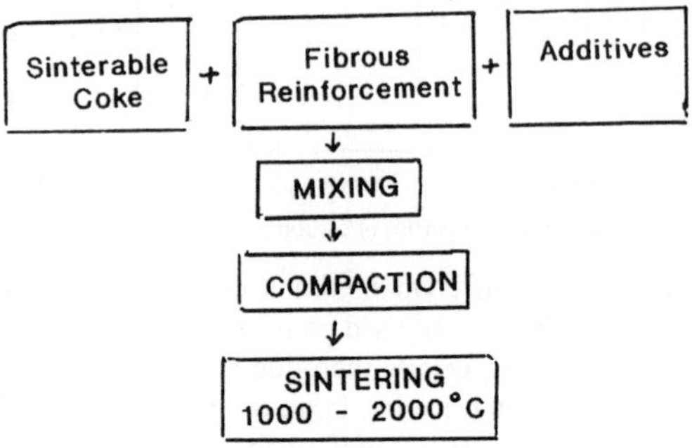

Fig. 3. Alternative route to make carbon/carbon composite (fabrication through solid/solid interactions).

2. Experimental

The present studies were made with petroleum-derived semi-coke carbon from Brauni Refinery, India. The coke was given a mild chemical treatment with toluene and mechanical treatment in a potmill. The time of mechanical treatment was varied. The coke carbon ultimately becomes very fine with particle size between 1–5 μm. The surface oxygen complexes (predominantly phenols) on dried samples were estimated using a chemical titrometric technique. The coke powder, after mechanical-chemical treatment as such or mixed with fine SiC powder and boron carbide or fibrous material, was compacted at room temperature under high pressure varying from 700 kg/cm^2 to 1500 kg/cm^2. The products were sintered to 1000°C and subsequently to between 2000–2700°C under nitrogen/argon atmosphere. The products were characterized for physical changes during sintering. Flexural strength of the products was measured using three point bending with a span to thickness ratio of 5. Electrical properties were measured using the four probe method.

3. Results and Discussion

3.1. Changes in Carbon Characteristics during Mechanical-Chemical Treatment

When the carbon is given mechanical treatment, it results in a decrease in size of the carbon particle, a change in shape, an increase in apparent density, and increases in volatile contents and surface complexes. Fig. 4 shows change in particle size and volatile content of

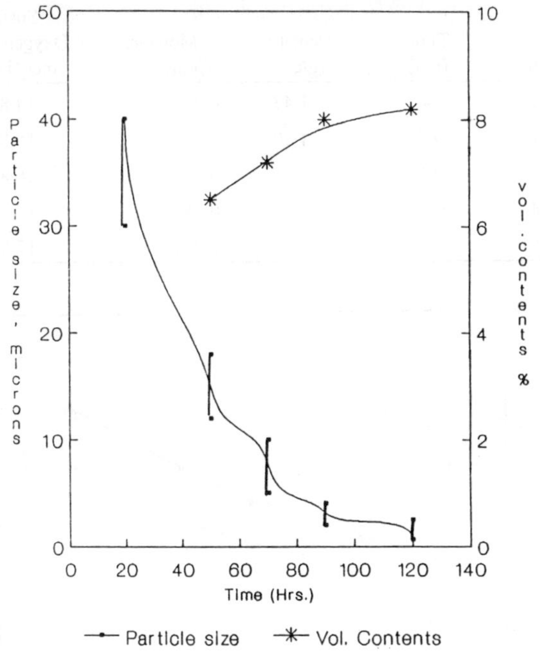

Fig. 4. Coke characteristics vs. treatment time.

carbon with treatment time. The surface oxygen complexes present on the powder are compiled in Table 1. As seen from the figure and the table, with increasing treatment time, the carbon becomes more reactive. The decrease in particle size results in increased surface area. This, along with the increase in beneficial volatiles and surface complexes, helps in making the carbon sinterable.

3.2. Solid Carbon from Sinterable Carbon

Sinterable carbon, when molded at room temperature and sintered at high temperatures, results in solid carbon. However, the ultimate density of the product will depend on the extent of solid-solid reactions between the particles or sinterability of carbon and molding conditions. Fig. 5 shows the density of solid carbon made from sinterable carbon mechanically treated for different times and sintered at 1000°C and 2000°C. As seen from the figure, the "green" density (molded) and the density of the sintered products increase with treatment time. This is a consequence of the decrease in particle size and increase in volatile contents and surface complexes of carbon which make carbon increasingly sinterable.

Table 1. Characteristics of Sinterable Carbons Treated for Different Times

Sr. No.	Time (hrs)	Density (g/cc)	Moisture Content (%)	Surface Oxygen Group (meq/100gm)
1.	--	1.47	0.208	14.836
2.	60	1.50	0.256	16.023
3.	90	1.51	0.327	16.847
4.	120	1.54	0.387	17.938
5.	150	1.55	0.391	17.976

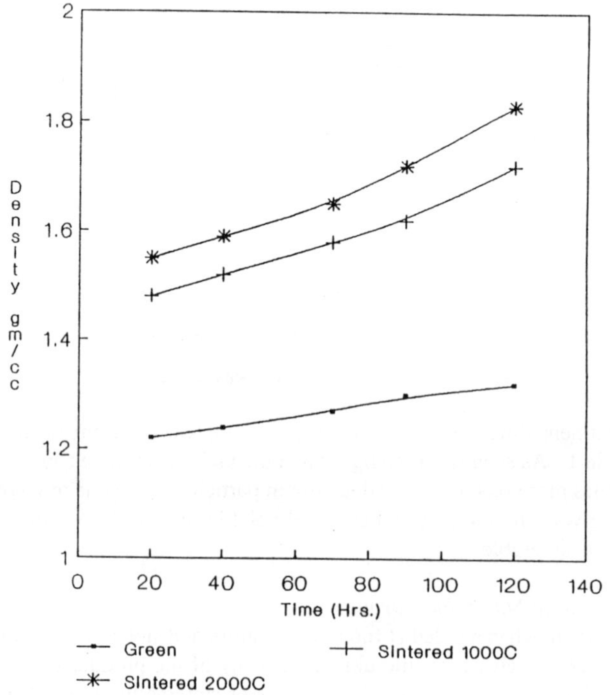

Fig. 5. Effect of treatment time of carbon powder on density of carbon product at different stages of sintering.

This is further evident from the SEM and optical micrographs of the sintered carbon products made from different carbon powders. As seen from Fig. 6a, carbon treated for 30 hours does not exhibit good sinterability. The particles are big enough to exhibit anisotropic structure (6b). On higher temperature sintering between 2000–2700 °C, these big particles

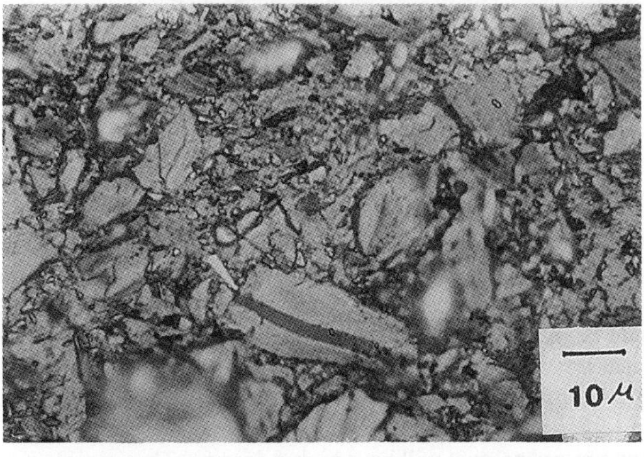

Fig. 6. Micrographs of sintered carbons sintered at 1000 °C made from carbon treated for 30 hours. (a) SEM, (b) optical.

suffer microcracks along the oriented planes. This results in a low density ultimate product. If the treatment is done for longer times, the carbon particles not only attain good sinterability, but are small enough that they lose the anisotropic character. Interparticle microcracking does not take place and the products show good sinterability and high density Likewise, compaction pressure is also quite significant in producing dense products.

Green carbon solids (compacted solids) when sintered to 1000°C undergo volume shrinkage of 20–30%. In order to follow the transformations during sintering, dimensional changes occurring during sintering to 1000°C were studied using a Mettler thermal analysis system TA4000. Fig. 7 shows the weight loss and linear shrinkage in the carbon products during sintering. As seen from the figure, weight loss begins around 400°C whereas the

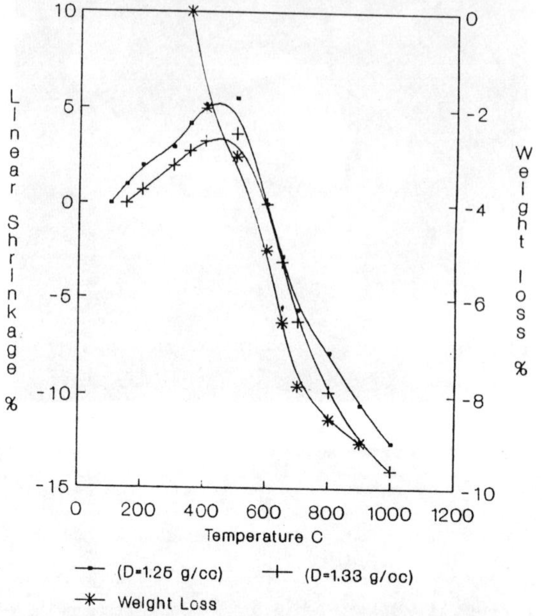

Fig. 7. Physical changes of carbon composites during sintering.

material exhibits expansion up to a temperature of 450°C followed by shrinkage up to 1000°C. The initial expansion is due to normal thermal expansion of the solid mass. Shrinkage is due to the sintering process. In order to get good sintering, the initial expansion up to 450°C should be minimal. This has been possible by having dense green solids either by long mechanical treatment of carbon and/or by using large compaction pressures. The ultimate aim is to get dense green carbon products. As seen from Fig. 7, if the green density is increased from 1.25 g/cc to 1.33 g/cc, the initial expansion is reduced, ultimate shrinkage is increased and the sintered product will have higher density. This is evident from SEM and optical micrographs (Fig. 8) of the sintered composites made with highly sinterable carbon.

3.3. Carbon/Ceramic Composites

After understanding the sinterability in carbon particles, carbon/ceramic particulate composites were prepared by adding SiC and boron carbide powder to sinterable carbons. The density of these carbon/SiC composites increased with increase in ceramic contents.

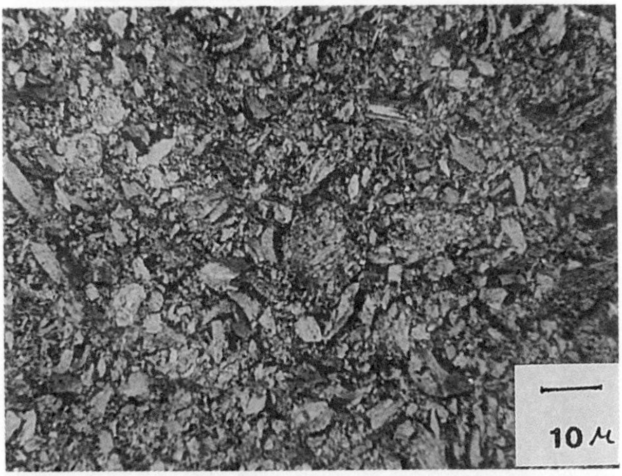

Fig. 8. SEM (top) and optical (bottom) micrographs of sintered composites made with highly sinterable carbon.

This is shown in Fig. 9 . Interestingly, the density of carbon/SiC composites having 30% SiC showed a larger increase in density after sintering to 2000°C than the other solids. Fig. 10 shows an optical micrograph of sintered composites. As seen from this micrograph, the silicon carbide is nicely distributed in the composites and carbon particles are nicely compacted around ceramic particles.

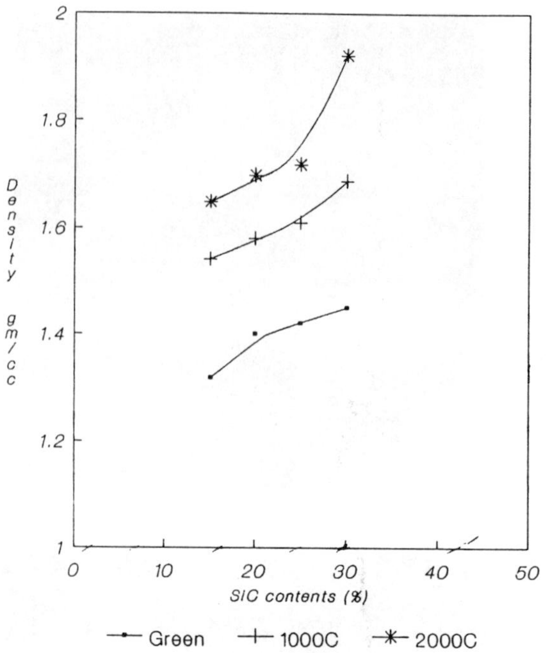

Fig. 9. Carbon/SiC composites.

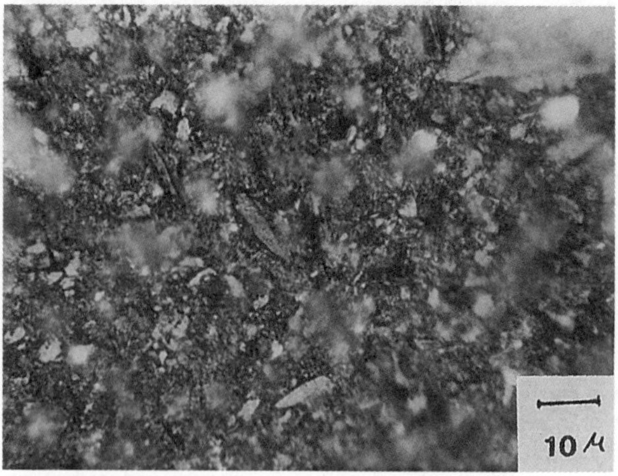

Fig. 10. Optical micrograph of sintered carbon/SiC composite.

3.4. Properties of Carbon and Carbon/Ceramic Composites

Like density, the mechanical, thermal and electrical properties of the carbon and carbon/ceramic materials are also dependent on the properties of the sinterable carbon and the processing parameters. The bending strength is found to increase with density while electrical resistivity decreases. As hypothesized in Fig. 1, the mechanical properties increase by about 50%, coefficient of thermal expansion and electrical resistivity increase to small extent and the values are still attractive for applications as a replacement for carbon materials. Typical properties of the materials developed in our Institute have been compiled in Table 2. The remarkable effect has been on the oxidation resistance of these materials. Fig. 11 shows the oxidation behavior of different materials. Of course, there is the further possibility of improving these properties by varying the constituents.

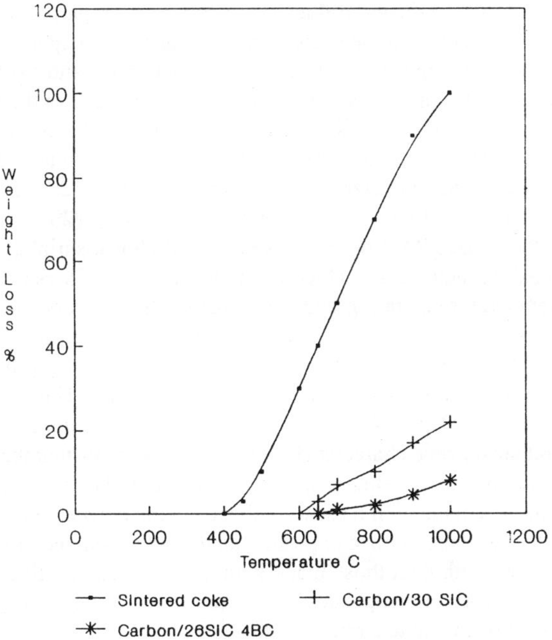

Fig. 11. Oxidation behavior of carbon/ceramics.

3.5. Carbon/Carbon Composites using Powder Technique

Properties of carbon materials, especially the structural properties, can be increased significantly by incorporating fibrous reinforcement. In previous studies by the author[6,7], it has been shown that fiber/matrix interactions in carbon/carbon composites, which are the consequence of the fiber/matrix bonding at the initial forming stages and are dependent on

Table 2. Properties of Carbon and Carbon-Ceramics

	Carbon		Carbon/Ceramic	
	1000°C	2000°C	1000°C	2000°C
Density (gm/cc)	1.60–1.64	1.82–1.85	1.71–1.73	1.94–2.1
Bending Strength MN/sq.m	96–102	84–95	117–122	146–154
Resistivity m cm	6.3–6.6	3.1–3.4	8.7–9.2	6.8–7.1
Shore Hardness	85–90	75–80	100–110	105–115
CTE (10E6)	4–5	3–5	6–8	5–7

the type of fibers, have significant influence on the end properties of carbon/carbon composites. Since the objective in the present studies was to develop a low cost technique for making carbon/carbon composites, high strength carbon fibers and oxidized PAN fibers were used as reinforcements with sinterable carbon discussed above as the matrix. The two reinforcing fibers showed different effects during sintering of the composites. As with thermosetting resins as matrix precursor[6], carbon fibers hindered the matrix shrinkage during sintering resulting in voids and big size cracks in the matrix (Fig. 12a). However, near the fibers, the carbon/matrix was found to be quite compacted (Fig. 12b). In contrast to this, composites made with oxidized PAN fibers showed 25–30% linear shrinkage during carbonization and did not exhibit any cracks. However, in this case, the fibers undergo carbonization during high temperature sintering and shrinkage of the fibers occurs. Moreover, no chemical bonding exists between fibers and matrix at the green stage. Therefore during sintering, the fibers shrink away from the matrix leading to a small gap between fiber and matrix (Fig. 13). At places the matrix is very close to the fibers. This is seen in the SEM micrograph of the composites (Fig. 14). On graphitization of the composites, the fibers have been found to exhibit anisotropic character (Fig. 14). This suggests that the fibers are under tension while the composite is being graphitized. Table 3 shows the characteristics of carbon/carbon composites made with these two fibers and the powder sintering method. As seen from this table, carbon/carbon composites made with oxidized fibers show higher density and bonding strength than those made with untreated carbon fibers. It shows that general purpose carbon/carbon composites can be fabricated at reasonably low cost using oxidized PAN fibers and sinterable carbon.

Table 3. Characteristics of Carbon/Carbon Composite

	Carbon Fibers		PAN OX Fibers	
	1000°C	2700°C	1000°C	2700°C
Density (g/cc)	1.48	1.42	1.45	1.66
Bending Strength MN/Sq.m	140	100	180	210

Fig. 12. Optical micrographs of C/C composites made with carbon fibers.

Fig. 13. Optical micrographs of C/C composites made with oxidized PAN fibers (sintered at 1000°C).

4. References

1. Y. Yamada et al., *Carbon* **12** (1974) 307.
2. G. Kanii, Y. Kyoji and H. Marsh, *Proc. Int. Symp. Carbon*, Tsukuba, Japan (1990) 714.
3. Y. Yamamoto and H. Izawa, *Proc. Int. Symp. Carbon*, Tsukuba, (1990) 730.
4. K. Kobayashi, T. Hagio and K. Miyazaki, *Proc. Int. Symp. on Densification and Sintering of Oxide and Non-oxide Ceramic*, Japan (1978) 309.
5. I. Ogawa et al., *Yogyo-Kyokai-shi*, **93(10)** (1985) 613.
6. L. M. Manocha, *Composites* **19** (1988) 311.
7. L. M. Manocha, *Carbon* **32** (1994) 213.

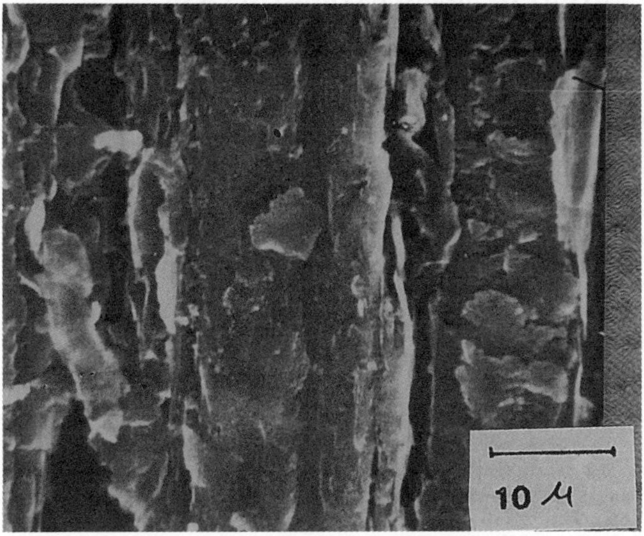

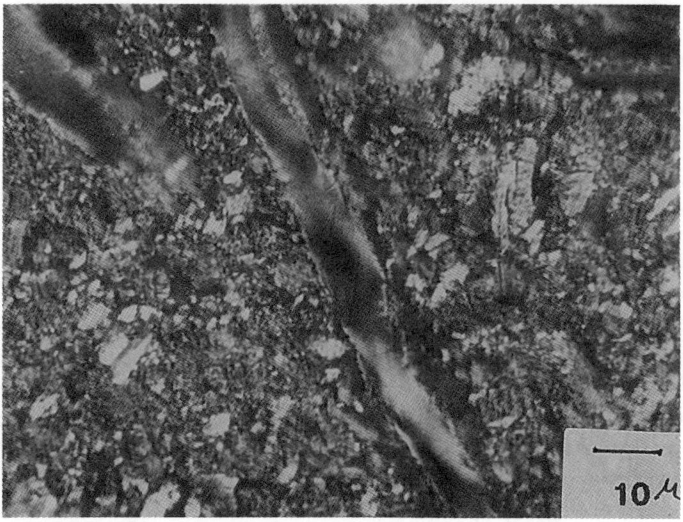

Fig. 14. SEM (top) and optical (bottom) micrograph of C/C composites made with oxidized PAN fibers (sintered at 2700 °C).

STUDY OF SIZING EXTRACTION AND AMMONIA PLASMA TREATMENT OF CARBON FIBERS BY COMBINED TIME-OF-FLIGHT SECONDARY ION MASS SPECTROMETRY AND X-RAY PHOTOELECTRON SPECTROSCOPY

V. B. WIERTZ, L. T. WENG, and P. BERTRAND
Unité de Physico-Chimie et de Physique des Matériaux,
Université Catholique de Louvain, Place Croix du Sud 1,
1348 Louvain-la-Neuve, Belgium

ABSTRACT

A two-step process has been carried out in order to increase the density of covalently bonded polar functionalities (mainly amines functionalities) at AU4 carbon fiber surface: sizing extraction and ammonia plasma treatment. At each step, surface modification has been followed by combined time-of-flight secondary ion mass spectrometry (ToF SIMS) and X-ray photoelectron spectroscopy (XPS). Quantitative and chemical information obtained by XPS combined with the molecular and the semi-quantitative information derived by ToF SIMS allows us to monitor the changes in functional groups on carbon surface. The results have shown that extraction in (1) CH_2Cl_2 and (2) H_2O is efficient to remove the sizing and to increase the density of functionalities at the surface. The effects of ammonia plasma treatment have first been studied on a model surface: highly ordered pyrolytic graphite (HOPG). Detection of aromatic amines at the HOPG surface as well as at the carbon fiber surface after plasma treatment and post-treatment washing let us suspect that amine functionalities have been introduced at both graphic surfaces. Moreover, applicability of combined ToF SIMS and XPS as useful tools for carbon fiber surface characterization has been illustrated in this work.

1. Introduction

The mechanical properties of carbon fiber (CF) composites depend not only on the properties of the fiber and the matrix, but also on the properties of their interface. The objective of this research is to modify the interface by creating a covalent bond between the carbon fiber and the matrix. Given the reactivity of the amines with anhydride functionalities as shown previously[1], a covalent bond may be created by reacting the amine functionalities present on the CF surface with the anhydride functionalities grafted on the matrix. This study aims at increasing the density of the useful functional groups on the fiber surface, mainly amine groups. In a first step, the fibers are washed in order to remove the sizing (in this paper, sizing means all compounds other than CF itself, which have been added voluntarily or involuntarily during the preparation of CF[1]). Indeed, this sizing forms a screen between the functionalities of the fiber and the polymer matrix. In a second step, washed fibers are treated by an ammonia plasma in order to introduce amine functionalities at the fiber surface. Different studies have reported cheering results in amine fixation to some surfaces by ammonia plasma treatment[2-5], notably at CF surface[5]. Poorly bonded fragments that could have been created by the plasma are removed by a post-treatment CH_2Cl_2 washing. The evaluation of each step of the surface processing was checked out by combined time-of-flight secondary ion mass spectrometry (ToF SIMS) and X-ray photoelectron spectroscopy (XPS).

XPS has been widely used to characterize CF surface modifications following different surface treatments[5-20]. Only a small number of papers have an interest in the N1s region[5,12,13,17,19], especially on CF after plasma treatment[5]. Different factors could justify this reserve: (1) there is some lack of correspondence between the binding energies of the N-containing functionalities reported in the literature[5,19,21-25], (2) on ex-PAN based fibers, a lot of N-containing functionalities are present before any treatment: N included at the end of thermally cyclised PAN sequence, as shown in Fig. 1[26], and different N-containing functionalities are present in the components of the sizing. Thus, in order to identify which N functionalities are introduced at a graphic surface following an ammonia plasma treatment, a graphite model surface for which no N was detected by XPS has first been treated: highly ordered pyrolitic graphite (HOPG); (3) after ammonia plasma treatment, it is likely that different N-containing functionalities are created at the fiber and HOPG surface. Moreover, the binding energy of each of those functionalities will depend on which structure it is linked, either on aliphatic hydrocarbon, on single aromatic cycle[23] or on a graphitic structure[27]. Therefore, for such samples, a large number of functionalities are present in N1s peak and it is not easy to obtain unambiguous information about the nature of N-functionalities from XPS. This major drawback can be overcome at least partially by ToF SIMS. This analytical technique offers qualitative information of molecular surface species with high mass resolution and high sensitivity[28], that is complementary to the quantitative XPS analysis. The combined use of the two techniques would be a reliable method to approach the surface chemistry of CF as will be shown in this paper.

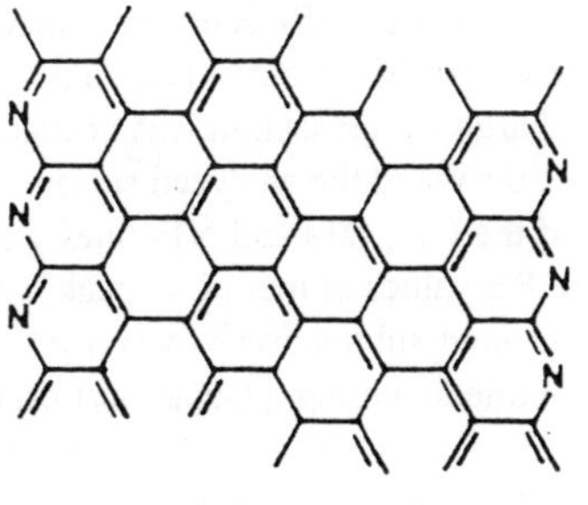

Fig. 1. N included at the end of thermally cyclised PAN sequence (reference 26).

2. Experimental

2.1. Materials

The CF studied in this work was a PAN-based CF: AU4 from Hercules (12K). The CF sample was first extracted with CH_2Cl_2 (dichloromethane) for 80h and then with H_2O for 48h (procedure 1) or with CH_3OH (methanol) for 48h (procedure 2). Ammonia plasma treatment was performed on HOPG samples and on CFs samples washed by procedure 1.

2.2. Surface Treatment

Ammonia plasma treatments were performed with a Chemprep 130 Barrel reactor from Chemex (capacitive coupling, 13.56 MHz). Reactor chambers were evacuated to $4\ 10^{-2}$ mbar prior to gas admission. The treatment time was 3 minutes and the power was 50W.

2.3. Surface Analysis

ToF SIMS measurements were carried out using a ToF SIMS spectrometer from Charles Evans & Associates[29]. The primary ion was generated from a pulsed ^{69}Ga ion source (15 kV) and rastered over a 97 x 97 μm^2 area. The secondary ions were extracted by a +/−3 kV voltage and then deflected by three electrostatic analyzers in order to compensate the initial energy distribution of ions with the same masses. The mass spectra of those ions (positive and negative separately) were recorded by measuring their time-of-flight distribution from the sample surface to the detector. The scanned mass range was from 0 to 5000 amu. The total ion fluence for the spectra presented in this work was about $1.5\ 10^{12}$ ions/cm^2, ensuring static conditions. The ToF SIMS data were treated using Cadence software from Charles Evans & Associates. This software calculates mass calibration from well-defined reference peaks. In this case, reference peaks were CH_3^+, $C_2H_5^+$, $C_3H_{11}^+$ and $C_4H_9^+$ in the positive mode and C^-, CH^-, C_2H^- and C_4H^- in the negative mode. The most accurate ion formula was assigned to each peak. Results were expressed in terms of relative intensities which are the ion peak integral normalized by the total secondary ion intensity (minus H+/H^- and inorganic ions, because of their high sensitivity and poor reproducibility).

XPS results were obtained with an SSI X-Probe (SSX-100/206 from Fisons)[30,31] equipped with an aluminum anode (10 kV, 11.5 MA) and a quartz monochromator. As CFs are sufficiently conducting, no charge compensation was needed for the measurements. The pass energy of the analyzer was 50 eV and the analyzed surface was a spot of 1000 μm. For each sample, a detailed scan of the C1s, O1s and N1s lines was recorded. For the CF and HOPG samples studied here, the BE values of main C1s peak was always 284.3 +/−0.05 eV. A Shirley type nonlinear background subtraction[32] was used, and the peaks were decomposed by using a least-square routine assuming a Gaussian/Lorenzien (85/15) function. The XPS atomic percentages were calculated from peak areas corrected by the sensitivity factors on the first principles model and assuming no hydrocarbon contamination[31]. The spectra differences which will be presented below were obtained by subtracting two spectra after the normalization of their intensity by the corresponding percentage.

2.4. Sample Preparation for Surface Analysis

A bundle of CF (1 cm) was cut from the fiber tow. This bundle was placed on a 1 x 1 cm^2 silicon wafer and mechanically tied on both sides using a scotch tape (for ToF SIMS) or a Cu foil (for XPS). For ToF SIMS analysis, a metallic grid was placed over the fibers in order to hold them flat. The sample was then fastened to the sample holder.

3. Results and Discussion

3.1. Effects of Solvent Extraction

Fig. 2 displays O and N percentages detected by XPS at the Au4 CFs as received, washed in dichloromethane then in H_2O (procedure 1) and washed in dichloromethane then in methanol (procedure 2). The number of analyzed samples is respectively 6, 4 and 4. The column height depicts the average value and the error bar is drawn between the minimum and the maximum detected values. The variability of O and N percentages decreases after both washing procedures. The high variability observed on "as received" fibers was related to external conditions (humidity, temperature, etc.) that could influence the sizing oxidation level. After procedure 1, %O and %N increase at the surface. Fig. 3 displays the overlay of representative N1s) XPS spectra. The N1s shape and intensity are similar for CFs as received and CFs washed following procedure 2. The main contribution is located around 400.6 eV. Table 1 summarizes the binding energies associated with different N-containing functionalities[21-23]. Most of them are taken from the literature and have been determined from polymeric samples. The others are logical proposals. This suggests that the contribution located around 400.6 eV can be attributed to amide and/or protonated nitrogen. The contribution located around 399.3 eV can be attributed to nitrogen included in aromatic cycle–N characteristic of ex-PAN fiber structure[26]–or possibly to amines or imines functionalities. The proper identification of this additional contribution appearing after procedure 1 is helpful because the increase of nitrogen included in aromatic structure would evidence the efficiency in the sizing removal.

Table 1. Binding Energy of Some N1s Species

Functional group	Eb (eV)	Ref.
R–C=NH; C–NH–C	~399.1	21
Aromatic N	~399.3	21
$R–NH_2$	~399.3	24
$\varphi–NH_2$	<399.9	
φ–NH–φ	~399.9	21
O=C–NH–(C,H)	~400.0	21
C–N+	~401.5	21
NO_2	~405.5	21

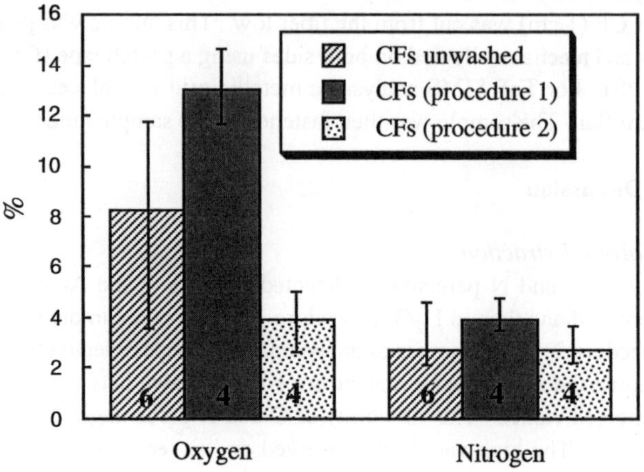

Fig. 2. %O and %N determined by XPS at unwashed CFs, CFs washed by procedure (1) and CFs washed by procedure (2).

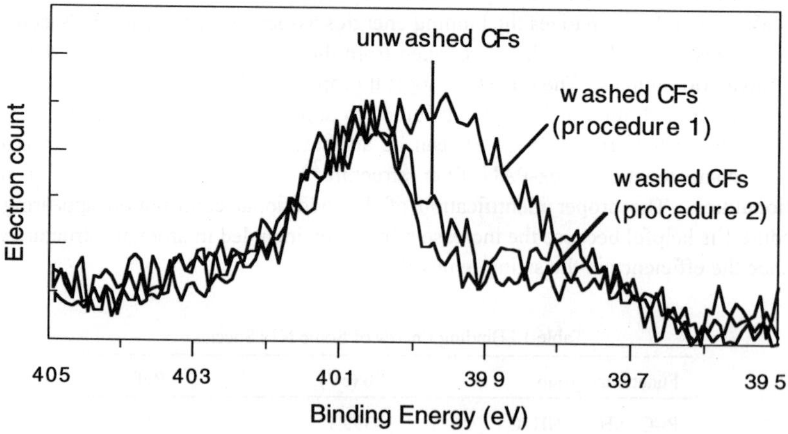

Fig. 3. N1s spectra overlay for unwashed CFs, CFs washed by procedure (1) and CFs washed by procedure (2).

Molecular information given by ToF SIMS analyses allows to specify surface functionalities identification. O^-/C^- and CN^-/C^- intensity ratios exhibit the same evolutions as %O and %N respectively. Fig. 4 shows intensity ratios for C^-/CH_2, C_2N^- and C_5N^-. The results have been normalized to the values of untreated CFs. C^-/CH_2^- indicates the level of insaturation of hydrocarbon components present at the surface. The increase of this ratio after procedure 2 indicates a more graphitic structure at the surface. C_2N^- and C_5N^- are believed to evidence the presence of nitrogen included in aromatic cycle. This relative intensity increases after procedure 1 and does not vary significantly after procedure 2. Bringing this together with XPS results, it can be deduced that the additional contribution at 399.3 eV (Fig. 3) is mostly due to an increase of nitrogen included in aromatic cycle at the surface. Hence, combined XPS and ToF SIMS information indicates that procedure 1 reveals a more ex-PAN fiber characteristic structure at the surface and thus demonstrates the efficiency of this washing procedure. Subsequent surface treatments will only be operated on CF washed by procedure 1 given: (a) the efficiency in the sizing removal, (b) the increase of functionalities at the surface, and (c) the more constant level of functionalisation observed.

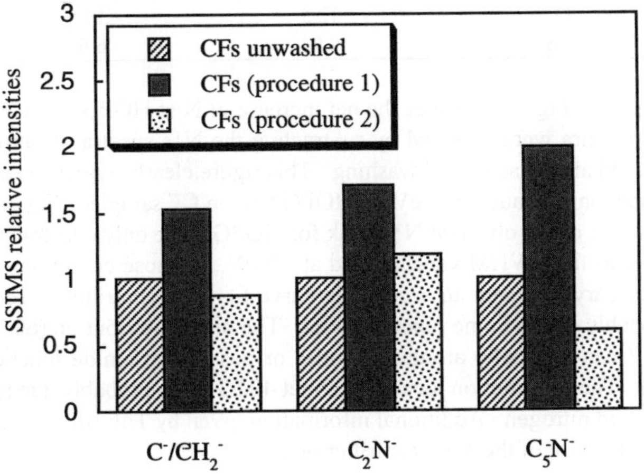

Fig. 4. ToF SIMS intensity ratios at unwashed CFs, CFs washed by procedure (1) and CFs washed by procedure (2) (the results are normalized to the unwashed CFs values).

3.2. *Effect of Plasma Treatment*

Surface modification by ammonia plasma was first studied on a HOPG sample, because of the complexity of CF surface chemistry. CF washed by procedure 1 were then treated under the same operating conditions. In order to remove all small fragments created by the plasma and that are not strongly bonded to the surface, CF and HOPG samples were washed with CH_2Cl_2 (20h) after plasma treatment.

Table 2 presents the %O and %N values evaluated by XPS at HOPG and CF surfaces before plasma, after plasma and after plasma and CH_2Cl_2 washing. No O or N was detected at the untreated HOPG surface. After plasma treatment, an important increase of %N is observed at HOPG and at CFs surface. A slight increase of %O is observed as well. After CH_2Cl_2 washing, a decrease of %N is observed. This is supposed to be related to the removal of adsorbed or poorly bonded N-containing degradation fragments created by the plasma. Nevertheless, %N after plasma and washing remains higher than on untreated samples. This net increase of N denotes that some N-containing functionalities were strongly

Table 2. %O and %N at HOPG and Procedure 1 Washed CFs before Plasma Treatment, after Plasma Treatment and after Plasma and CH_2Cl_2 Washing

	HOPG	HOPG plasma	HOPG plasma + washing
%O	< detection limit	2.7	6.9
%N	< detection limit	7.4	2.9
	CFs (procedure 1)	CFs (procedure 1) plasma	CFs (procedure 1) plasma + washing
%O	12.0	12.5	13.5
%N	3.8	8.9	6.6

fixed at the surface. Fig. 5 illustrates the net increase of N at HOPG and CF surface. The displayed N1s spectra were obtained by subtracting the N1s spectra of untreated samples from those obtained after plasma and washing. This figure clearly evidences the presence of one main contribution at about 399.8 eV on HOPG and on CF samples. Fig. 6 illustrates the deconvolution result of the obtained N1s peak for HOPG. The only constraint fixed for this deconvolution was the FWHM values, fixed at 1.8 eV. In those conditions, three components were necessary to obtain an acceptable curve fit. The contribution located at about 398.7 eV is probably due to imine functionalities. The main contribution (68.7%) located at about 399.8 eV can be due to aromatic amines or possibly to amide functionalities or to aliphatic amines. The contribution located at about 402.3 eV is probably due to the presence of some protonated nitrogen. Additional information given by ToF SIMS will be helpful in the proper identification of the main contributions.

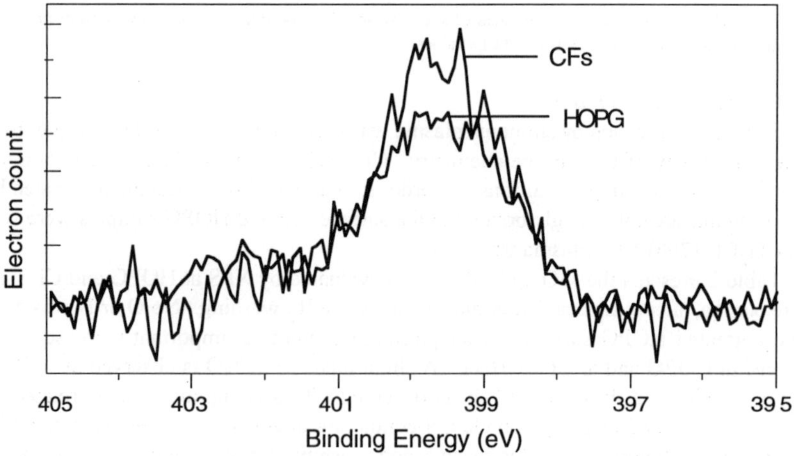

Fig. 5. N1s XPS difference spectra between samples before and after plasma treatment and CH_2Cl_2 washing for CFs and HOPG.

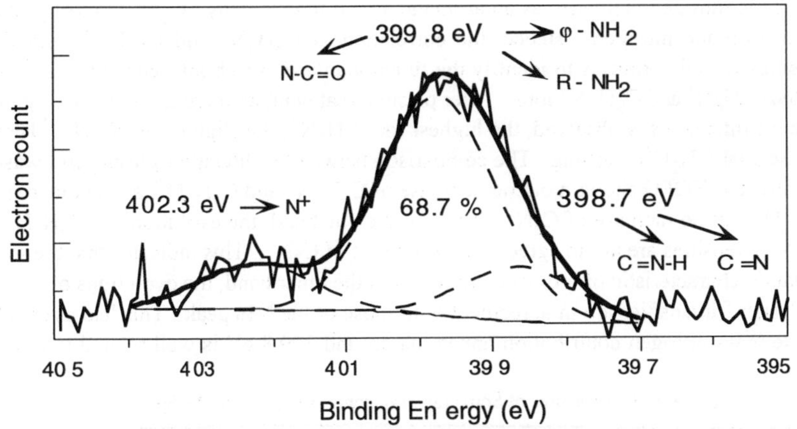

Fig. 6. Curve fitting of N1s difference spectrum between HOPG before and after plasma treatment and CH_2Cl_2 washing.

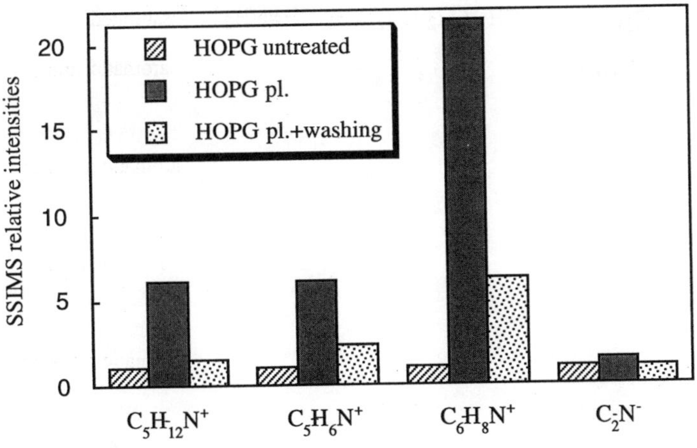

Fig. 7. Relative ToF SIMS intensities of untreated HOPG, plasma treated HOPG and HOPG after plasma and washed (the results are normalized to the untreated HOPG values).

Fig. 7 displays some ToF SIMS relative intensities on untreated HOPG, plasma treated HOPG and HOPG after plasma treatment and washing. In this histogram, relative intensities have been normalized to the values of untreated HOPG. In Table 3, some observed fragments that are representative of different hydrocarbon environment for nitrogen are shown with their possible structures. $C_5H_{12}N^+$ (m/z = 86) is characteristic of amine on a saturated hydrocarbon chain. $C_5H_6N^+$ (m/z = 80) and $C_6H_8N^+$ (m/z = 94) can be associated to two different structures. In the first, the nitrogen is included in an aromatic cycle. In the second, the nitrogen is present as an aromatic amine functionality. C_2N^- is most probably due to aromatic nitrogen. Hence, the correlation of $C_5H_6N^+$ and $C_6H_8N^+$ with C_2N^- intensities would permit us to identify the functionality in which nitrogen is present in the detected $C_5H_6N^+$ and $C_6H_8N^+$ ions. After plasma treatment, an increase of all N-containing fragment intensities is observed, the highest for $C_6H_8N^+$, the slightest for C_2N^-. They all decrease after CH_2Cl_2 washing. The comparison between HOPG after plasma and washing and untreated HOPG shows a big net increase of $C_6H_8N^+$ and $C_5H_6N^+$, a slight increase of $C_5H_{12}N^+$ but no increase of C_2N^-. Thus on the one hand, the evolutions of $C_5H_6N^+$ and $C_6H_8N^+$ intensities are not in agreement with those of C_2N^-. This indicates that the former are mainly characteristic of aromatic amines. On the other hand, the evolutions of $C_5H_6N^+$ and $C_6H_8N^+$ intensities are in agreement with those of the N1s peak. This lets us conclude that the main nitrogen contribution appearing around 399.8 eV is well related to aromatic amines.

Table 3. **Structures of Some Ion Fragments in Positive SIMS Spectra**

Mass	Ion formula	Proposed structure	N functionality	Ref.
86+	$C_5H_{12}N^+$	CH_2-CH_2-CH_2-CH_2-CH=NH_2^+	saturated amine	(33)
80+	$C_5H_6N^+$	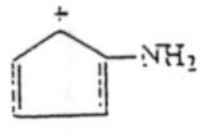	aromatic amine	(33)
			aromatic nitrogen	(33)
96+	$C_6H_8N^+$	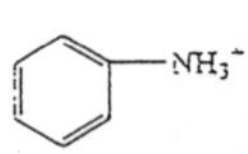	aromatic amine	(33)
		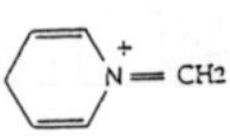	aromatic nitrogen	(33)
38-	C_2N^-		aromatic nitrogen	

Fig. 8 displays relative intensities for the same ions on untreated CFs, CFs treated by plasma and CFs treated by plasma and washed. In this case, relative intensities measured on untreated CFs are taken for the normalization. The lack of correlation between trends of $C_5H_6N^+$ and $C_6H_8N^+$ intensities and C_2N^- intensity confirms their identification as aromatic amines. The net increase of $C_5H_6N^+$ and $C_6H_8N^+$ with plasma treatment evidences the fixation of aromatic amines at the fiber surface, which is supported by the position of the main contribution in the net increase N1s spectra (Fig. 5). Thus, the comparison of Fig. 8 with Fig.

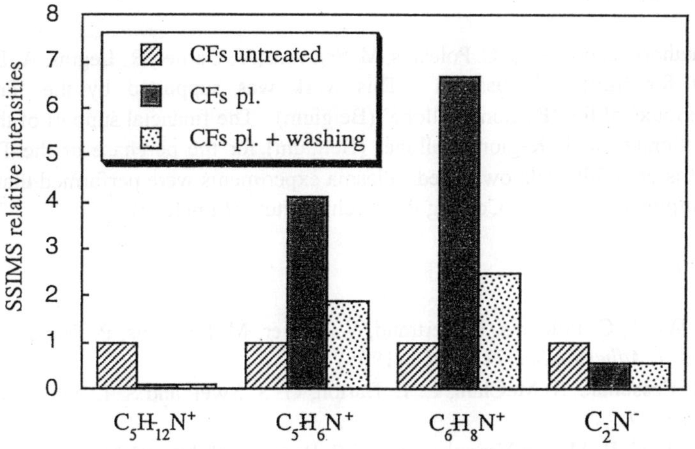

Fig. 8. Relative differences of ToF SIMS ions at untreated CFs, plasma treated CFs and CFs after plasma and washing (the results are normalized to the untreated CFs values).

7 gives mostly the same information. The major difference is the decrease, after plasma treatment on CFs, of characteristic fragments of aliphatic amines ($C_5H_{12}N^+$) and characteristic fragments of nitrogen included in aromatic structure (C_2N^-). This is due to the almost non-existence of N-containing species at untreated HOPG surface unlike at CFs surface. These results show that ammonia plasma treatment induces nitrogen fixation directly on the graphite structure of CFs while eliminating some fragments present at the surface.

4. Conclusions

Combined use of XPS and ToF SIMS was shown to be fruitful for the surface characterization of CFs and HOPG. Changes in functional groups at the surface after surface treatments were monitored in a qualitative and semi-quantitative manner by XPS and ToF SIMS simultaneously.

The results have shown that extraction in (1) CH_2C1_2 and (2) H_2O is efficient to remove the sizing and to increase the density of functionalities at the surface. The use of a model surface HOPG facilitates the identification of functionalities introduced after ammonia plasma treatments. Detection of aromatic amines at the HOPG surface as well as at the carbon fiber surface after plasma treatment and washing allows to support that amine functionalities have been introduced at both graphitic surfaces.

5. Acknowledgments

The authors acknowledge C. Poleunis, M. Sclavons, V. Carlier, R. Legras, A. Doren and M. Genet for fruitful discussions. This work was supported by the 'Programme Multimatériaux' of the 'Région Walloné' (Belgium). The financial support of the FNRS-Loterie Nationale' and 'Région Walloné' (Belgium) for the purchase of the ToF SIMS equipment is gratefully acknowledged. Plasma experiments were performed thanks to the plasma equipment lent by the Coating Research Institute (Limelette).

6. References

1. L. T. Weng, C. Poleunis, P. Bertrand, V. Carlier, M. Sclavons, P. Franquinet and R. Legras, *J. Adhesion Sci. Technol* **9** (1995) 859.
2. M. G. Fessehaie, S. McClain, C. L. Barton, G. S. Swei, and S. L. Suib, *Langmuir* **9** (1993) 3077.
3. J. Lub, F. C. B. M. van Vroonhoven and E. Bruninx, *Polymer* **30** (1989) 40.
4. J. R. Hollahan, B. B. Stafford, R. D. Falb and S. T. Payne, *Journal of Applied Polymer Science* **13** (1969) 807.
5. C. Jones and E. Sammann, *Carbon*, **28** (1990) 509.
6. I. H. Loh, R. E. Cohen and R. F. Baddour, *Journal of Materials Science* **22** (1987) 2937.
7. R. J. Smiley and W. N. Delgass, *Journal of Materials Science* **28** (1993) 3601.
8. M. Morra, E. Occhiello, F. Garbassi and L. Nicolais, *Composites Science and Technology* **42** (1991) 361.
9. Okhusen, R. C. Cochran, R. E. Allred, R. Sposili and T. M. Donnellan, *J. Adhesion* **45** (1994) 3.
10. G. J. Farrow, K. E. Atkinson, N. Fluck and C. Jones, *Surface and Interface Analysis* **23** (1995) 313.
11. G. Krekel, K. J. Hüttinger and W. P. Hoffman, *Journal of Materials Science* **29** (1994) 3461.
12. C. Kozlowski and P. M. A. Sherwood, *Carbon* **24** (1986) 357.
13. M. R. Alexander and F. R. Jones, *Surface and Interface Analysis* **22** (1994) 230.
14. J. Mahy, L. W. Jenneskens, O. Grabandt, A. Venema and G. D. B. van Houwelingen, *Surface Interface Anal.* **21** (1994) 1.
15. B-W. Chun, C. R. Davis, Q. He and R. R. Gustafson, *Carbon* **30** (1992) 177.
16. R. H. Bradley, X. Ling and I. Sutherland, *Carbon* **31** (1993) 1115.
17. M. R. Alexander and F. R. Fones, *Carbon* **32** (1994) 785.
18. S. T. Gardner, C. S. K. Singamsetty, G. L. Booth, G-R. He and C. U. Pittman, *Carbon* **33** (1995) 587.
19. M. J. Reis, A. M. Botelho do Rego, J. D. Lopes da Silva and M. N. Soares, *Journal of Material Science* **30** (1995) 118.

20. E. Desimoni, A. M. Salvi, U. B. Ceipidor and I. G. Casella, *Journal of Electron Spectroscopy and Related Phenomena* **70** (1994) 1.
21. G. Beamson and D. Briggs (Eds.), *High Resolution XPS of Organic Polymers: The Scienta ESCA 300 Database.* (John Wiley, Chichester, 1992).
22. E. Desimoni, G. I. Casella, T. R. I. Cataldi, A. M. Salvi, T. Rotunno and E. Di Croce, *Surface Interface Anal.* **18** (1992) 623.
23. C. D. Wagner, W. M. Riggs, L. E. Davis, J. F. Moulder and G. E. Muilenberg, *Handbook of X-Ray Photoelectron Spectroscopy* (Perkin-Elmer Corp., Eden Prairie, 1979) 40.
24. P. A. Gerin, P. B. Dengis and P. G. Rouxhet, *J. Chim. Phys.* **92** (1993) 1043.
25. S. O'Kell, T. Henshaw, G. Farrow, M. Aindow and C. Jones, *Surface Interface Anal.* **23** (1995) 319.
26. J. B. Donnet, R. C. Bansal, *Carbon Fibers* (Marcel Dekker, Ney-York and Basel, 1990) 37.
27. D. W. Davis and Shirley, *J. Elec. Spec. And Rel. Phenom.* **3** (1974) 137.
28. F. M. Petrat, D. Wolany, B. C. Scwede, L. Wiedmann and A. Benninghoven, *Surface Interface Anal.* **21** (1994) 274.
29. B. W. Schueler, *Microsc. Microanal. Microstruct.* **3** (1992) 119.
30. L. T. Weng, G. Vereecke, M. J. Genet, P. Bertrand and W. E. E. Stone, *Surface Interface Anal.* **20** (1993) 179.
31. L. T. Weng, G. Vereecke, M. J. Genet, P. Rouxhet, J. H. Stone-Masui, P. Bertrand and W. E. E. Stone, *Surface Interface Anal.* **20** (1993) 193.
32. D. A. Shirley, *Phys. Rev.* **B5** (1972) 4709.
33. J. G. Newman, B. A. Carlson, R. S. Michael, J. F. Moulder and T. A. Hohlt (Eds.), Statis SIMS Handbook of Polymer Analysis (Perkin-Elmer, Minnesota, 1991).

Mechanical, Chemical

and

Tribological Properties

CHEMICAL AND PHYSICAL PROPERTIES OF CARBON AS RELATED TO BRAKE PERFORMANCE

C. BLANCO, J. BERMEJO, H. MARSH, and R. MENÉNDEZ
Instituto Nacional del Carbón, Apartado 73,
33080 Oviedo, Spain

ABSTRACT

Objectives are to review the essentials of carbon science as related to carbon disc brake performance, and to relate the known trends in the variations of aircraft brake performance with relevant chemical and physical properties of carbon materials. Relevant literature is reviewed. Structures in carbon materials are described followed by a list of the properties of carbon materials which are important to the functioning of carbon composite disc brakes in aircraft. There are two dominant mechanisms of wear, or carbon loss from the surface of the carbon brakes, viz, losses by gasification resulting from the generation of high frictional temperatures, in air, and by abrasion losses. In the former, the carbon is lost as atoms of carbon in the form of oxides of carbon, carbon monoxide and carbon dioxide. In the latter, the carbon is lost as wear dust (brake debris). However, as well as forming oxides of carbon, the gasification (oxidation) processes produce the chemisorbed species of oxygen known as surface oxygen complexes. These are described further and their relevance to the process of physical adsorption of water molecules is enlarged upon. This leads to a discussion of the role of adsorbed water in terms of braking efficiency. Wear dust also has a role in the braking process, probably with the role of a lubricant between surfaces of the disc brakes. This wear dust is lost by physical displacement from between the discs and by oxidation at the high temperatures of brake operation. By being fragmented, the chemical reactivity of the edges of wear dust to molecular oxygen is high. As is often the case in the manufacture of carbon artifacts, it is not imperfections in the design model which cause performance problems, rather it is imperfections in the manufacturing process which limit performance. The review concludes with a summary of relevant definitions of carbon materials.

1. Literature Review

In this area of carbon-carbon-composite brake materials there are a limited number of references to published materials[1-27]. A reason for this is that much of the research and development is carried out by industry, where patent protection is necessary.

Longley et al.[1], in 1963, presented a discussion of frictional behavior. Friction coefficients of carbons can be low ($\mu = 0.05 - 0.1$) in dry carbon dioxide (5-50 ppm of water), as well as in laboratory air (50% of relative humidity). It is considered that all the evidence available is consistent with the theory that the low friction of highly graphitic carbon is due to low surface forces associated with the presence of adsorbed gases, not to low shear strength, nor to the separation of the surfaces by rolled-up wear products. Adsorbed gases also account for the low friction of low and non-graphitic carbons. After running in, the friction of low and non-graphitic carbons becomes low in dried carbon dioxide and laboratory air, but then subsequently increases to a high value. In air, the high value persists, but in carbon dioxide, the friction cycles between the low and high values. It is considered that the increase in friction is due to an increase in true contact area. The increase in friction can be delayed by encouraging the wear dust to escape. The carbon dioxide is associated with the formation and subsequent break-up of the ultra-flat compacted surface layer which is not formed in air owing to the lower wear rate. *In vacuo*, low graphitic carbon has a high friction and wears rapidly due to desorption, as does highly graphitic carbon. Surface friction is strongly dependent on the orientation of the surface material.

Low friction is associated with adsorbed gases

Weaver[2] discusses the advantages of using carbon as a structural material for disc brakes. Although beryllium has a higher thermal conductivity and specific heat, beryllium oxide is toxic and this property automatically excludes its use. Carbon materials can be used at temperatures in excess of 2000°C, being an integral structural member and frictional surface, at the same time.

Carbon materials, non-toxic, operate at high temperatures

Hill[3] made caliper-type simulation experiments on seven carbon-graphite material formulations, against a steel disc and a carbon-graphite disc. Reductions in binder levels and additions of B_4C each resulted in increased wear. The wear rate was not affected by the addition of graphitic fibers. Transition to severe wear and high friction was observed in the case of graphite-cloth reinforced carbon sliding against a disc of similar composition. This transition was related to the disruption of a continuous graphite shear film that must form on the sliding surfaces if low wear is to occur. The exposure of the fiber structure of the cloth constituent is believed to play a role in the shear film disruption.

Low to high friction coefficients relate to a graphite film

Chang and Rusnak[4] consider the contribution of oxidation to the wear of carbon-carbon composites. Wear tests were performed on samples under various oxygen partial pressures, and measurements were made of wear rates in weight and thickness. Wear rate was found to increase with increasing oxygen partial pressure. The wear rate in air was more than 4.5 times the wear rate in an inert atmosphere under the conditions of the experiment. At least 60% of the total weight loss was caused by oxidation, that is, to direct removal by conversion of the solid to carbon dioxide and carbon monoxide. Furthermore, the oxidation weakened the surface and sub-surface, which in turn facilitated removal by mechanical action at the rubbing surface.

60% of weight loss due to oxidation, 40% to wear

Liu[5] made a study of wear rates and transfer material films of frictional materials. The wear rates of two commercial friction materials–drum lining and a disc pad–were measured using a Chase dynamometer at a constant temperature of 150°C. Doubling the load (45 to 90 kg) and the speed (150 to 300 rpm) simultaneously increased the drum lining wear by 70%, and the disc pad wear by 250%. Specimens cut from the drums were examined by scanning electron microscopy (SEM) and scanning Auger microscopy (SAM). Transfer films were observed in all cases ranging from 1 to 7 μm in thickness. Increasing severity of running conditions increased both the thickness and continuity of the film, the films from the drum being thicker than those from the discs. The microscopy studies showed that frictional materials and their decomposition products were found in the brake drum surface and that oxides and carbides from the drum were detected in the transfer films.

Increasing severity of brake operation increases film thickness

Stimson and Fisher[6] discuss the design and engineering of carbon brakes. The use of carbon brake disks offered a 60% weight saving, for Concorde brakes, compared with steel. Orientation of the fibers on account of stress and heat flow requirements is vital to the achievement of a successful design. The material is anisotropic and represents a compromise between strength and thermal properties, and manufacturing costs. Dunlop selected the chemical vapor deposition of carbon into the carbon fiber lay-up as the method of manufacture of the composite for Concorde brakes.

Li and Sheehan[7] studied the friction and wear properties of graphite- and a carbon-composite in air and in helium, over a range of temperature. It was found, in general, that friction and wear decreased with increasing temperature, 20 to 950°C. This is considered to be due to the chemisorption of impurities to form lubricating films and to oxidation at higher temperatures, both of which reduce friction and wear.

Chemisorption and oxidation reduce friction and wear

In a study of correlations of wear with oxidation of carbon-carbon composites, Chang[8] used an inertial dynamometer. Wear rates were measured in terms of weight loss and thickness reduction. Wear rates in terms of weight loss were always greater than those in terms of thickness reduction over a wide range of braking conditions. This type of result is explained in terms of oxidation of carbon beneath the porous friction surfaces. Good agreement was obtained between the activation energies of oxidation of carbon (122 kJ mol^{-1}) and for material loss on non-frictional surfaces (113 kJ mol^{-1}). Such good agreement suggests that the rate-controlling step for oxidative weight loss is the diffusion of oxygen through the pores.

Oxidative weight loss is controlled by in-pore oxygen diffusion

Kimura et al.[9] measured friction and wear of carbon-carbon composites by sliding surfaces against each other in air using an inertial braking system. Specimens, of different HTT, were prepared from carbon fiber textile as a filler and phenolic resin-derived carbon as a matrix. The composites showed flat torque curves during 50 braking tests. The frictional coefficient ('μ') increased in the range 0.3 to 0.5 with increasing HTT and Young's modulus. Worn surfaces were covered with a graphite film in the composite with HTT > 2000°C. The minimum wear rate of the composite was about 3.5 x 10^{-5} mm^3 Nm^{-1}. The wear of the composites with a 'μ' value above 0.4 is considered to be controlled by an abrasion mechanism and that below 0.4 by an oxidation mechanism.

Wear, with a value of 'μ' < 0.4, is due to oxidation

Energies involved with a Boeing 767 aircraft are set out by Awasthi and Wood[10]. At take-off, the kinetic energy of a Boeing 767 aircraft, weighing ~ 170,000 kg (170 t) and moving at 320 km h^{-1} is 670 MJ. In the event of an aborted take-off, this energy must be dissipated in 30 s by the eight brakes of the aircraft. The total weight of steel brakes is 1,085 kg, and the total weight of the carbon-carbon composite brakes is 690 kg. Carbon brakes are adequate for about 3,000 landings, compared with 1,500 for steel brakes. These authors report on the variation of coefficient of dry friction as a function of sliding speed over the entire range of aircraft materials. Commencing with a value of 'μ' of about 0.2, this value increases to a maximum of about 0.55 with a speed of 6 m s^{-1}, subsequently decreasing to about 0.2 at the take-off speed of ~30 m s^{-1}. This is due to the varying conditions at the interfaces of the discs.

Kinetic energy of a Boeing 767 is 670 MJ at take-off

Lhymn and Lhymn[12] investigated the effects of sliding velocity on wear-friction behavior to test a mathematical model. The microstructure of worn surfaces was examined and correlated with wear mechanisms.

The microstructure of worn pitch/resin/CVI carbon-carbon composites has been examined by Murdie et al.[14] Optical microscopy, scanning and transmission electron microscopy were used. Friction and wear were tested under different energy conditions using a laboratory dynamometer. Results indicated that both a particulate-type (type I) and a film-type (Type II) of wear debris co-oxidized on the worn surfaces. Under low energy conditions, type I debris predominates. This dominance of the particulate debris and the lack of the film debris (type II) is thought to be responsible for the observed higher coefficients of friction and wear rates. Under high energy conditions, the predominance of the type II debris serves as a self lubricant, yielding a lower coefficient of friction and wear rate than under low energy conditions. Type II debris is composed of plastically deformed and fractured components of the composite, including fibers, CVI carbon and resin char. A thin, non-uniformly distributed amorphous film, identified by selected area electron diffraction, is also formed during the wear process.

Particulate and film type wear dust (debris) are formed

The wear properties of carbon-carbon composites, prepared by chemical vapor deposition, have been studied by Kim et al.[15] Friction and wear tests were carried out using a pin-on-disc-type apparatus at sliding speeds ranging from 0.14 to 0.62 m s^{-1}, and normal loads ranging from 5 to 22 x 10^4 kg m^{-2}. At a constant sliding speed of 0.62 m s^{-1} the coefficient of friction ('μ' = 0.2) was constant to a normal loading of about 18 x 10^4 kg m^{-2}, and then decreasing to about 'μ' = 0.13. Under these conditions, wear rates were constant at about 0.5 x 10^{-9} kg m^{-1}, to rise steeply to about 8.5 x 10^{-9} kg m^{-1}, at a normal loading of 22 x 10^4 kg m^{-2}. With increasing sliding speed, at a constant normal load of 22 x 10^4 kg m^{-2}, values of 'μ' decreased from about 0.25 to 0.12, whilst the wear rate increased from about 0.25 to 8.5 x 10^{-9} kg m^{-1}. As normal work load and sliding speed increase the friction force on the fiber increases. If the friction force on the fiber exceeds the fiber-matrix bond strength, and the transverse strength of the fibers, this causes the breakage and loss of fibers. More breakage and loss of fibers means that more frictional energy is expended as mechanical energy. The transverse strength of fibers is an intrinsic property of fibers. Kim et al.[15] confirmed the effect of the fiber-matrix bond strength on the wear behavior of carbon-carbon composites made with a different matrix carbon. Tests were performed at a normal load of 17 x 10^4 kg m^{-2} and a sliding speed of 0.44 m s^{-1}, when it was found that the wear rate of matrix of smooth laminar structure is higher than for those with a matrix of isotropic structure. The higher degree of preferred orientation of the matrix leads to a weaker bonding between fiber and matrix, a conclusion supported by the studies of Oh and Lee[11,13].

An isotropic matrix decreases wear rate

Constructional and operational aspects of carbon-carbon composites and their applications are reviewed by Savage[16]. Carbon-carbon composites exhibit a good thermal shock resistance, a high thermal conductivity and a very low coefficient of thermal expansion. They have a heat capacity 2.5 times that of steel and a high strength at elevated temperatures. Energy dissipation during actuation causes a rise in temperature. When undertaking analysis of the performance of a device of this type, the factors of concern are (1) the actuating force, (2) the torque transmitted, (3) the energy loss (dissipation) and (4) the rise in temperature. The friction surface of the thermoset resin-based brakes heats up relatively quickly because of the essentially two-dimensional nature of the fiber reinforcement. Thermal conductivity is far greater along the fibers than through the glassy carbon matrix; thus the dissipation of heat energy away from the rubbing surface is relatively inefficient. The matrix is very brittle and only weakly attached to the fiber reinforcement. The friable isotropic carbon forms a characteristic 'powdery' debris which is subsequently compacted into the surface. The resulting glazed surface thus loses a degree of braking efficiency, but as a direct result wear is reduced. Using a CVD system, the surface is dominated by large pores because CVD fills the smallest of pores. The bond between fibers and matrix is much stronger than for a resin based composite. Debris (wear dust) is much larger and more irregularly shaped, so providing a much rougher friction surface. The fibers in this matrix are able to conduct more heat away from the rubbing surface as it is being heated during operation. This rougher surface topography accounts for the more constant braking behavior but also increases wear. Values of thermal conductivity[6] ($J\ m^{-1}\ s^{-1}\ K^{-1}$) are in the order of 10 to 150 for carbon-carbon composites, about 59 for steel and 346 for copper. For a high strength polycrystalline graphite, the value is about 85. For single crystal graphite, thermal conductivities increase with temperature, (about 50 to 100 K) to a maximum value of about 680 $J\ m^{-1}\ s^{-1}\ K^{-1}$ to decrease to about 100 $J\ m^{-1}\ s^{-1}\ K^{-1}$ at about 2000 K. This is because of the semi-conductor properties of graphite, when the initial increase in temperature elevates electrons into the conduction band. For carbon composite disc brakes, the thermal conductivities may be higher at lower temperatures of initial operation.

Fiber thermal conductance is crucially important

During oxidation conditions, it is the matrix carbon which is oxidized preferentially and this results in loss of performance. The use of antioxidants is not really possible, because any painting of a surface results in loss of performance. Attempts of enhanced ventilation only result in enhanced oxidation. Limited access to air and accumulation of oxides of carbon can be considered as optimizing measures. Development objectives, according to Savage[16], to improve carbon-carbon composite materials include (a) an increase in thermal conductivity especially perpendicular to the friction surface (that is control over matrix structure), (b) an increase in the static coefficient of friction, (c) a greater strain to failure ratio, (d) improved specific strength, and (e) a reduction in cost.

Matrix carbon is oxidized preferentially

Chen and Ju studied friction and wear of PAN/pitch-, PAN/CVI- and pitch/resin/CVI-based 2-D and 3-D carbon-carbon composites[17], and the low energy tribological behavior of carbon-carbon composites[20]. Under high speed conditions of 2,000 rev min^{-1} and a pressure of 1.7 MPa, the 2-D composites were suitable for application whereas the 3-D composite underwent extensive structural damage. The composites from 2-D PAN/pitch and PAN/CVI exhibited lower friction coefficients (about 0.4) and lower wear rates than the 2-D pitch/resin/CVI composite and the 3-D PAN/pitch composite. For some of the composites, a transition in friction occurred causing the friction coefficient to rise abruptly from about 0.4 to 0.5–0.9. During the transition, the initially formed thin, smooth lubricative film was disrupted to become a thick powdery debris layer. For all composites, a flat and smooth self-lubricative debris film was generated on the worn surfaces after sliding for a few meters. Debris covering the surface of the 2-D PAN/pitch composite developed at a lower rate. For the composites of this study, the development of a steady-state friction coefficient value generally occurred earlier than the development of a steady-state wear rate.

2-D composites have lower wear rates

Low energy wear behavior of polyacrylonitrile, fiber-reinforced pitch-matrix, carbon-carbon composites has been studied by Ju et al.[18] Essentially, these authors studied two dimensional (2-D) and three-dimensional composites (3-D), and always noted that the 2-D composite, initially, had a much larger weight loss than the 3-D materials, but eventually decreased to levels comparable to the 3-D material and to graphite. SEM indicated that the formation of wear dust film is critical to wear behavior. Once a smooth, adherent, and uniform wear dust film is formed on the sliding surfaces, there is a drop in wear rates. The larger wear rate in the wear-in period of the 2-D composite is explained by the rapid formation of the loose, powdery debris, which can escape from between the surfaces more easily.

Powdery debris can escape from the braking system

Yen and Ishihara[19] also examined surface morphology and structure of carbon-carbon composites in high-energy sliding contact. These authors used chopped carbon fiber felt impregnated with phenolic resin and pitch. A ring-on-ring specimen configuration with fiber randomly oriented in the plane of sliding was used to simulate aircraft brakes. The relative sliding speed between two composite rings decelerates from an initial speed of 23 m s^{-1} to a complete stop under a load of 3,100 N to simulate a high energy aircraft braking system. Two types of surface morphology were observed: a dull-looking grey surface with machined-type of finish, and a lustrous black area with a mirror-like finish. The lustrous area is composed of wear dust, about 1 μm thick, composed of aggregates of equiaxial particles with no preferred crystalline orientation at the surface. The two types of surface morphology are the result of differences in local contact pressures. In the grey area the contact pressure is higher, leading to a rougher surface without continuous wear-dust-film coverage. In the lustrous area, the contact pressure is lower, which allows the maintenance of a debris film. Differences in contact pressures are due to non-uniform frictional heat generation, causing unequal thermal expansion of the contact area, a phenomenon experienced in tribological tests involving high energy dissipation rates.

Two types of surface debris are generated

Trefilov[21] reports that the strongest impact on the frictional-wear properties of carbon-carbon materials is exerted by the value of the elastic modulus of reinforcing fibers. Changes in the elastic modulus result in change in dissipative properties of the materials and hence also in temperature and force conditions at the contact spots. The interrelation of friction and wear characteristics with the elastic modulus of fibers also stem from the influence of other parameters of fibers, such as density and thermal conductivity, which change as modulus changes. The structure of surface layers of various carbons after friction tests were studied by scanning electron microscopy. For samples with a relatively small wear, a dense smooth film is formed in the course of friction on the surface of the materials, heterogeneous in the initial state. With greater wear, the surface film is loose, consisting of particles weakly bonded to each other. With catastrophic wear, there is formed a compacted conglomerate of particles not bonded to one another. By being elastically and plastically deforming, the formed films take up the frictional load and protect the base material from wear. When the deformability of the film has been exhausted, it breaks down or separates from the base. Fragments of the film get partly dispersed and participate once again in its formation, and are partly thrown out by centrifugal forces beyond the friction unit. It can be assumed that the wear resistance of a material is associated with the efficiency of the protective action of the surface film, which depends on its deformation-and-strength properties and adhesion to the base material. These characteristics appear to be governed by the size, shape and surface properties of wear particles, which depend on the composition and structure of the material, as well as on the structure and properties of the components, the carbon fibers and matrix carbon.

Friction and wear rates relate to elastic modulus of fibers

Murdie et al.[22] examined the frictional performance of a series of pitch matrix C-C composites. Four types of composite were used: (TD and STD) from continuous rayon fibers from a pitch matrix, and heated to 2000 and 1500°C respectively, (DNV) from continuous PAN fibers from a pitch with an initial pitch matrix, and (DF) similar to DNV but made with discontinuous chopped filaments. Results indicate that there is a general trend of decreasing friction coefficient with increasing kinetic energy for all materials except for the DF material. The friction coefficient for this material was stable, at about 0.26, in the range of 2,000 to 8,500 J cm^{-2}. Phosphate-containing detergents when applied to the discs cause an unrecoverable loss of friction. Park et al.[23] studied the effect of carbon fiber heat treatment on the frictional properties of C-C composites. Pregraphitization treatment of carbon fiber on the frictional surface was found to be very effective in increasing the thermal conductivity, lowering friction coefficients with well developed frictional films. Low density samples with lower temperature treatment carbon fibers showed higher friction coefficients, abrasive wear, and oxidation induced wear.

Hutton and McEnaney[24,25] discuss wear mechanisms in carbon-carbon composite aircraft brakes. These authors studied wear debris and structural changes at wear surfaces using optical microscopy, scanning electron microscopy and X-ray diffraction. The carbon composite discs had high wear rates during taxiing operations, associated with dull, unpolished wear surfaces and wear debris containing less well-ordered carbon than the base composite. Under landing conditions, low wear rates occur due to the deformation of the abraded material into a friction film that protects the wear surfaces. Low wear rate debris is more structured than high wear rate debris, and is more closely related to that of the base composite.

Carbon composite disc brakes are manufactured from a carbon fiber weave with pyrolytic carbon as the matrix carbon. Pyrolytic carbon is graphitic with its characteristic columnar structures. Discs made from such materials, during use, generate wear dust (debris) from the pyrolytic graphite carbon, initially as groups of clusters of graphene sheets. But immediately, these clusters become subjected to the comminution effects of brake use, which essentially is a form of amorphization of graphite by mechanical milling. In 1995, Zhou et al.[26] published a transmission electron microscopy study of amorphization of graphite by mechanical milling. Starting with a Johnson Matthey Co. graphite of high purity, the authors subjected this material to 5, 10, 30, 90 and 206 h of ball milling in a Spex 8000 mixer/mill. Resultant materials were examined by high resolution transmission electron microscopy and X-ray diffraction. At all times, the samples were protected from air by the use of argon as the containing gas. After 30 hour of ball milling, the graphite layers became crumpled lamellae with nanocrystal size. The folded lamellae were very distorted with a large number of defects. In summary, the results indicate that during long ball milling, the graphite undergoes a crystal-nanocrystal-amorphous transition. It is anticipated that such highly defective material would be extremely reactive to molecular oxygen. Earlier, in 1978, Nakamizo et al.[27] subjected a natural graphite to similar comminution treatments. Raman spectra of resultant specimens indicated that they were very similar to those of glassy carbon heated to above 2000°C.

2. Introduction

The objectives of this review article are (a) to introduce the essentials of carbon science associated with carbon disc brake performance, and (b) to relate known trends in the variations of aircraft brake performance with relevant chemical and physical properties of carbon materials. This review does not make use of references to the literature of carbon science, mainly because these are too extensive. It is felt that the review should be read continuously. However, further references and discussion are to found in the journal CARBON, in the series of Chemistry and Physics of Carbon, and in the book "Introduction to Carbon Science"[28]. There exists the possibility to have carbon brakes made from carbons with quite different structural characteristics. Accordingly, a working experience of the concepts of structure and graphitizability of carbon materials must be understood. Further, it is necessary to know something about the mechanisms (kinetics) of gasification or ablation of carbon surfaces by molecular oxygen. Further, one should be able to model the processes of formation, by a chemi-sorption process, of surface oxygen complexes and the formation of clusters/multilayers of physically adsorbed water on carbon surfaces as brake discs cool to ambient temperatures in moist atmospheres. The processes of wear, mechanical and physical should be established, as far as is possible. Further, there is discussion of how chemical and physical properties of carbon materials may be used to improve aircraft brake performance. Fig. 1 is a diagram showing applications of carbon composite disc brakes in a civilian aircraft, a military aircraft and a racing car.

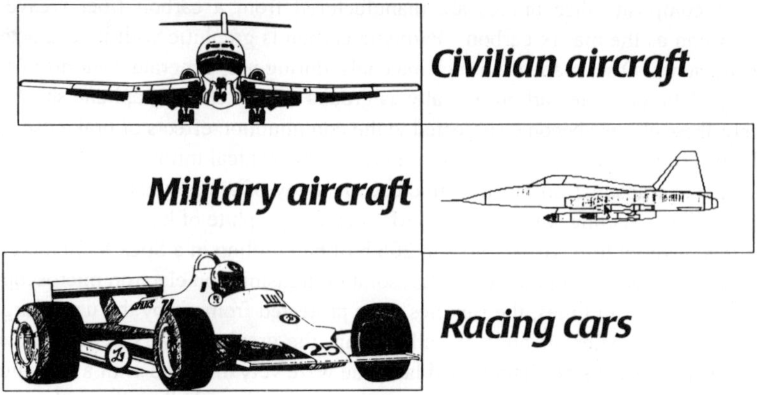

Fig. 1. Applications of carbon-carbon composites in braking systems.

3. Structures in Carbons

The element **CARBON** exists, in crystalline form dominantly as graphite (see Section 12) or diamond. The term '**CARBON**' should be reserved to describe the element **CARBON**, only. In terms of 'carbon materials' these can be described, quite accurately, as defective forms of graphite, some being very defective. However, it is best to use the term 'carbon' in association with a descriptive word, such as 'activated', 'pyrolytic', etc. Carbon materials can be grouped into the following main groups (See Section 12). There are those carbon materials which, on heat treatment to ~3000°C, ARE NOT transformed into a three-dimensional graphite lattice. Such carbon materials are described as being **NON-GRAPHITIC** and **NON-GRAPHITIZABLE**. There are those carbon materials which, on heat treatment to ~3000°C, ARE transformed into a three-dimensional graphite lattice. Such carbon materials are described as being initially **NON-GRAPHITIC** but are **GRAPHITIZABLE**. A **GRAPHITIZABLE** carbon material, after heat treatment to 3000°C, is then **GRAPHITIC**. An X-ray study of a non-graphitizable, non-graphitic carbon material, Heat Treatment Temperature (HTT) of 900°C, and a non-graphitic but graphitizable carbon material (HTT = 900°C) would not distinguish between these two carbons materials, even though their structures are very different.

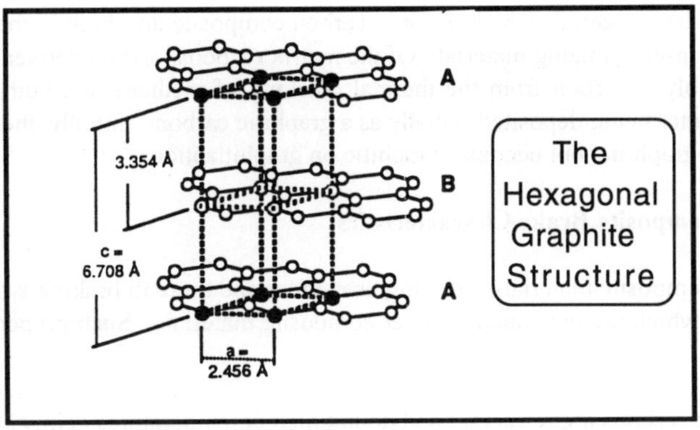

Use is often made of the optical microscope to distinguish between these two structures. Using plane polarized light, reflected from a polished surface of the carbon material, the non-graphitizable carbon is seen to be **ISOTROPIC** in structure, whereas the graphitizable carbon material is **ANISOTROPIC** in structure, exhibiting optical activity relative to the plane of polarization of the incident light. Generally, carbon materials made from wood or other materials which do not fuse, yield **ISOTROPIC** carbon materials. Carbon materials made from pitch or other materials which fuse, yield **ANISOTROPIC** carbons. The essential difference between these two types of carbon is the mode of relative alignment of the constituent **GRAPHENE** layers (always defective relative to single crystal graphite). In **ISOTROPIC, NON-GRAPHITIZABLE** carbon materials the graphene layers are smaller and arranged relative to each other, and bonded to each other, in all directions. In **ANISO-TROPIC, GRAPHITIZABLE** carbon materials the graphene layers are larger and arranged relative to each other, and bonded to each other, essentially parallel to each other. The **ISOTROPIC** carbon materials have structures which relate in some way to the parent (wood or resin or PAN fibers, for example). The **ANISOTROPIC** carbon materials derive their structure from an aromatic, discotic, nematic liquid crystal system, called **MESOPHASE**, which is generated from within the pyrolysing and carbonizing pitch.

Structures in Carbons

Isotropic Anisotropic

Of the carbon materials which occur in carbon composite disc brakes, the PAN fiber is essentially a non-graphitizing material. Of the matrix carbons, carbon derived from coal-tar pitch and pyrolytic carbon from the thermal cracking of methane, are both graphitizable carbons, the latter being deposited initially as a graphitic carbon. Initially, the coal-tar pitch carbon is non-graphitic, but becomes graphitic on graphitization.

4. Carbon Composite Brake Characteristics

Carbon composite materials are being used more and more in braking systems because of properties which are not found in other composite materials. Such properties are listed below:

(a) Graphitic carbons are refractory and do not melt at temperatures below 3000°C.
(b) Graphitic carbons have excellent coefficients of thermal conductance (to get frictional heat away from the brake surfaces).
(c) Graphitic carbons have coefficients of friction which change little with surface temperature (when dry).
(d) Graphitic carbons have high thermal capacities (storage of energy generated at minimum temperatures).
(e) Graphitic carbons provide interfacial "debris" to create a frictional surface.
(f) Graphitic carbons do not "*fatigue*" as is understood in metals.
(g) Graphitic carbon composites provide both frictional and structural material, at the same time, within the disc brake.

Dominant Characteristics
Of Graphite

THERMAL CONDUCTANCE
THERMAL CAPACITY

For some time now, it has been recognized that carbon-carbon fiber composite systems provide minimum recession rates, on re-entry of space vehicles, for example, in comparison with say tungsten and fiber reinforced plastics. Carbon fibers offer high specific stiffnesses and strengths in comparison with say silicon nitride and silicon carbide. The high thermal capacities continue to be available at temperatures above 1200°C, where other materials cease to have the required physical properties. The marked anisotropic physical properties of graphite result in the high thermal conductances in directions parallel to the plane of the graphene sheets. Thus, in terms of thermal management, graphitic materials, in the correct orientation can 'pipe-out' thermal energy very efficiently.

As discussed above, carbon composite systems, as used in disc brakes, are made of carbon fibers within a matrix carbon. The matrix carbon can be isotropic (non-graphitizable) as made from a resin matrix precursor, or anisotropic (graphitizable) as made from, e.g., a pitch precursor such as coal-tar pitch, or the matrix carbon can be pyrolytic carbon (anisotropic and graphitizable) as created within the carbon fiber systems by the thermal cracking of, usually, methane. The isotropic carbons, on heat treatment, will never exhibit graphitization and so their frictional properties will never approach those of graphite.

The anisotropic properties of the pitch coke are an immediate function of the properties of the pitch feedstock used. It must never be assumed that all coal-tar pitches are identical, and that all petroleum pitches are identical, and that one pitch type is as good as another. There exists an immense variability in the specifications of coal-tar pitches, often in terms of their Quinoline-Insoluble (QI) contents, as well as in chemical composition, those pitches originating from lower rank coals used in the cokeries being richer in oxygen and hence being less graphitizable. Thus, the structures exhibited by matrix carbons originating from pitch sources are variable.

The structures of the matrix carbons can also be a function of the arrangements and types of carbon fibers used in the composite. All this adds up to a situation in which control, in any detail, of the structures found within carbon matrices from pitch, becomes very tenuous. Similarly, the detail of structure within pyrolytic carbons, e.g., the growth of the columnar structures, must be difficult to control. Structure is dependent upon the feedstock gas used, its pressure and temperature, the rate of flow of the feedstock gas, the presence of a diluent gas, e.g., hydrogen or helium and the flow conditions of the cracking gas within the furnace, at temperature.

Accordingly, following the above discussion, only indications can be given of the variation of say the coefficient of friction (COF = 'μ') of the composite disc with the degree of graphitization of the matrix carbon. Fig. 2 is a qualitative diagram indicating how COF may vary with degree of graphitization. Another aspect to this discussion relates as to how the degree of graphitization can be measured. Are X-ray data to be used, or thermal conductance, for example? Whatever, the result will be the same. Instead of there being a narrow line, of high correlation coefficient, there will be a broad band into which the data will fit.

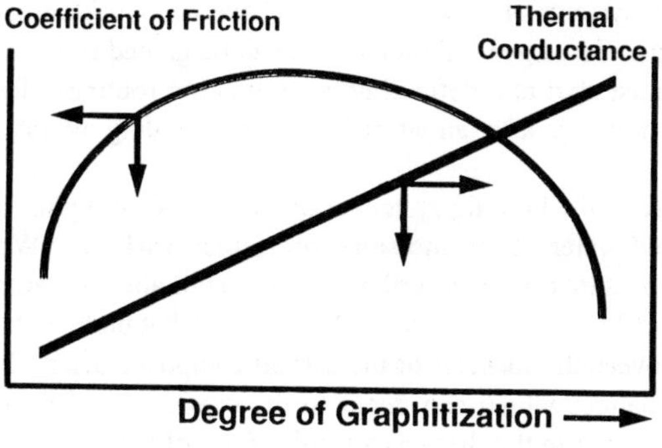

Fig. 2. The variation of coefficient of friction and thermal conductance of carbon disc brakes with degree of graphitization of matrix carbon.

As Fig. 2 indicates, the COF of carbonaceous materials will not increase linearly with increasing degree of graphitization. Carbons prepared at below about 1000°C will become harder with increasing heat treatment temperature (HTT), as is a common experience with green delayed cokes and calcined delayed (filler) cokes. As the HTT of the carbon increases so the surfaces become more graphitic and the frictional properties of the surface improve. But there has to be a maximum, because of the enhanced levels of perfection of graphitic structures, when the surface will disintegrate because of the ease of pulling out of graphene layers from the surface. At this stage the surface assumes significant lubricating properties (Fig. 2). In terms of creating a precise Fig. 2, for one type of system only, a great deal of careful and detailed, but empirical study is required, no doubt being expensive to carry out.

CARBON DISC BRAKES CAN
'ACCEPT'
THE ENERGIES
OF AN
ABORTED TAKE-OFF

Fig. 2 also indicates how the thermal conductance of carbons increases with increasing graphitization. This physical property is of major importance in brake use, as heat generated at the carbon disc interfaces has to be transmitted into the bulk of the disc as rapidly as possible. In this way, excessive surface temperatures can be avoided, so preventing the ignition and actual burning of the carbon disc. Another aspect to this technology is that, in the case of an aborted take off, the temperatures reached must not, if at all possible, cause damage to the non-carbonaceous metallic structures within the braking system.

5. Trends in Braking Performances

Very little exists in the way of precise information concerning what exactly is happening to braking systems of aircraft, during an average working day, involving several take-offs and landings. Some general indications can be gained from considerations of Fig. 3. The discussion must start at a defined point in braking routines of an aircraft. A good point is to start in the morning, after an aircraft has been standing overnight, say in a docking area.

What is the state of the braking system at this point? Overnight, it is assumed that the brakes have adsorbed water vapor, moisture, on to their surfaces. Why this is so will be explained below. The situation is described, in diagrammatic format, of an exploded disc brake system (cold) in Fig. 4. Also, it has to be assumed that brake debris or wear dust will have accumulated between the surfaces of the carbon composite discs. The curve ABC, Fig. 3, is indicative of changes which can occur to the coefficient of friction of the discs, as temperature changes occur in the discs as a result of use of the brakes, during taxiing to the runway and subsequent use.

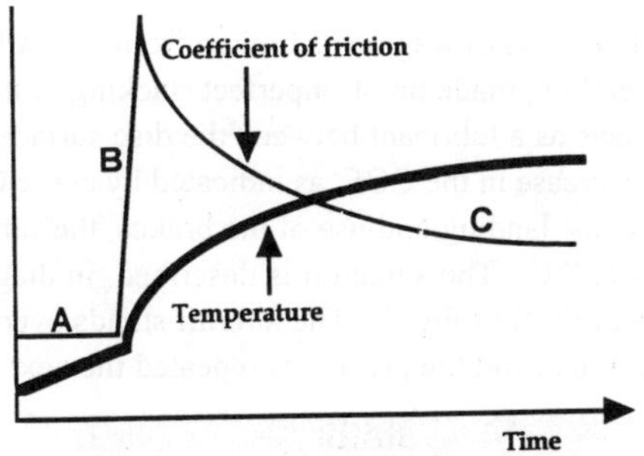

Fig. 3. Trends in braking performance with time, during the course of a typical 'flying' day.

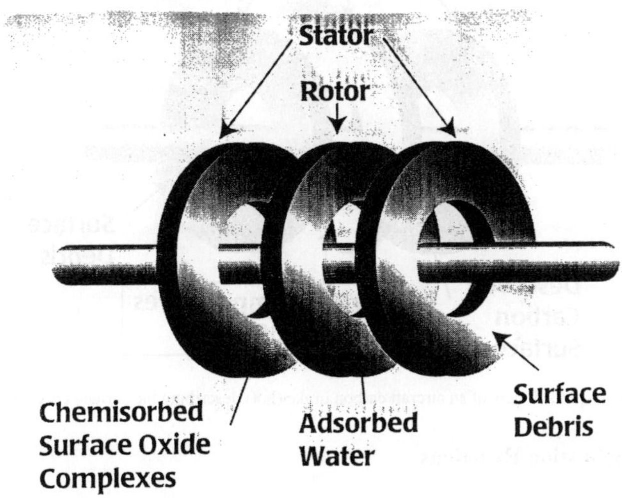

Fig. 4. Exploded view of an aircraft carbon brake (cold) describing the surface conditions.

Initially, curve A, the braking efficiency is low. This is because on the surfaces of the discs are the wear dust particles, associated with adsorbed moisture to an extent such that it begins to resemble a colloidally suspended aqueous system. Thus, frictional properties are not going to be that good. However, following the initial application of the brakes, the surface temperatures will be sufficiently high to cause evaporation of the moisture, perhaps so rapid that the expulsion of the water vapor carries with it the wear dust particles leaving carbon disc surfaces free from wear dust and adsorbed moisture. During this process, the COF rises very sharply along the curve B.

Subsequent use of the braking systems, during the course of the day, never allows the brakes to cool completely to ambient temperature so that there is no further adsorption of water at this stage. But there is abrasion at the operating surfaces and this generates further wear dust. This material consists of fragments broken from the carbon fiber system and of small particles of matrix carbon, made up of imperfect stacking of the graphene layers from the matrix carbon. This acts as a lubricant between the disc surfaces, and the consequence of this is the progressive decrease in the COF, as indicated by curve C, Fig. 3. At the end of the working day, after the last landing and use of the brakes, the surface temperature of the discs can be as high as 1800°C. The situation is described, in diagrammatic format of an exploded disc brake system (hot) in Fig. 5. The aircraft stands overnight and the COF will decrease to the level of curve A and the process is repeated the next day.

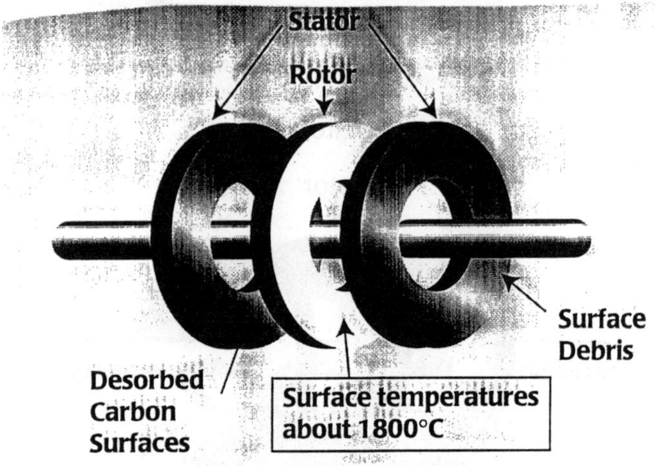

Fig. 5. Exploded view of an aircraft carbon brake (hot) describing the surface conditions.

6. Carbon Gasification Reactions

The dominant advantage to the use of braking systems composed of carbon fiber composites is the stability of structure at high temperatures. These materials do not melt or undergo any type of plastic deformation. Their COF will vary little with operating temperatures. Relative to other composite materials their oxidation resistance is extremely high, but not total. No carbon material is totally inert to the molecular oxygen of the air at temperatures of about 1800°C. These are white heat temperatures. Some aspects of the chemistry of oxidation reactions, relevant to this paper, will be considered below.

The gasification reactions of carbon, in stoichiometric format, look very simple:

$$C + O_2 = CO_2$$
$$2C + O_2 = 2\,CO$$

These equations appear to be indicating that the approach of an oxygen molecule to the surface of a carbon material results directly in the removal of either (a) one carbon atom as carbon dioxide, or (b) the removal of two carbon atoms as carbon monoxide. The equations give no indication as to why or how the carbon dioxide, say, is formed at a particular site, rather than the carbon monoxide. In fact, the reactions of molecular oxygen with carbon surfaces are much more complex. A further reaction product, relevant greatly to this discussion, are the surface oxygen complexes.

CARBON-CARBON COMPOSITE MATERIALS ARE NOT INERT TO OXYGEN IN THE TEMPERATURE RANGE 0°C TO 1800°C

Not all of the oxygen molecules approaching the carbon surface return to the gas phase as carbon dioxide or carbon monoxide. Some stay, chemically bonded to the carbon surface in a variety of bondings, some of which are set out in Fig. 6. These chemically bonded oxygen atoms constitute what are called '*surface oxygen complexes*'.

ATOMS OF OXYGEN AND CARBON

ARE MOBILE

OVER THE SURFACE OF CARBON DURING THE COURSE OF THE CARBON MOLECULAR OXYGEN REACTION

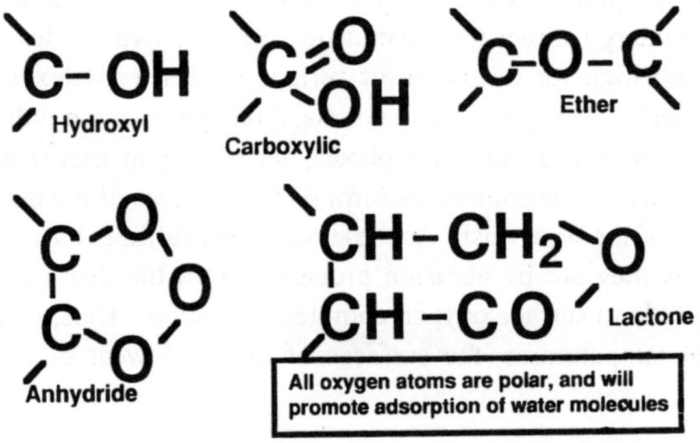

Fig. 6. Indicative structures of carbon-oxygen surface complexes.

The gasification of carbon, by molecular oxygen, does not occur as a single chemical stage. Rather there is a sequence of surface reactions, involving mobility of oxygen atoms over the carbon surface as reactions proceed, and possibly also with the mobility of carbon atoms as the surfaces attempt to anneal out defects introduced by the removal of carbon atoms by the oxygen. Fig. 7 indicates that the formation of carbon dioxide and carbon monoxide occurs via the interactions of mobile surface oxygen complexes at suitable sites of the carbon surface. Suitable sites may be defective sites, relative to the perfection of the graphene layer, or sites with the required stereo-capabilities to permit the oxygen atoms to orientate around a suitable carbon atom to give the carbon dioxide molecule (O-C-O) which is then subsequently desorbed from the surface.

The Carbon/Molecular Oxygen Reaction

$C + O_2 = C(O_2)_{localized}$ and $C(O_2)_{mobile}$ 1

$C(O_2)_{mobile} = C(O)_{localized}$ and $C(O)_{mobile}$ 2

$C(O)_{mobile}$ and $C(O)_{mobile}$ 3

$C(O)_{localized}$ and $C(O)_{localized}$ 4

$C(O)_{localized} = CO$ gas 5

$C(O)_{mobile} = CO$ gas 6

$C(O)_{mobile} + C(O)_{mobile} = CO_2$ gas + C 7

$C(O)_{mobile} + C(O)_{localized} = CO_2$ gas + C 8

$CO + C(O)_{localized} = CO_2$ gas 9

$CO + C(O)_{mobile} = CO_2$ gas + C 10

$O_2 + 2C(O)_{localized} = 2CO_2$ gas 11

Fig. 7. A summary of intermediate stages in the reaction of carbon with molecular oxygen.

Looking again at Fig. 3, at the stage when an aircraft has just landed and the brake surfaces could have temperatures as high as 1000°C, under these conditions, the surfaces of the carbon discs undergo direct gasification to form the two oxides of carbon. These temperatures are too high for formation of surface oxygen complexes. At temperatures below 1000°C, particularly for graphitic materials, direct gasification will be replaced by the formation of these surface oxygen complexes, increasing in extent as the temperature decreases. Not only will these complexes form on the surfaces of the discs, themselves, but also on the surfaces of the wear dust. As these wear dust surfaces have extremely defective surfaces, formed as they are by abrasion processes from the disc surfaces, they will be particularly susceptible to surface oxygen complex formation. Thus, when the brakes have cooled to ambient temperatures, the surfaces of disc and wear dust will be essentially oxidized.

THERE IS A
ROLE
FOR SURFACE OXYGEN COMPLEXES FORMED
ON SURFACES OF
WEAR DUST (DEBRIS)
AND ON THE
CARBON DISC SURFACES

That this is so can be demonstrated very easily, as indicated in Fig. 8. In the experiment, take a pristine brake disc and a disc that has seen service in an aircraft. Place a drop of water on the surface of each disc. For the pristine disc, the drop of water will not wet the surface, but will remain as a globule. For the used disc, the water wets the surface and spreads quickly over the surface. This is the result of the hydrophobic nature of the surface in the first place, and the hydrophilic nature of the surface in the second case. The hydrophilicity arises from the interactions between the highly polar water molecules and the polarity of the surface oxygen complexes. The importance of this to brake operation is critical. It is the presence of the surface oxygen complexes on the surfaces of the discs, after cooling, which cause the layers, or films, or even bulk water to be established in the disc systems. And, as indicated in Fig. 3, curves A to B, the water has a crucial role in brake operation.

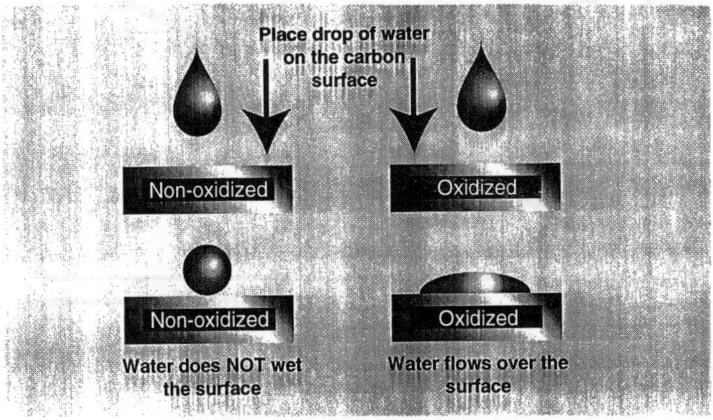

Fig. 8. An experiment describing the wetting by water of non-oxidized and oxidized carbon/graphite surfaces.

7. Structure and Surface Oxygen Complex Formation

In terms of brake operation the presence of water in the systems can only be considered as a liability. To minimize effects, extents of water adsorption have to be minimized. Fig. 9, in qualitative diagram format, indicates that as the degree of graphitization within a carbon increases, so extents of formation of surface oxygen complexes decrease. There are two principal reasons for this. First, with increasing graphitization, usually with increasing heat treatment temperature (HTT) there occurs a decrease in the surface area of the material. Hence, there is a reduced facility for complex formation on a reduced area. Second, with increasing extent of graphitization the graphene constituent entities become more perfect in terms of structure and planarity, and again, the material becomes less susceptible to formation of surface oxygen complexes. However, there may be some compensation for these effects. As Fig. 9 indicates, increasing extents of graphitization result in an increase in extent of formation of debris, or wear dust. This is the result of the graphite becoming softer, and more susceptible to wear, with increasing extent of graphitization. Wear dust is composed of extremely finely divided particles of carbon, from the carbon fiber and the matrix carbon, and as a result, it presents to the molecular oxygen of the ambient air, an extremely defective, reactive and available surface. It is established that if graphitic materials are comminuted, finely, < 1 mm, in an inert atmosphere, than on exposure to air, the graphite particles will spontaneously ignite, such is the exothermicity of the chemisorption of oxygen on the reactive sites of the newly generated particles.

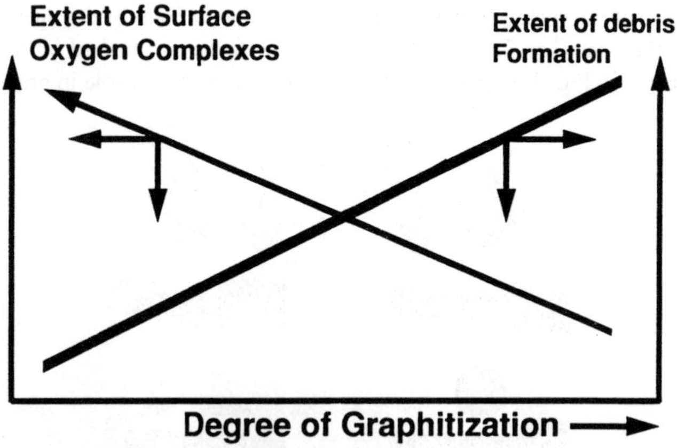

Fig. 9. The variation of extent (mmol g^{-1}) of formation of surface oxygen complexes, and extent (g g^{-1}) of debris (wear dust) formation, with degree of graphitization of matrix carbon.

8. Properties of Surface Oxygen Complexes

Fig. 10 attempts to describe the changes occurring on surfaces of carbon discs as temperatures are increased, that is, as the brake surface is heated by use, after standing say overnight at ambient temperatures. Initially, the surface will contain physically adsorbed water and chemi-sorbed oxygen. The water will be attached principally at the sites of oxygen bonding to the surface. The first species to come away from the surface will be the water molecules. The phenomena of physical adsorption approximate to evaporation and condensation processes of liquids, and for water, temperatures not much higher than 100°C are involved. Thus, by the time the disc surface has reached about 150°C almost all of the water molecules will have been desorbed, or boiled out. A few water molecules may, relatively, be quite strongly attached to the carbon surface, and would come off above 150°C. But this is not significant. Being chemisorbed, the oxygen requires higher temperatures for desorption. These surface oxygen complexes must not be looked upon as a single compound. There exists a spectrum of compound formation with a wide range of enthalpies of formation. Thus, these complexes will desorb over a range of temperatures, starting at about 100°C, reaching a maximum at about 300–350°C, with amounts being desorbed with increasing temperature, then falling away. Fig. 10 indicates that at temperatures of about 800°C little in the way of oxygen remains on the carbon surface. In terms of brake operation, effects which can be related to water on the disc surfaces are quickly removed on initial brake application. At higher temperatures the removal of the surface oxygen complexes is not as molecular oxygen, but is as oxides of carbon. Thus, carbon atoms are removed form the disc surface, and this can be looked upon as a form of gasification or ablation, even as a form of wear.

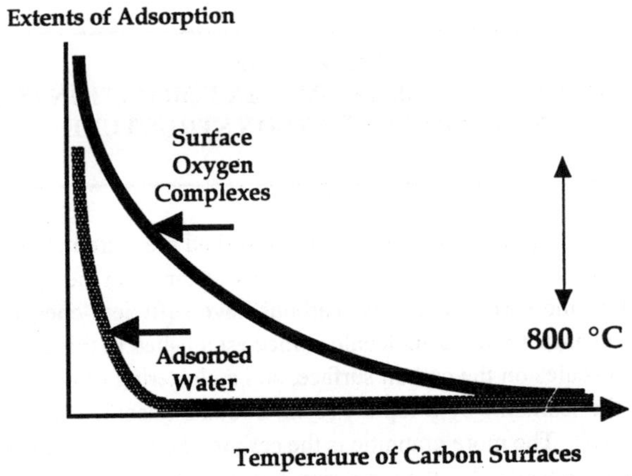

Fig. 10. The variation of extent (mmol g^{-1}) of formation of surface oxygen complexes, and adsorption (mmol g^{-1}) of water molecules with temperature of the surface of the carbon disc brake.

Fig. 11, in diagrammatic format, looks at the water isotherms of carbons, non-graphitized and graphitized. An isotherm is a graph indicating the variation of amount of adsorbate (in this case water vapor) on the surface of the adsorbent (in this case the carbon surface). Carbon surfaces, outgassed to about 800°C and subsequently exposed to air, essentially are hydrophobic, meaning that, at low relative pressures, p/p^o, a molecule of water (vapor) prefers to remain in the vapor phase where the molecule-molecule interactions are stronger than water molecule-carbon interactions.

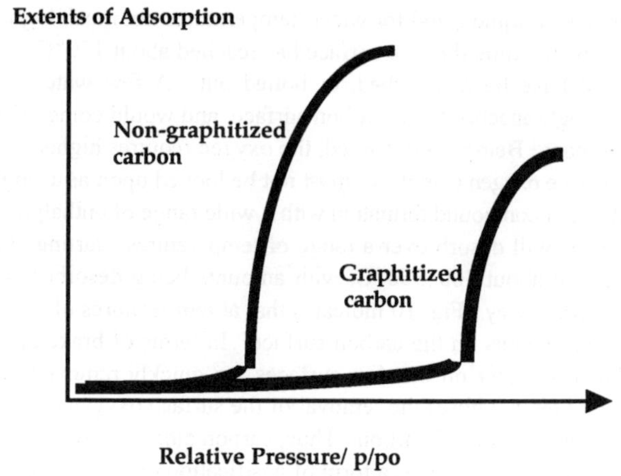

Fig. 11. The variation of extent (mmol g^{-1}) of adsorption of water vapor on non-graphitic and graphitic carbon, with increasing relative pressure of water vapor (adsorption isotherms).

EXTENT OF SURFACE OXYGEN COMPLEX FORMATION IS A FUNCTION OF CARBON STRUCTURE

Thus, looking at Fig. 11, amounts of water adsorbed are extremely limited at the lower p/p^o values. However, structure in these carbons is far from being perfect, and some locations (sites) on the surfaces of these carbons have sufficient energy, because of their defect structures, to attract a water molecule. Once established in this way, water molecules attract further molecules on the carbon surface, and a clustering effect begins to take place. This is seen in the isotherm by a rapid rise in amounts adsorbed at the higher levels of relative pressure, p/p^o. The more graphitic is the carbon, the less susceptible is the carbon to the adsorption of water vapor. So, the rapid rise in water uptake is displaced to higher values of p/p^o with increasing extents of graphitization, that is of increasing heat treatment temperature. The implication of this for brake properties is that the more graphitic is the carbon, the less will be the tendency for water adsorption.

Whereas Fig. 10 describes the changes occurring on a brake surface as it is heated, that is, as it comes into service, Fig. 12 describes the changes occurring on a brake surface as it cools, after a landing, with the aircraft standing in a docking area. Initially, the brake temperature is high, in the region of about 1800°C. It will cool, but in the temperature range 1800 to about 800°C, it will gasify or ablate, that is, carbon atoms are removed from the surface directly, by the molecular oxygen, as carbon monoxide and carbon dioxide. However, at about 800°C, these ablations reactions will have slowed down appreciably, and now the formation of the surface oxygen complexes occurs. These complexes are thermally unstable above about 800°C and, hence, will not form until the brake temperatures are below this temperature. So, as Fig. 12 is indicating, the surface becomes initially associated with the complexes. As is characteristic of chemisorption processes, these surface oxygen complexes are essentially in the form of a monolayer. However, this monolayer of chemi-sorbed oxygen atoms attracts water molecules to such an extent, that appreciable multilayer formation occurs by cluster formation of water molecules, perhaps leading to the formation of a 'bulk' water layer between the discs of the brakes. This is indicated by the asymptotic rise in the water uptake for the heavily oxidized carbon surface, Fig. 13.

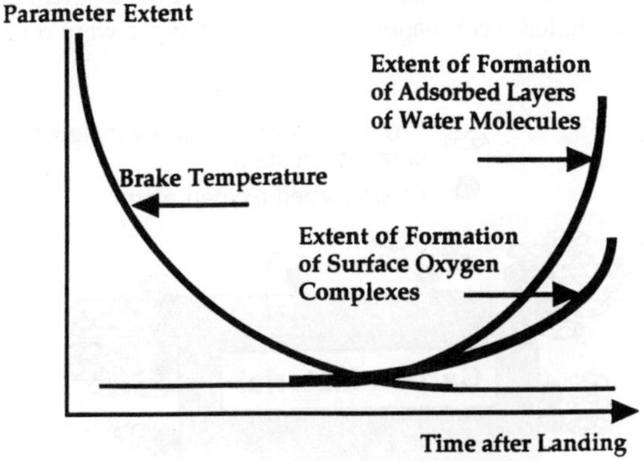

Fig. 12. The variation of extent (mmol g^{-1}) of formation of surface oxygen complexes, and adsorption (mmol g^{-1}) of water molecules, and temperature (°C) of the surface of the carbon disc brake, with time, after an aircraft landing.

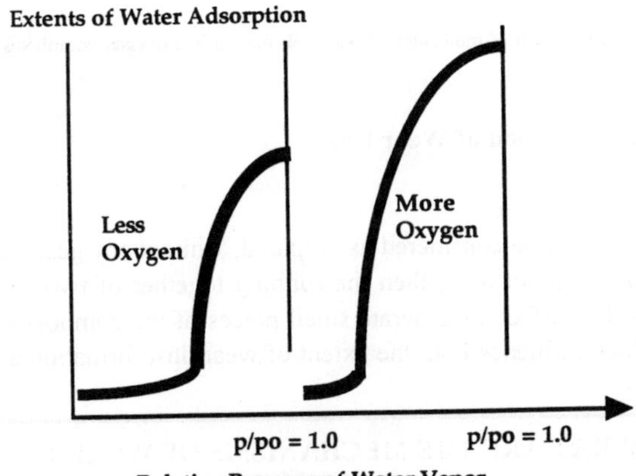

Fig. 13. The variation of extent (mmol g^{-1}) of adsorption of water vapor with relative pressure of the water vapor, on a carbon surface in terms of extents (mmol g^{-1}) of pre-formed surface oxygen complexes.

A model of the adsorption of molecules of water on the surface oxygen complexes, chemisorbed on the surface of carbon is reproduced in Fig. 14. This model gives some idea of the import role that surface oxygen complexes have in controlling the extent and nature of physically adsorbed material on the carbon surface. It is the adsorption of water molecules, which can be extensive, which has the major influence on braking characteristics. It is the high dipole moment of the water molecule which creates a 'hydrogen-bonding' between the chemi-sorbed oxygen and the water molecule. In the review, Section 2, mention was made of the role of adsorbed carbon dioxide[1]. This gas is not adsorbed to the same extent as is water vapor at ambient temperatures, nor is it held as strongly. The experiments of Longley et al.[1] may not be that conclusive because the replacement of the ambient gasses with carbon dioxide could have excluded water vapor, so that two changes were operating at the same time.

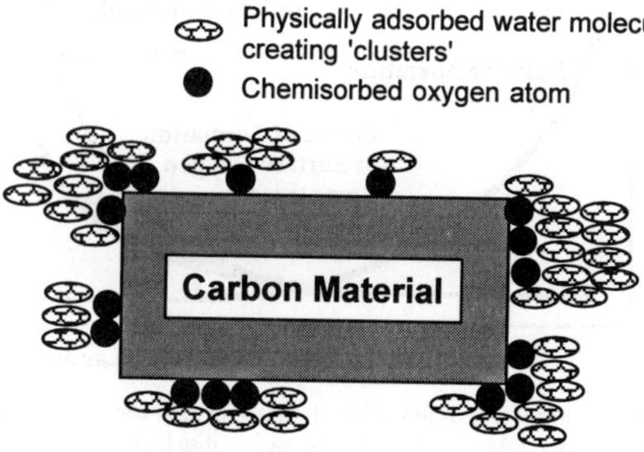

Fig. 14. A model of the adsorption of molecules of water on the surface oxygen complexes, chemisorbed on the surface of carbon.

9. Mechanisms of Generation of Wear Dust

9.1. Without Gasification Effects

Initially, surfaces can be considered as prepared, without any gasification effects. As carbonaceous surfaces do not melt, then the rubbing together of two surfaces will cause fracture of parts of the surface to generate small pieces of the composite. The list below contains factors which influence both the extent of wear dust formation and the shape and size of the wear dust.

WHAT ARE THE MECHANISMS OF WEAR IN CARBON COMPOSITE DISC BRAKES IN AIRCRAFT ??

(a) The relative speeds of rotation of the surfaces of the brakes.
(b) The pressure being applied to the surfaces.
(c) The state of final preparation of the surfaces of the brakes.
(d) The 'type' of carbon fiber used in the carbon-carbon composite.
(e) The 'type' of matrix carbon used in the carbon-carbon composite.
(f) The bonding mechanisms between the carbon fiber and the matrix carbon.

9.2. With Gasification Effects

When considering the processes of gasification of carbon surfaces, always, two aspects have to be considered at the same time, and never independently. These two aspects are: (a) gas/solid reaction kinetics as traditionally applied to gas/solid interactions, and (b) surface reaction anisotropy, that is, changes in surface structure as a result of gasification of the surface.

9.2.1. Gas/Solid Reaction Kinetics

This includes the traditional kinetics of gas/solid phase reactions, as used, e.g., in studies of catalysis. Thus, there are measured rates of reaction, the order of reaction and the temperature dependence of reaction rate to obtain activation energies. Mechanisms are derived in terms of Langmuir-Hinshelwood kinetics to obtain overall rate equations. Extreme care has to be taken to monitor if the design of the experimental system is allowing diffusion control (partial or total) of reaction rate. If there is some influence of diffusion then the kinetic data pertain only to the reaction system of the study, and not to the fundamentals of gasification kinetics as found in the scientific literature. An ideal in design of gasification experimentation is to 'mimic' as far as possible the situation in which a braking system finds itself.

9.2.2. Reactivity

In studies of traditional kinetics, the units used to express rates of gasification (that is the so-called *reactivity*) are important. Because gasification reactions are occurring on a carbon surface, the area available to the gasification agent, certainly, will influence the overall rates of gasification, on a per-gram basis. Thus, the most meaningful of reactivity measurements of gasification use as units: weight lost, per gram, per square meter of surface, per second, that is $g\ m^{-2}\ s^{-1}$.

9.2.3. Surface Areas

Thus, not only must the kinetic parameters be obtained, but also the surface area of the carbon composite material must be assessed. Again, this is a tricky situation and the value obtained for a surface area value in the experimentation may be a function of the state of preparation of the carbon sample. For example, for a non-porous solid, the surface area which a sample presents to a reacting gas, is a function of its state of comminution. The smaller it is sub-divided, the higher the surface area, per gram.

9.2.4. Rates of Gasification

For such a surface as is presented in a braking system, the requirement is to know the rate of 'ablation', that is gasification, from the surface only. It is probably a reasonable assumption to make that, during brake system use, the effects that we are interested in, and which are in fact occurring, are surface effects, not bulk effects. Of course, the next question concerns the definition we can give to 'surface effects'. Probably, we are thinking of effects which penetrate into the bulk of the carbon to no more than 100 μm, maybe less.

A reasonable experimental approach in such circumstances is NOT to attempt to obtain fundamental kinetic data from brake artifacts (discs) themselves, but from the components which make up the discs. By experimenting with different forms of matrix carbon, with variations in structure and possibly with and without additives, such carbons can be placed in a list of say decreasing 'reactivity'. Such a list of 'pecking-orders' can be then compared with the 'pecking-orders' obtained of brake wear data of controlled experimentation, elsewhere.

However, all of the above must be associated with what is called 'surface reaction anisotropy', see below.

9.3. Surface Reaction Anisotropy

As far as we are concerned, the carbons of the fiber and the matrix carbons are essentially polycrystalline. This means that at any prepared surface of such carbons there is a continuum of presentation of surface structures, from basal plane presentation to prismatic edge presentation, with all other possible intermediate presentations. This shows up in optical micrography of polished surfaces to obtain optical textures of carbons.

It is established within carbon science that the basal planes of carbons are much less reactive to a gasifying agent than are the prismatic edges. Put into other words, this means that not all parts of a surface which are being presented to a gasifying agent have identical reactivities. Parts of *an area* of a surface, and *areas* can vary from some μm^2 to mm^2, will therefore gasify much more quickly than other areas.

The essentials of this discussion are seen in Fig. 15. In this model, the various presentations of the graphene layers are seen. Those graphene layers with basal plane presentation (parallel to the plane of the photograph, and to the right side of the photograph) possess a minimum in chemical reactivity to gasifying agents. Hence, this part of the carbon would be more resistant to gasification (ablation) and would appear to be unaffected during the initial stages of gasification. On the other hand, those graphene layers with basal plane presentation (at right angles to the plane of the photograph, and to the left side of the photograph) possess a maximum in chemical reactivity to gasifying agents. Hence, this part of the carbon would be less resistant to gasification (ablation) and channeling and pitting can occur. In this way, the phenomena of reaction anisotropy can be explained. Because of these effects, with deep and localized channeling, for example, significant changes in mechanical strength of a carbon artifact are brought about, but with only levels of gasification being recorded.

Fig. 15. Model of the structure of surfaces of carbons to explain the causes of reaction anisotropy.

9.3.1. Selective Surface Gasification: Highest Reaction Temperatures

This is the phenomenon of surface etching. The outcome of this is very serious in terms of the mechanical stability of the carbon materials comprising the surface which constitutes a braking surface.

Depending on the conditions of gasification, the gas used, its partial pressure and rate of flow, the temperature of gasification, etc., this surface etching can be minimized or maximized.

Usually, at the highest of temperatures of gasification experienced by carbonized materials, say < 2000°C, this differential, selective gasification mechanism (another way of saying surface etching) is minimized. Basal plane and all edge presentations of structure will gasify at the same rate (controlled by rates of diffusion of gasifying molecules to the surface). This is because, under these conditions, the rate controlling processes are not associated with the structures of the carbons, but only with the availability of gasifying molecules at the surface. Reactivities are so high, and the gasifying molecules are so reactive at these temperatures that gasification occurs at the collision site. There is no penetration of the gasifying molecule into the porosity of the bulk material. Under these conditions, which represent one extreme, a limit, in the study of topographical gasification effects, the surface may appear to be essentially unchanged after the gasification process. Note that I say '*may appear*' because I believe that not too much exists in the literature of studies of topographical changes following gasification at these highest of temperatures, and so we are a little in a grey area.

9.3.2. Selective Surface Gasification: Lowest Reaction Temperatures

This effect of differential selective gasification (also called reaction anisotropy) will be maximized at the lowest of temperatures where gasification process rates can be detected. Under these conditions there will occur selective gasification to depths commensurate with the creation of a "macroporosity" into the surface (by surface I now mean a layer, probably some micro-meters thick) and this gasification weakens the strength of the surface layer. Again, by weakening of a surface, I am meaning a decrease in its abrasion resistance. The rubbing of such a gasified surface should produce more 'wear dust' in comparison with a surface gasified at the highest of temperatures.

10. The Role of Wear Dust During Braking

During braking operations it is inevitable that material fragments come away from the surfaces of the carbon discs. The space between the surfaces is very confined and the wear dust, as these fragments are called, stay between the surfaces and in the surface irregularities or cavities, or porosity, whatever, in the surface. They are not then inert materials but have a role, a significant role in fact, in controlling the performances of braking systems. The wear dust can originate from the following sources:

Carbon fibers, isotropic as in PAN,
Carbon fibers, isotropic as from pitch fiber,
Carbon fibers, anisotropic as from mesophase,
Matrix carbon, as from a resin-based carbon,
Matrix carbon, as from a pitch-based carbon,
Matrix carbon, as from a CVD process.

Scanning electron microscopy and transmission electron microscopy studies of these wear dust particles indicate that they are essentially lamellar, are very thin, and have sizes from about 100 nm to 10 μm. Examination of the edges indicates significant serration which is indicative of gasification or ablation from the edges. Thus, in such particles, the edge atom to in-plane carbon atom ratio must be quite high. This means that there is adequate facility for formation of surface oxygen complexes, the relevance of which is discussed above.

11. Conclusions

Section 1 of this article reviews aspects of the literature which are available. The review indicates that no overall understanding of the mechanisms of brake wear, brake operation and the influence of structure of the composite, including the effects of oxidation and adsorption of gases and vapors, exists. The modeling of such systems is multi-dimensional. The parameters to a model must include the physical properties of the fibers, strength and modulus, hardness, friability, as well as mechanisms of adhesion to the matrix carbon.

Similarly, the comparable physical parameters of the matrix carbon must be included in the model. Then, there must be considered the ways in which the wear dust is generated, and how this is a function of slide velocity and applied pressure. A question not considered in the review is the nature and extent, if any, of interactions between the wear dusts generated from the fibers and from the matrix carbon.

The review indicates that adsorption of gases/vapors, plays a role in brake operation. However, further experimentation must indicate that if studies of brake properties are made, say in an inert atmosphere, not only will oxygen be excluded but also water vapor will be excluded. Here, it has to be stressed that oxygen is chemi-sorbed, only removed at high temperatures of brake operation, whereas water vapor is physically adsorbed and is removed, essentially at temperatures of about 100°C.

The review is of studies which are difficult to compare because of differences in the composites studied, and the modes of testing. But they indicate, quite clearly, the interactive roles of fiber, matrix carbon, wear dust, gasification processes of surfaces and debris and the importance of adsorption of species on to the carbon surfaces.

In more detail, the observed trends in braking performances (friction) could relate to:

(1) Extent of debris formation,
(2) Structure of carbon debris,
(3) The shape, size, and surface area of the debris particles, as related to initial carbon structures (fiber and matrix).
(4) The planarity of the debris particles as related to initial carbon structure (frictional properties both dry and in the water medium).
(5) Ability of the carbon debris to form surface oxygen complexes.
(6) Porosity, cracks or other defects in the disc brake surface. As well as promoting the destruction of the surface, the porosity, etc., can act as a "sink" for the removal of lubricating debris from the surface. Thus, frictional properties will vary as "sinks" are created and destroyed, perhaps contributing to the *juddering* or *grabbiness* phenomena observed during braking.
(7) There appears to be preferential removal, by oxidation, of the fiber component within debris from carbon disc brakes.
(8) As is often the case in the manufacture of carbon artifacts, it is not imperfections in the design model which cause performance problems, it is either
    (a) the absence of a realistic model, or
    (b) it is imperfections in the manufacturing process which limit performance.

12. Definitions.

The International Union of Pure and Applied Chemistry (IUPAC) recently published a terminology to characterize carbon materials, *Pure and Applied Chemistry, 67, pp. 473-506,*

1995). It is imperative that there should be uniformity in such descriptions, and therefore those of relevance to this paper are reproduced below.

BAKING: The process in which carbonaceous BINDER, usually COAL-TAR PITCH or PETROLEUM PITCH, as part of a shaped CARBON MIX is converted to CARBON yielding a rigid carbon body by the slow application of heat.

BINDER: A BINDER is usually a COAL-TAR PITCH or PETROLEUM PITCH (but may include thermosetting resins or MESOPHASE PITCH powders) which, when mixed with a BINDER COKE or a FILLER, constitutes a CARBON MIX.

CALCINED COKE: CALCINED COKE is a PETROLEUM COKE or COAL-TAR PITCH COKE obtained by heat treatment of GREEN COKE to about 1600K (1300°C). It will normally have a hydrogen content of less than 0.1 wt.%.

CARBON ELECTRODE: A CARBON ELECTRODE is an electrode for an electrical application. In its green state it comprises granular material bound with pitch.

CARBONIZATION: CARBONIZATION is a process by which solid residues with increasing content of the element carbon are formed from organic material usually by pyrolysis in an inert atmosphere.

COAL-TAR PITCH: COAL-TAR PITCH is a residue produced by distillation or heat treatment of coal-tar. It is a solid at room temperature, consists of a complex mixture of numerous predominantly aromatic hydrocarbons and heterocyclics, and exhibits a broad softening range instead of a defined melting point.

COKE: COKE is a solid high in content of the element carbon and structurally in the NON-GRAPHITIC state. It is produced by pyrolysis of organic material which has passed, at least in part, through a liquid or liquid-crystalline state during the CARBONIZATION process. COKE can contain mineral matter.

DELAYED COKE: DELAYED COKE is a commonly used term for a primary CARBONIZATION product (GREEN COKE or RAW COKE) from high-boiling hydrocarbon fractions (heavy residues of petroleum or coal processing) produced by the DELAYED COKING PROCESS.

FILLER COKE: FILLER COKE is the main constituent of a carbon artifact, introduced as a solid component (predominantly in the form of particulate CARBON) into the CARBON MIX from which POLYGRANULAR AND GRAPHITE materials are obtained by heat treatment.

GRAPHENE LAYER: GRAPHENE is a single carbon layer of the graphite structure, describing its nature by analogy to a polycyclic aromatic hydrocarbon of quasi infinite size.

GRAPHITE: GRAPHITE is an allotropic form of the element carbon consisting of layers of hexagonally arranged carbon atoms in a planar condensed ring system (GRAPHENE LAYER). The layers are stacked parallel to each other in a three-dimensional crystalline long-range order. The chemical bonds within the layers are covalent with sp^2 hybridization and with a C-C distance of 141,7 pm. The weak bonds between the layers are metallic with a strength comparable to VAN DER WAALS bonding only.

GRAPHITIC CARBON: GRAPHITIC CARBONS are all varieties of substances consisting of the element carbon in the allotropic form of GRAPHITE irrespective of the presence of structural defects.

GRAPHITIZABLE CARBON: GRAPHITIZABLE CARBON is a NON-GRAPHITIZABLE CARBON which upon GRAPHITIZATION HEAT TREATMENT converts into GRAPHITIC CARBON.

GRAPHITIZATION: GRAPHITIZATION is a solid-state transformation of thermodynamically unstable NON-GRAPHITIC CARBON into GRAPHITE by means of heat treatment.

ISOTROPIC CARBON: ISOTROPIC CARBON is a monolithic CARBON MATERIAL without preferred crystallographic orientation of the microstructure.

MESOPHASE PITCH: MESOPHASE PITCH is a PITCH with a complex mixture of numerous essentially aromatic hydrocarbons containing anisotropic liquid-crystalline particles (CARBONACEOUS MESOPHASE) detectable by optical microscopy and capable of coalescence into the BULK MESOPHASE.

METALLURGICAL COKE: METALLURGICAL COKE is produced by CARBONIZATION of coals or coal blends at temperatures up to 1400K (1100°C) to produce a macroporous CARBON MATERIAL of high strength and relatively large lump size.

NON-GRAPHITIC CARBON: NON-GRAPHITIC CARBONS are all varieties of solids consisting mainly of the element carbon with two-dimensional long-range order of the carbon atoms in planar hexagonal networks, but without any measurable crystallographic order in the third direction (*c*-direction) apart from more or less parallel stacking.

NON-GRAPHITIZABLE CARBON: NON-GRAPHITIZABLE CARBON is a NON-GRAPHITIC CARBON which cannot be transformed into GRAPHITIC CARBON solely by high-temperature treatment up to 3300K (3000°C) under atmospheric pressure or lower pressure.

PETROLEUM PITCH: PETROLEUM PITCH is a residue from heat treatment and distillation of petroleum fractions. It is a solid at room temperature, consists of a complex mixture of numerous predominantly aromatic and alkyl-substituted aromatic molecules, and exhibits a broad softening range instead of a defined melting temperature.

REGULAR COKE: REGULAR COKE is a PETROLEUM COKE with good graphitizability and is characterized by a combination of properties which differ significantly from those of METALLURGICAL COKE but do not reach the quality level of PREMIUM (NEEDLE) COKE: These properties are: optical anisotropy, medium reversible thermal expansion, and low ash content.

13. Acknowledgments

This research is supported by the CICYT, project MAT95-0206, and the Ministerio de Educación y Ciencia, Madrid, Spain.

14. References

1. R. I. Longley, J. W. Midgley, A. Strang and D. G. Teer, Mechanism of the frictional behavior of high, low and non-graphitic carbon, Lubrication and Wear Convention, (1963).
2. J. V. Weaver, Advanced materials for aircraft brakes, *Aeronautical Journal* **76** (1972) pp. 659-698.
3. R. C. Hill, Friction and wear of carbon-graphite materials for high-energy brakes, NASA Technical Note, (1975) NASA TN D-8006.
4. H .W. Chang and R .M. Rusnak, Contribution of oxidation to the wear of carbon-carbon composites *Carbon* **16** (1978) pp. 309-312.
5. T. Liu, S .K. Rhee and K. L. Lawson, A study of wear rates and transfer films of frictional materials, Bendix Research laboratories, Southfield-Michigan, (1980).
6. I. L. Stimson and R. Fisher, Design and Engineering of Carbon Brakes: New fibers and their composites, *Phil. Trans. R. Soc. Lond.* **A294** (1980) pp. 583-590.
7. C. C. Li and J. E. Sheehan, Frictional and wear studies of graphite and a carbon composite in air and helium, Intern. Conference, ASME, N.Y. (1981) pp. 525-533.
8. H. W. Chang, Correlation of wear with oxidation of carbon-carbon composites, *Wear* **80** (1982) pp. 7-14.
9. S. Kimura, E. Yasuda and N. Narrate, Friction and wear of carbon-carbon composite, *J.S.L.E., International Ed.* **5** (1984) pp.11-16.

10. S. Awasthi and J. L. Wood, C/C Composite Materials for Aircraft Brakes, *Advanced Ceramic Materials* **3** (1988) 449-456.
11. S.-M. Oh and J. Y. Lee, Structure of pyrolytic carbon matrices in carbon-carbon composites, *Carbon* **26** (1988) pp.763-768.
12. C. Lhymn and Y. Lhymn, On the wear of short carbon fiber reinforced carbon matrix composite, *Adv. in Polymer Technology* **9** (1989) pp. 33-39.
13. S.-M. Oh and J. Y. Lee, Fracture behavior of two-dimensional carbon-carbon composites, *Carbon* **27** (1989) pp.423-430.
14. N. Murdie, C. P. Ju, J. Don and F. A. Fortunato, Microstructure of worn pitch-resin-CVI C-C composites, *Carbon* **29** (1991) pp. 335-342.
15. D.-G. Kim, D.-W. Kweon and J.-Y. Lee, The wear properties of C-C composites prepared by chemical vapor deposition, *J. Mat. Sci., Letters* **12** (1993) pp. 8-10.
16. G. Savage, Carbon-Carbon Composites, Applications of carbon-carbon composites, The Properties of Carbon-Carbon Composites: Chapman and Hall, 1993, pp. 323-346.
17. J. D. Chen and C. P. Ju, Friction and wear of PAN/pitch-, PAN/CVI- and pitch/resin/CVI-based carbon-carbon composites, *Wear* **174** (1994) pp. 129-135.
18. C. P. Ju, K. J. Lee, H. D. Wu and C. I. Chen, Low-energy wear behavior of polyacrilonitrile, fiber-reinforced, pitch-matrix, carbon-carbon composites, *Carbon* **22** (1994) pp. 971-977.
19. B. K. Yen and T. Ishihara, The surface morphology and structure of carbon-carbon composites in high energy sliding contact, *Wear* **174** (1994) pp. 111-117.
20. J. D. Chen and C. P. Ju, Low energy tribological behavior of carbon-carbon composites, *Carbon* **33** (1995) pp. 57-62.
21. Trefilov, V. I., (Ed.) Ceramic- and Carbon-Matrix Composites, Chapter 3, Carbon-based composites, Chapman and Hall, London, (1995).
22. N. Murdie, D. E. Hayes, V. I. Kostikov and V. V. Kulakov, Extended Abstracts and Program, Carbon '95, San Diego, University of San Diego, U.S.A., pp. 154-155.
23. H. S. Park, J. H. Park and K. S. Kim, Extended Abstracts and Program, Carbon '95, San Diego, University of San Diego, U.S.A., pp. 156-157.
24. T. J. Hutton and B. McEnaney, Extended Abstracts and Program, Carbon '95, San Diego, University of San Diego, U.S.A., pp. 158-159.
25. T. J. Hutton and B. McEnaney, Abstracts "Carbon and Carbonaceous Composite Materials" conference, Malenovice, Czech Republic, October 1995, Materials Technology Center, Southern Illinois University, Carbondale, IL 62901, U.S.A.
26. W. L. Zhou, Y. Ikuhara, W. Ahao and J. Tang, *Carbon* **33** (1995) pp. 1177-1180.
27. M. Nakamizo, H. Honda and M. Inagaki, *Carbon* **16** (1978) pp. 281-283.
28. Introduction to Carbon Science, Ed. Harry Marsh, Butterworths, London, (1989).

PORES AND CRACKS ANALYSIS OF CARBON-CARBON COMPOSITES

Z. WEISHAUPTOVÁ and K. BALÍK
Institute of Rock Structure and Mechanics
Academy of Sciences of the Czech Republic
182 09 Prague 8, Czech Republic

J. MEDEK
Carbochem
Na Hubálce 8, 169 00 Prague 6, Czech Republic

ABSTRACT

Determination of the extent and character of voids in composite materials and their resolution into pores and cracks is significant not only from the viewpoint of a continuously followed technological parameter, but also for the qualification of the final product. The attempt at a complex texture analysis of the sample 1-D composite by combination of sorption, porometric, pycnometric and optical methods and SAXS method was performed. In the individual classes were determined the volumes and percentage representations of open voids, open + closed voids and closed voids. It was identified that the highest share of the voids is formed by the coarse pores and cracks above 0.5 μm, where a broad spectrum of information on their arrangement, shape and size was provided by image analysis.

1. Introduction

The strength of composite materials is affected not only by the kind of fibers used, their treatment, and the kind of matrix (especially its microstructure), but also by the number and shape of the voids consisting of pores and cracks[1]. The presence of voids influences not only the resulting mechanical properties and the conductivity of the composite but also the possibility of its application as a biomaterial. Whereas pores are formed in the course of the release of volatile substances from the matrix precursor during carbonization, the cracks are formed by the shrinkage of the matrix proper and difference volume contraction of the matrix and the fibers in the course of cooling of the carbonized material[2].

The knowledge of volume, size and shape of voids is important also for the further treatment of the composite by impregnation. As there is a wide range of sizes of voids, for their evaluation besides the total porosity above all the distribution of the pore sizes in the particular size classes (micropores $r < 1.6$ nm, mesopores $1.6 < r < 35$ nm, macropores $35 < r < 7500$ nm, coarse pores $r > 7500$ nm) is of importance. The cracks caused by shrinkage may have a width of the order of 10 μm, the cracks formed by the difference in the contraction of the matrix and of the fibers have a width of only 1 μm[3]. This broad sizes range of pores and cracks leads necessarily to the application of a number of methods of their determination.

This study performed with a 1-D composite is an attempt at a complex texture analysis by the combination of various methods applied to the characterization of pores and cracks.

2. Experimental

2.1. Material

The high-strength carbon fibers Torayca T 800 (6000 filaments in tow) were employed. 1-D composites with a resin-based matrix were manufactured using the wet winding (prepreg) technique by impregnating the fibers with the phenolic resin (Umaform). The prepregs were stacked in seven layers and cured at 0.5 MPa and 170°C. The obtained 1.3 – 1.5 mm thick carbon-polymer plates were cut to specimens and carbonized at the heating rate of 100°C/h up to 1000°C in the atmosphere of nitrogen. No impregnation step was applied.

2.2. Methods

The real density d_r was determined pycnometrically with methanol after a drift of seven hours, the apparent density d_a was determined stereometrically. To determine the real densities of the fiber and matrix samples, they were separately heat-treated at 1000°C and then pulverized into grain size below 0.06 mm.

The volume of the open micropores $V_{micro,o}$ and their surface area $S_{micro,o}$ were determined from the adsorption isotherm of CO_2 at 25°C (Sorptomatic 1800, Carlo Erba). The volume of the voids with their distribution in the range of meso- and macropores V_{mm} and the mercury density d_{Hg} were determined by high-pressure mercury porosimetry (Porosimeter 2000, Carlo Erba).

The total volume of the micropores $V_{micro,t}$ was determined by the SAXS method (CuK_{α}).

The texture parameters of the voids with the effective radii r > 0.5 μm were determined by image analysis using the LUCIA M system. The analysis was performed with a section oriented perpendicularly to the fibers with the magnification of 140x.

3. Results

For the calculation of the volume of the voids the following basic data were used: real density $d_r = 1.8224$ g/cm^3, apparent density $d_a = 1.3813$ g/cm^3, mercury density $d_{Hg} = 1.5557$ g/cm^3, volume of a mass unit of the composite $V_k = 1/d_a = 724$ mm^3/g. The ratio of the fibers and the matrix was calculated from their respective real densities: $d_f = 1.9467$, $d_{matrix} = 1.5893$ and that of the composite, $d_{comp} = d_r = 1.8224$ g/cm^3.

For the description of the open porous texture the following parameters were used:

- total volume of the voids: $V_{p,o} = (d_r - d_a) / d_r\, d_a$,
- volume of the coarse voids above 0.5 μm: $V_{c+m,o} = V_{c,o} + V_{m,o}$, including:
  a) volume of the voids above 7.5 μm: $V_{c,o} = (d_{Hg} - d_a) / d_{Hg} \cdot d_a$,

b) volume of the voids in the range $0.5 < r < 7.5$ μm: $V_{m,o}$ determined from the distribution curve obtained by Hg porosimetry

- volume of the voids in the range of meso- and macropores below 0.5 μm: $V_{mm,o}$
- volume of the voids with the size in the range $1.6 < r < 3.5$ nm:

$$\Delta V_{1.6-3.5} = V_{p,o} - (V_{c+m,o} + V_{mm,o} + V_{micro,o})$$

- volume of micropores: V_{micro}.

The values of the particular partial volumes and their respective percentage are presented in Table 1, where $[\%] = 100(V_{x,o} / V_k)$.

Table 1. Volumes and Percentages of the Particular Size Classes of Open Voids

	$V_{p,o}$	$V_{c+m,o}$ >0.5μm	$V_{mm,o}$ <0.5μm	$\Delta V_{1.6-3.5}$	$V_{micro,o}$
[mm³/g]	175.2	94.0	6.4	11.9	62.9
[%]	24.2	13.0	0.9	1.6	8.7

The total volume of open and closed voids was determined by using the accessible experimental techniques only in two size ranges:

- total volume of micropores $V_{micro,t}$ by the SAXS method, and
- total volume $V_{c+m,t}$ in the range of macrovoids and coarse voids above 0.5 μm by image analysis where their percentage in the surface was 14.7 %, corresponding to 106.4 mm^3/g, supposing that the surface and volume relative ratios are equivalent. From mercury porosimetry the volume of macrovoids V_m with r in the range of 0.5 – 7.5 μm was determined.

The results are presented in Table 2.

Table 2. Volumes and Percentages of Open + Closed Voids

	$V_{p,t}$	$V_{c+m,t}$ >0.5μm	$V_{mm,t}$ <0.5μm	$\Delta V_{1.6-3.5}$	$V_{micro,t}$
[mm³/g]	240.6	106.4	6.4	11.9	115.9
[%]	33.2	14.7	0.9	1.6	16.0

The volumes of closed voids calculated from the difference $V_{x,t} - V_{x,o}$ are presented in Table 3.

Table 3. Volumes and Percentages of Closed Voids

	$V_{p,cl}$	$V_{c+m,cl}$ >0.5μm	$V_{mm,cl}$ <0.5μm	$\Delta V_{1.6-3.5}$	$V_{micro,cl}$
[mm³/g]	65.4	12.4	–	–	53.0
[%]	9.0	1.7	–	–	7.3

By image analysis the parameters describing in detail the geometric character and arrangement of the voids were evaluated. Figs. 1 through 4 show the distribution of voids by width, length, circularity and orientation, respectively.

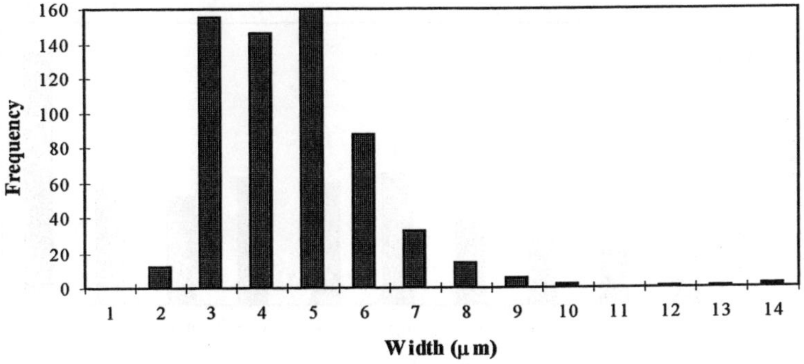

Fig. 1. Distribution of voids in dependence on their width.

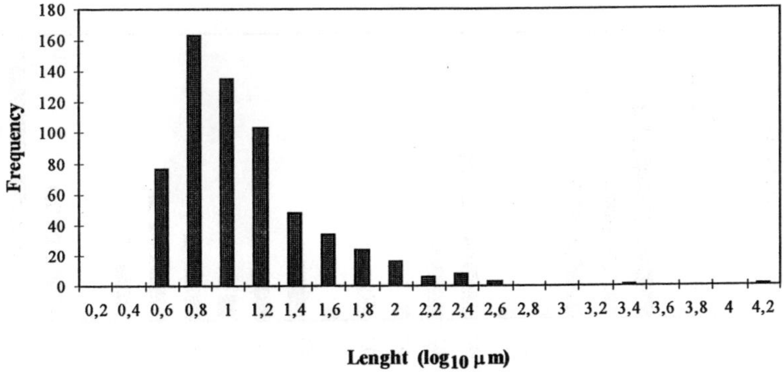

Fig. 2. Distribution of voids in dependence on their length

4. Discussion

From the knowledge of the real density of the composite and the densities of the fibers with the matrix thermally treated in the same way, the volume and mass relative ratios may be determined by using Eq. (1) and Eq. (2):

$$m_f = d_f(d_{comp} - d_{matrix})100 / d_{comp}(d_f - d_{matrix}) \qquad [\text{wt }\%] \qquad (1)$$

$$v_f = m_f d_{matrix} 100 / [m_f d_{matrix} + (100 - m_f)d_f] \qquad [\text{vol }\%] \qquad (2)$$

where m_f and v_f are the mass and volume percentage of the fibers and $m_{matrix} = (100 - m_f)$ and $v_{matrix} = (100 - v_f)$ the mass and volume percentages of the matrix, respectively.

With the investigated sample, the content in fibers was $m_f = 69.7$ wt % and $v_f = 65.3$ vol.%, and thus for the matrix, $m_{matrix} = 30.3$ wt % and $v_{matrix} = 34.7$ vol.%.

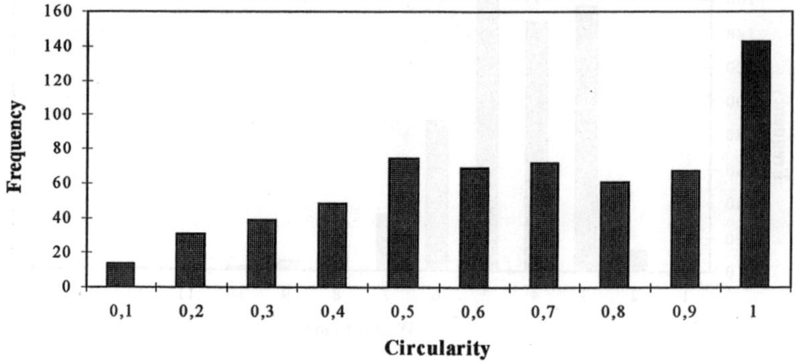

Fig. 3. Distribution of voids according to their circularity.

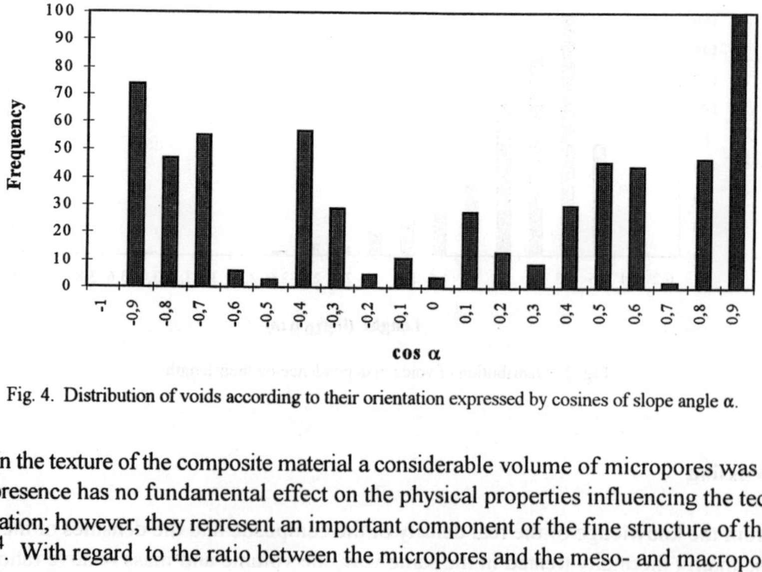

Fig. 4. Distribution of voids according to their orientation expressed by cosines of slope angle α.

In the texture of the composite material a considerable volume of micropores was found; their presence has no fundamental effect on the physical properties influencing the technical application; however, they represent an important component of the fine structure of the solid phase[4]. With regard to the ratio between the micropores and the meso- and macropores the investigated composite may be considered as a microporous material. From the CO_2 isotherm no presence of micropores in the carbonized fibers was proved, whereas in the matrix carbonized alone the volume of micropores of 63.5 mm^3/g and the surface area S_{micro} of 152 m^2/g were determined. In the sample of the composite the volume of micropores of 62.9

mm^3/g and the surface area of 164 m^2/g were determined. As the content in the matrix is only 30 wt %, which corresponds to the surface area of about 46 m^2, the remaining part of the total surface area obviously corresponds to the microvoids which results from the microroughness on the contact surfaces between the matrix and the fibers.

The highest share of the voids is formed by the coarse pores and the cracks above 0.5 μm which are of importance to the final properties of the composite. The majority of voids is contained in the open system, the volume of which may be determined by conventional methods of texture analysis. However, these methods do not give the possibility of investigating the arrangement, shape and size of the voids, represented in this size range mostly by cracks. The parameters in question may be obtained by optical methods, among which a broad spectrum of information is given above all by image analysis. Its function is irreplaceable in the determination and treatment of the data concerning the shape, length and orientation of the voids.

From the evidence of the frequency of the voids on their shape factor (Fig. 3) it is evident that the considerable occurrence is observed with voids of approximately isometric shape with a circularity value of 1 which in the ideal concept corresponds to a circular cross-section. On the other hand, the elongated voids with the circularity of 0.1, where the ratio of the width to the length is approximately 1:30, exhibit the lowest relative content. For the assessment of the mechanical properties, the orientation of the cracks in the section expressed by their angle of inclination is of importance. In the case that no direction of orientation prevails, we may suppose that the material is isotropic and in no direction is there a preference to disintegration. With the occurrence of cracks mainly in one direction, a complex failure is formed which leads to the weakening of the inner coherence of the material. As can be seen in Fig. 4, the largest number of cracks in the investigated material exhibit a nearly horizontal orientation and their effect on the strength of the material may not be assessed unambiguously without assignment to the size.

5. Conclusion

It was identified that the greatest share of the voids in a 1-D composite is formed by coarse pores and the cracks above 0.5 μm where the majority of voids is contained in the open system. According to volume of micropores, the investigated composite may be considered as a microporous material. The predominant part of total surface area of micropores obviously corresponds to the microroughness on the contact surfaces between the matrix and the fibers.

This study demonstrates the possibility of the complex analysis of the porous texture of the voids present in the composite material with the application of sorption, porometric, pycnometric and optical methods. According to the needs of technology or application, some of the methods mentioned here may be stressed. The study represents also a suggestion for the further elaboration of the individual methods with the aim of obtaining new mutual dependencies with a higher information capacity.

6. Acknowledgments

The financial support of this work by the Grant Agency of the Czech Republic, grant no. 104/94/1789, is gratefully acknowledged.

7. References

1. S. Kimura, S. Kasuya and E. Yasuda, in *Proc. Int. Symp. of Factors in Dens. and Scritering of Ceramics* (Japan, 1978), p. 229.
2. *Essentials of Carbon-Carbon Composites* (C. R. Thomas, The Royal Society of Chemistry, 1993), p.150-151.
3. J. Jortner, *Carbon* **24** (1986) 603.
4. J. Medek and Z. Weishauptová, *Uhlí (Coal)* **35** (1987) 213.

MECHANICAL PROPERTIES OF RESIN-BASED CARBON-CARBON COMPOSITES MODIFIED WITH NEEDLE COKE

K. BALÍK, J. GREGOR, M. ČERNÝ, and M. KLUČÁKOVÁ
Institute of Rock Structure and Mechanics
Academy of Sciences of the Czech Republic
V. Holešovičkách 41, 182 09 Praha 8, Czech Republic

ABSTRACT

Uni- and bi-directional (plain-weave) C/C composites from Torayca carbon fibers and phenol-formaldehyde resin, containing 1.5 wt % of finely ground needle coke were prepared. The "green" samples were heat treated at 1000 and 2200°C. Flexural strength, modulus, microscopical character and fracture pattern of the composites with and without needle coke were tested and compared.

1. Introduction

During the last twenty years, several research centers have paid continuous attention to carbon-carbon composite materials. The mechanical properties of these composites at high temperatures, low density and low thermal expansivity make them especially suitable for advanced applications.

These materials consist of uni-, bi- or multidirectional carbon fiber reinforcement embedded in a carbon matrix. The resulting mechanical properties are determined not only by the above-mentioned macrostructure but also by the type of carbon fibers, the initial precursor of the carbon matrix, and the process conditions of carbonization or possibly graphitization[1].

As matrix precursors, synthetic polymers or coal-tar pitches are used most frequently. The carbonization of synthetic resins yields a glass-like carbon matrix. Its microstructure and thus also the mechanical properties of the composite can be changed by further thermal processing, by alteration of the adhesion bonds at the fiber-matrix interface, or by the addition of finely ground fillers. The last possibility seems to be simple and smart.

Yasuda et al.[2] modified the furfuryl alcohol-based matrix by the addition of finely ground (< 5 μm) graphite in quantities up to 45 wt %. The prepared unidirectional composites attained the highest mechanical strength values and revealed a pseudoplastic character of failure with the addition of 5 wt % of graphite.

Joo[3] modified the phenolic-resin based matrix by the addition of fine graphite powder (< 5 μm) to a maximum amount of 5.3 vol %. The lowest values of volume shrinkage were obtained with unidirectional composites with the addition of about 2 vol % of graphite. These samples were also characterized by the highest mechanical strength and Young's modulus values.

In our previous paper[4] we verified the effect of the addition of the fine fraction of needle coke on the mechanical properties and the character of failure of the 1-D components.

By adding a very small amount of needle coke (1.5 wt %) with particle sizes of < 5 μm phenol-formaldehyde matrix precursor the flexural strength of composites may be increased to attain at the same time a pseudoplastic character of failure.

The aim of this investigation was to assess the effect of the coarser fraction of needle coke (10 μm) added in the quantity of 1.5 wt % to the phenol-formaldehyde precursor of the matrix on the mechanical properties of 1-D and 2-D carbon-carbon composites.

2. Experimental

For all materials: 1-D (material A), plain-weave (material B), the high-strength carbon fiber Torayca T800 (6000 filaments in a tow) with sizing was employed. Material B had a fill-yarn count of about 430 yarns per meter; the warp count was about 430 yarns per meter.

In weaves, the warp and fill yarns are nominally perpendicular to each other. The fractions of needle coke (Shell) powder (~ 10μm) were prepared by grinding in a vibratory mill and fractionated by centrifugal force and by sedimentation. All materials (A, B) were manufactured by using the prepreg technique by impregnating the fibers and cloths with a mixture of phenolic resin (Umaform) and the needle-coke powder (1.5 wt %). The prepregs were stacked in 7 layers and cured at 0.5 MPa and 170°C. The resulting 1.3 to 1.5 mm thick carbon-polymer plates were cut to specimens and carbonized at the heating rate of 50°C/hr up to 1000°C in a nitrogen atmosphere. In the case of cloth composites, one-step impregnation with pure phenolic resin was employed.

Heat treatment to 2200°C in argon, called "graphitization," was applied as the final process step.

The flexural strength and Young's modulus (1-D composites in fiber direction, cloth composites in warp direction) were determined with groups containing 8 specimens processed under identical conditions. A three-point bending arrangement was used (height-to-span ratio approximately 1:25, cross-head speed of 0.5 mm/min) on the material tester Minimat with a 1000 N load cell.

The open porosity and density were measured by the water penetration method according to the ASTM-C20 standard.

Structure and crimp angle were studied and measured with an optical microscope.

3. Results and Discussion

3.1. Material A

The results of the tests of the 1-D composites are presented in Table 1. From the table, it is evident that the addition of needle coke has no significant effect on the values of open porosity, apparent density, flexural strength and modulus. The effect of graphitization, which

Table 1. Properties of Material A - 1D Composites

Sample	V_f (%)	Porosity (%)	Density (g/cm^3)	Flexural Strength (MPa)	Flexural Modulus (GPa)
Carbonized 146	55	8.5	1.61	484 ± 61	146 ± 2
Carbonized 145 with coke	55	7.1	1.63	453 ± 40	129 ± 5
Graphitized 146	58	3.2	1.67	1074 ± 60	261 ± 11
Graphitized 145 with coke	59	3	1.68	1037 ± 63	247 ± 13

reduces the value of open porosity down to 3% and increases the flexural strength to the value of more than 1000 MPa, is very significant. The results of the flexural strength tests are graphically presented in Fig. 1 where the values of F.S.T. are compared. They were obtained from the equation (1):

$$F.S.T. = \frac{R_{mc} \times 100}{R_{mf} \times V_f}$$

where F.S.T. is the fiber strength translation, R_{mc} the flexural strength of the composite, R_{mf} the tensile strength of the fiber, and V_f the volume fraction of the fibers in the composite.

From Fig. 1 it is clear that the value of F.S.T. (15%), which is practically equal for samples with and without coke, is increased by graphitization with both series again to the equal value (about 30%).

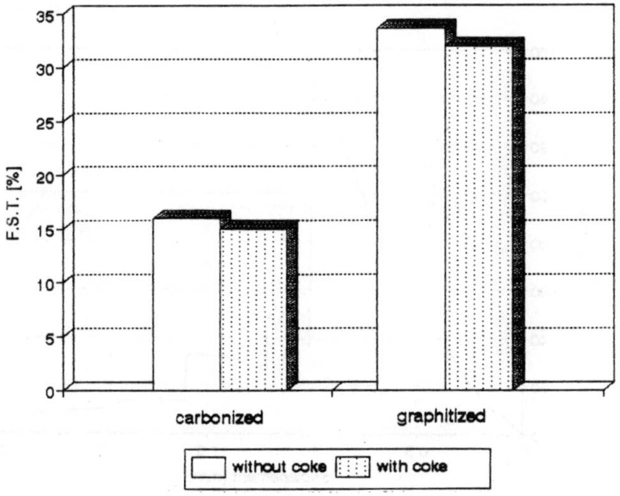

Fig. 1. F.S.T. of 1-D composites with and without needle coke after carbonization and graphitization.

In contrast to the particle size of ~ 5 μm, the effect of which we have tested in our previous publication[4] and which caused a pseudoplastic character of the fracture, the current particle size of ~ 10μm does not change the catastrophic failure of the 1-D composites after carbonization or after graphitization – see Fig. 2.

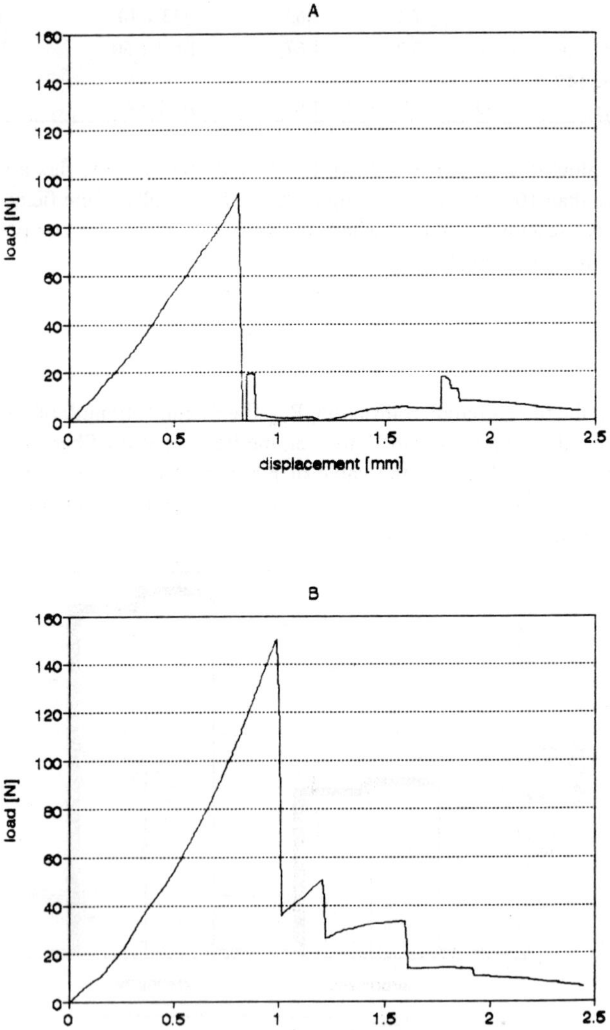

Fig. 2. Load-displacement curves for 1-D composites with needle coke, A–Carbonized, B –Graphitized.

3.2. *Material B*

The results of the tests of the 2-D composites with plain weave are presented in Table 2. Again, there is no significant difference in the values of open porosity and apparent density of the samples with and without coke, after carbonization or after graphitization. With regard to the different value of V_f in the two series in Table 2, for comparison the flexural strength and the modulus were expressed as their ratios to V_f, i.e., as R_{mc}/V_f and E/V_f, respectively–see Figs. 3 and 4.

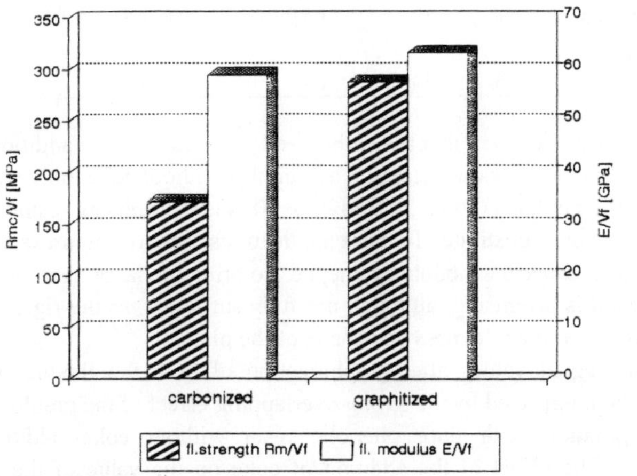

Fig. 3. R_{mc}/V_f and E/V_f of carbonized and graphitized plain-woven composites without needle coke.

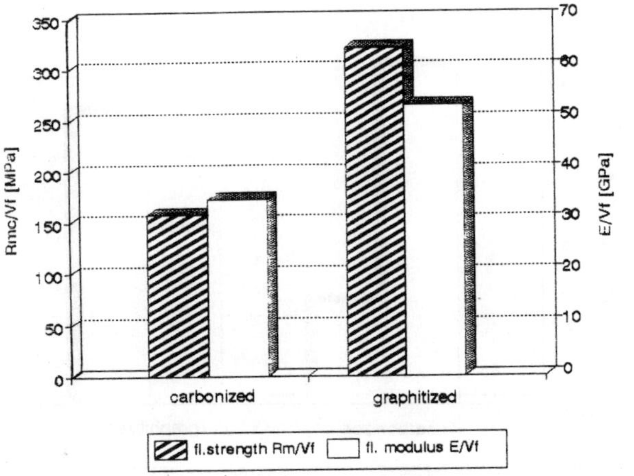

Fig. 4. R_{mc}/V_f and E/V_f of carbonized and graphitized plain-woven composites with needle coke.

Table 2. Properties of Material B–
Plain-woven Composites, Crimp Angle in Warp Direction 11°

Sample	V_f (%)	Porosity (%)	Density (g/cm^3)	Flexural Strength (MPa)	Flexural Modulus (GPa)
Carbonized 155	61	23	1.42	105 ± 22	36 ± 5
Carbonized 141 with coke	52	24	1.37	82 ± 5	18 ± 3
Graphitized 155	68	12	1.51	196 ± 21	43 ± 3
Graphitized 141 with coke	53	14	1.45	170 ± 31	28 ± 8

From Fig. 4 it is evident that the carbonized samples with the addition of coke exhibit nearly the same value of flexural strength as samples without coke, only with a significantly lower value of the modulus (about 20% lower). This is a promising result for application of this material as a bone substitute. In these applications, one tries to preserve the mechanical strength value and reduce the modulus value, i.e., to bring the latter near to that of the human bone. The result is promising, although not fully sufficient, as the rigidity of the material depends above all on the thickness and shape of the plate.

Practically equal values after graphitization of both the flexural strength and the modulus may be interpreted by the strong overlapping effect of the graphitization as well as by the impregnation with pure phenolic resin without coke addition prior to the graphitization. The effect of the addition of coke on the value of the relative standard deviation of the flexural strength with carbonized samples is also interesting (see Fig. 5). In

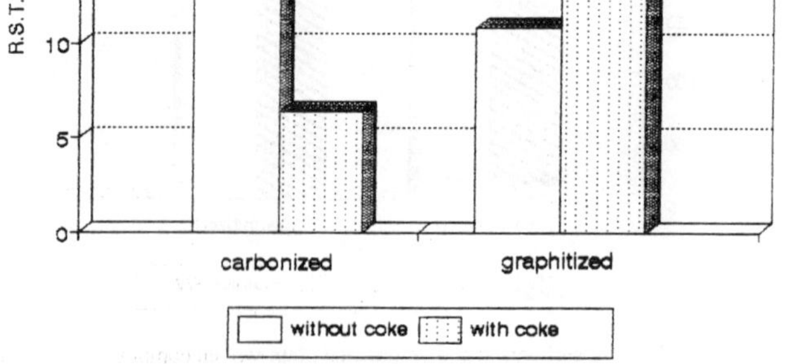

Fig. 5. The RSTD of carbonized and graphitized plain-woven composites with and without needle coke.

this case, the addition of filler causes a decrease of the value of the relative standard deviation, i.e., the equalization of the values of the mechanical properties. This may be caused by the effect of the presence of particles of coke, especially in the regions of pure matrix where coke in fact forms a particulate composite with glassy carbon (see Fig. 6).

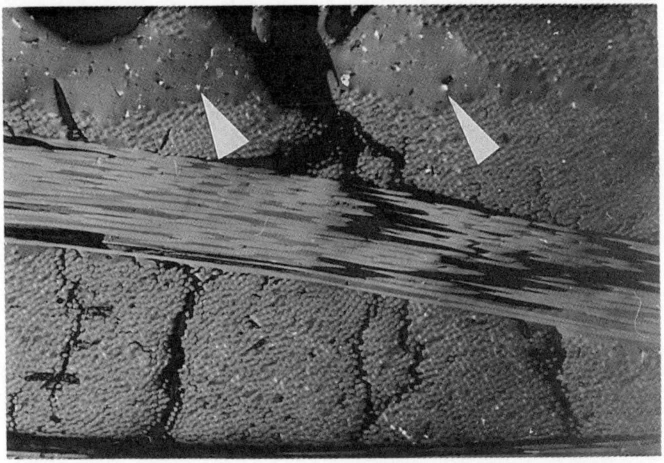

Fig. 6. Optical micrograph of polished section of plain-woven composite after carbonization, arrows–needle coke particles, 200x.

The character of the failure of the samples differs again only after carbonization. From the interdependence of load and displacement shown in Fig. 7 it follows that composites with the addition of coke are less susceptible to delamination. It is also documented in Fig. 8, which shows the SEM fractographies of the composites with and without needle coke. The debonding of fiber/matrix interfaces and fiber pullout are observed, first of all in the composite with coke (Fig. 8, B).

4. Conclusion

Material A: The fraction of needle coke of ~ 10 μm size has no substantial influence on the mechanical properties of 1-D composite in contrast to the results for ~ 5 μm size.

Material B: The addition of the needle coke fraction has an influence on mechanical properties of plain-woven composite. It reduces the value of modulus, but not the value of flexural strength after carbonization. It seems that this filler influences the uniformity of flexural strength of this material after carbonization.

5. Acknowledgment

The financial support of this work by the Grant Agency of the Czech Republic, grant no. 104/94/1789, is gratefully acknowledged.

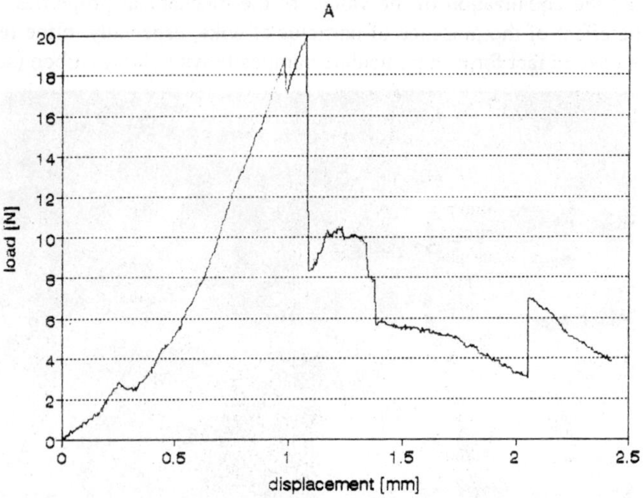

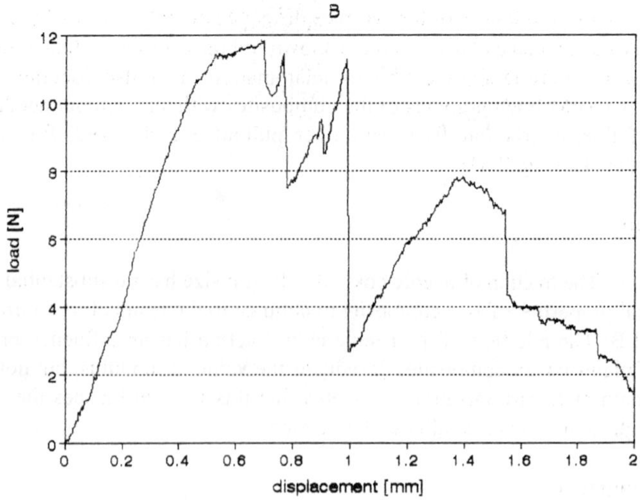

Fig. 7. Load-displacement curves for plain-woven carbonized composites, A–without needle coke, B– with needle coke.

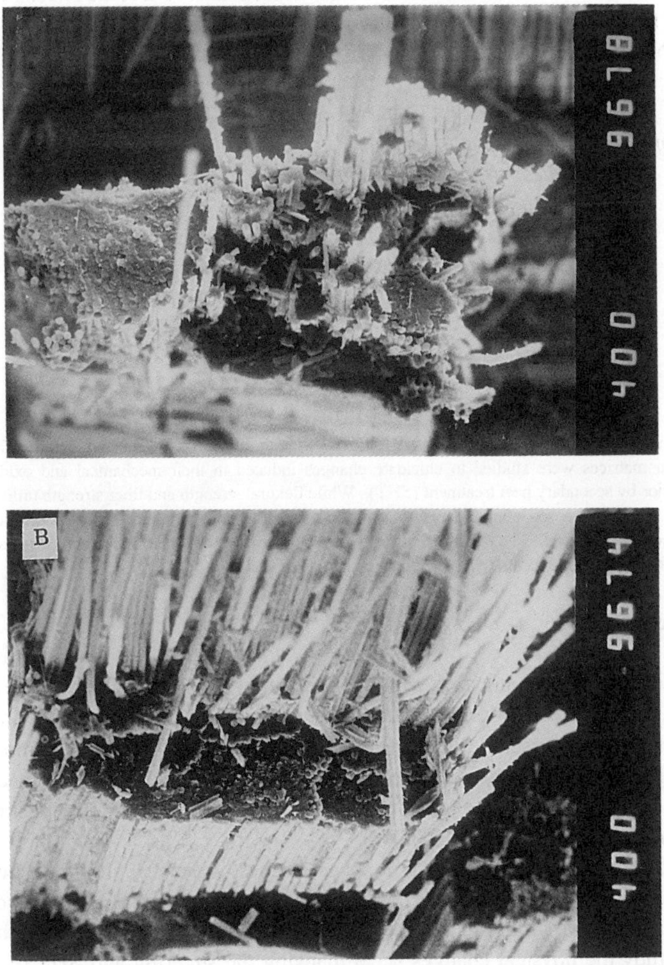

Fig. 8. SEM fractographics of plain-woven carbonized composites, A –without needle coke, B–with needle coke, 400x.

6. References

1. E. Fitzer, *Carbon*, **25** (1987) 163.
2. E. Yasuda at al., *Carbon*, **26** (1988) 255.
3. H.-J. Joo, *High Temperatures-High Pressures*, **22** (1990) 649.
4. K. Balík at al., *CFI Beihefte (Fortschrittsberichte der DKG)*, in press.

THE INFLUENCE OF A SECONDARY HEAT TREATMENT ON MECHANICAL AND OXIDATION PROPERTIES OF UNIDIRECTIONAL CARBON/CARBON AND CARBON/FURFURAL - SILOXANE COMPOSITES

P. GLOGAR and P. FOTT
Institute of Rock Structure and Mechanics, Academy of Sciences of the Czech Republic, V Holešovičkách 41, 182 09 Prague 8, Czech Republic

L. M. MANOCHA and A. PATEL
G. H. Patel Institute of Material Science, Sardar Patel University, Vallabh Vidyanagar, 388 120, Gujarat, India

ABSTRACT

Four series of carbon fiber reinforced composites with furfural-siloxane and polymer based carbon matrices were studied to elucidate changes induced in their mechanical and oxidation behavior by secondary heat treatment (SHT). While flexural strength and fiber strength utilization of all series increased after the SHT, development of the pore structure reflected differences between the matrices. The enhanced porosity of secondary treated furfural-siloxane composite is probably responsible for its relatively poor oxidation resistance.

1. Introduction

Processing of C/C composites with phenolic and polyfurfuryl alcohol resins as matrix precursors[1] though possessing lower graphitizability, has the advantages of time and cost savings over pitch precursors. Pitch precursors, on the other hand, produce more graphitic structures[2]. However, there has not been enough literature on the changes in macrostructure of the composites, especially the porosity during pyrolysis of the resin composites and their secondary heat treatment (SHT) to 2800°C.

Recently, preceramic polymers (siloxanes, silanes, silazanes, etc.) have been used as matrix precursors for ceramic matrix composites[3]. Their processing through sol-gel route may result in different structures depending on the reaction conditions. It is, therefore, of interest to study the physical changes accompanying the sintering and SHT of the fibrous composites made with mixed ceramics as matrices.

2. Experimental

Four series (A – D) of unidirectional composites made by the wet winding technique were used in the present study. After pyrolysis at 1000°C in nitrogen at ambient pressure selected composites were impregnated and repyrolyzed (Table 1). Subsequent SHT was done under argon.

Table 1. Parameters Characterizing the Four Series of Investigated Composites

Series A	Series B	Series C	Series D
HT fiber, phenolic resin	HM fiber, phenolic resin	HM fiber, polyfurfuryl alcohol	HT fiber, furfuryl siloxane
carbonized	carbonized, impregnated, re-carbonized	carbonized	pyrolyzed, impregnated, re-pyrolyzed
SHT at 1450°C	SHT at 1450°C	SHT at 2800°C	SHT at 1450°C

The composites were characterized by their weight loss, density and porosity changes during pyrolysis and SHT, flexural strength (three-point bending), and failure behavior (SEM of the fractured surfaces).

Oxidation behavior of the composites was studied by a weight loss test in air under a constant heating rate and air flow.

3. Results and Discussion

3.1. Physical Changes During Pyrolysis and Secondary Heat Treatment of Composites

The volumetric shrinkage during pyrolysis of the Series A composites ($\Delta V_r = 14.9\%$) is higher than that of the Series B ($\Delta V_r = 8.2\%$) which agrees with observations of other authors on shrinkage of resin composites made with HT and HM carbon fibers. During SHT the composites A exhibit a very low volumetric shrinkage ($\approx$ 0.9%) and an increase in density while the Series B composites exhibit a slight expansion ($\approx -3\%$). The composites made with polyfurfuryl alcohol resin (Series C) though having a low fiber volume content possess a high density and exhibit a remarkable shrinkage ($\approx$ 15%) after SHT. The composites made with furfural and siloxane mixed matrix system (Series D) undergo a decrease of density during the SHT though they possess quite high densities at the pyrolyzed level. These composites also exhibit an appreciable weight loss ($\approx$ 14%) during the SHT but no substantial volume shrinkage occurs. This suggests that in these composites the chemical transformation is not completed at 1000°C but continues even at higher temperatures during the SHT. Fig. 1 gives a graphical representation of the change of the bulk and geometrical densities of the composites on SHT.

3.2. Porosity

Porosity and pore size distribution are dependent on the processing of composites, volume fractions of the constituents, and heat treatment temperatures[4]. From water penetration data and porosimetry, the porosities determined by using different fluids were calculated enabling us to compare porous structures of different specimens. The mercury porosity P_1 corresponds to the cumulative specific volume of pores with orifice radius 7.5 to

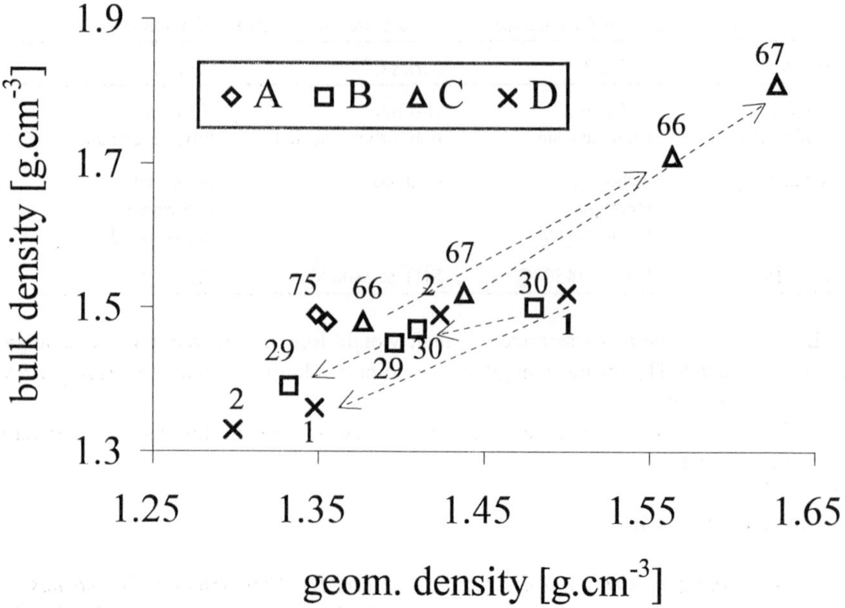

Fig. 1. Variations of bulk and geometric densities during the secondary heat treatment (the arrows direct from the initial towards the heat-treated state).

0.00375 μm (macro- and mesopores). P_2 was determined by taking into account the sum of the macro-, meso-, and coarse pore volumes. The coarse pore volumes are those of radius greater than 7.5 μm and were determined by taking the difference of the reciprocal values of the bulk and mercury densities. The total open porosity P_3 was determined according to ASTM C-20 by water penetration. The mean values of P_1, P_2, and P_3 for individual series are plotted in Fig. 2.

Before SHT, the open porosity $\boldsymbol{P_3}$ in composites made with phenolic resin (Series A and B) is higher than for those made with polyfurfuryl alcohol resin (Series C). However, in Series C the mercury porosity $\boldsymbol{P_1}$ contributes more to the overall porosity than in Series A and B. The difference $\boldsymbol{P_2} - \boldsymbol{P_1}$ (amount of coarse pores) for Series A and B composites is larger for the Series C samples. After SHT the porosity $\boldsymbol{P_1}$ in the Series A and B composites remains almost unchanged but their total open porosity $\boldsymbol{P_3}$ (in contrast to the Series C and D composites) decreases. This may be due to closing of the micropores created in the cross-linked polymer matrix during evolution of pyrolysis gases. In Series C the porosities in all size ranges increase after SHT to 2800°C. This is partly due to the creation of new microporosity as a result of reorganization of graphitic layers which prevails over the high-temperature closing of the matrix microporosity.

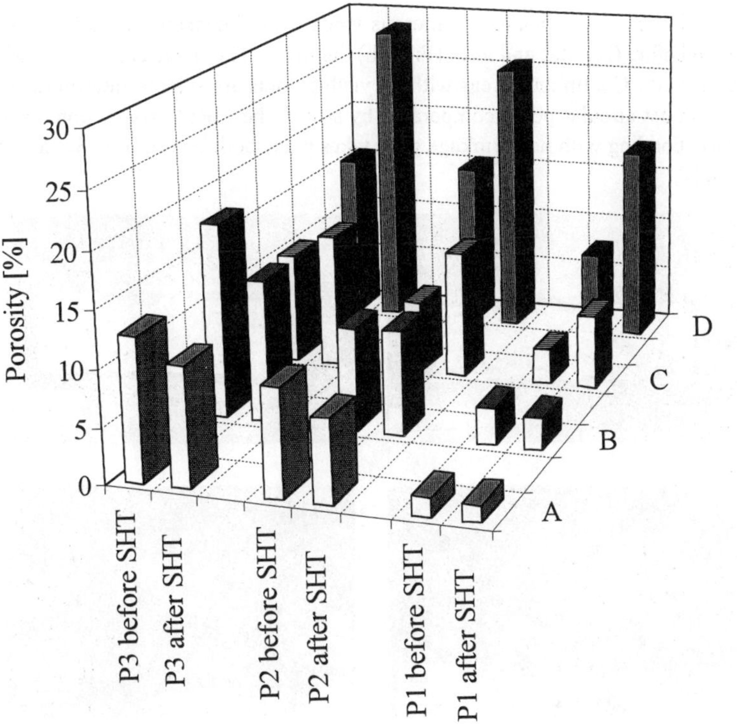

Fig. 2. Mean values of porosity P_1-P_3 of four series of composite specimens (A – D).

Composites belonging to Series D possess the highest total porosity. They contain coarse open pores as well as meso- and macropores. The porosity P_1 of Series D is approximately four times greater than for Series A. It is due mainly to the cross-linked stage of the polymer matrix in the A composites which undergoes solid state pyrolysis resulting in a condensed structure while in the D composites the matrix is in gel state and undergoes chemical reaction between different species in the matrix which results in a higher porosity. Accordingly, all the porosities increase almost two-fold after the SHT and are accompanied by a remarkable reduction of bulk density.

3.3. Optical Microscopy

A visual characterization of the polished cross-sections of the investigated composite specimens was performed at magnifications of approximately 16, 250, 600, and 1600 times.

Series A and B. Composites from Series A and B possess very similar pore structures despite the (slight) densification of the Series B composites. No significant changes were observed after SHT to 1450°C.

In the lowest magnification, numerous voids with dimensions up to 0.2 – 0.5 mm are visible both before (Fig. 3a) and after (Fig. 3b) the SHT. The fibers are evenly and relatively densely packed. Continuous areas without visible voids are seldom interrupted by slit-like voids or almost circular pores comparable by size to the fiber cross-section. A very good fiber/matrix bonding with no shrinkage gaps takes place both before (Fig. 3c) and after (Fig. 3d) the SHT.

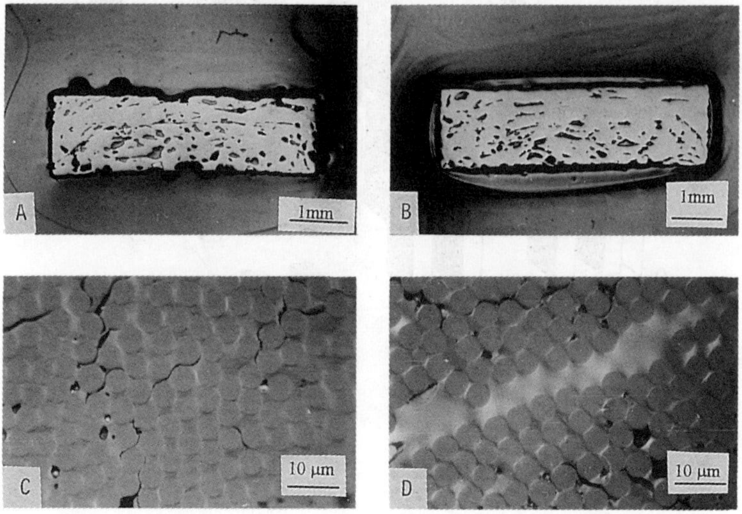

Fig. 3. Optical micrographs of the Series B composites before (a, c) and after (b, d) the secondary heat treatment to 1450°C. Magnification: a) 16 x, b) 16 x, c) 1600 x, d) 1600 x.

Series C. The initial composite from Series C has a lower population of coarse pores than the Series A and B specimens which is consistent with the observed lower porosity $\boldsymbol{P}_2$. An uneven distribution of the fibers takes place. Some areas densely occupied by fibers are visible and other areas that are extended "lakes" of matrix without fibers are visible (Fig. 4a). After the SHT (2800°C in this case) the matrix becomes anisotropic and its "lakes" are broken down to variously oriented domains (Fig. 4b). Relatively weak fiber/matrix bonding is observed in the initial specimens (Fig. 4c) with a large number of shrinkage gaps. A columnar texture of the secondary heat-treated matrix is documented in Fig. 4d.

Series D. The secondary heat-treated composites differ significantly from initial ones. Before SHT a dense network of interconnected, randomly oriented slit voids (up to 1 mm length and 0.1 mm width) was observed (Fig. 5a). After the SHT the minor voids almost disappear and larger voids are arranged primarily along the cross-section breadth (Fig. 5b). Simultaneously, the initially irregular distribution of fibers becomes more uniform while irregularly shaped voids of the initial specimen become smaller and more regular in shape. The heterogeneity of the matrix containing sub-micron brighter inclusions is clearly seen (Figs. 5c and 5d). Semi-annular shrinkage gaps are seldom detected.

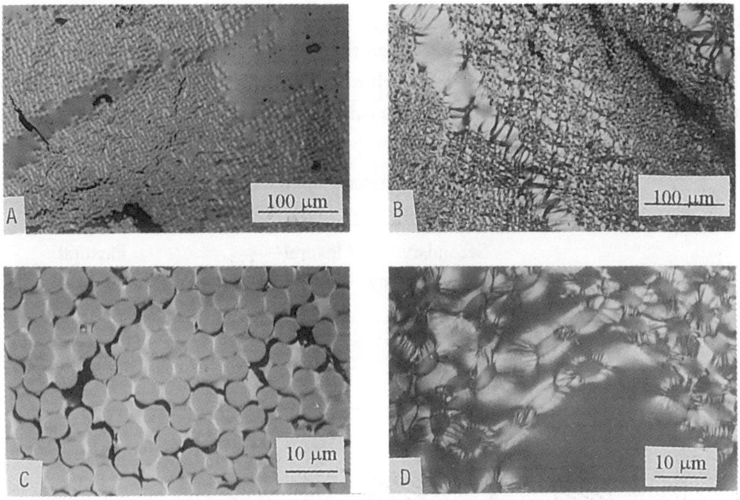

Fig. 4. Optical micrographs of the Series C composites before (a, c) and after (b, d) the secondary heat treatment to 2800°C. Magnification: a) 250 x, b) 250 x, c) 1600 x, d) 1600 x.

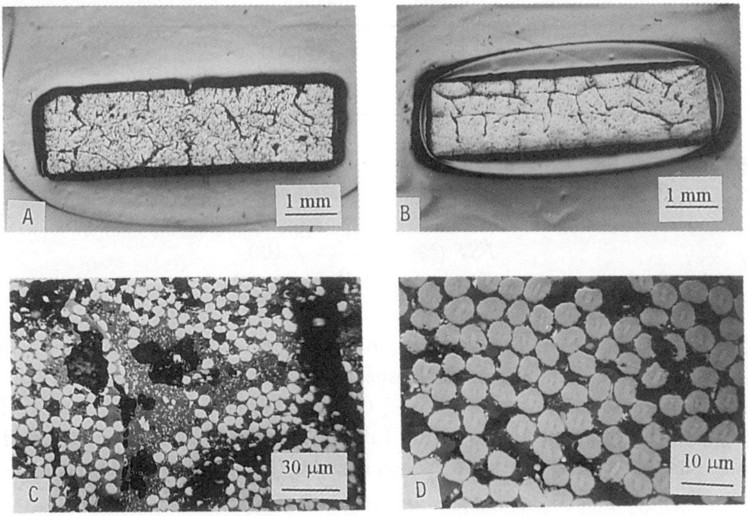

Fig. 5. Optical micrographs of the Series D composites before (a, c) and after (b, d) the secondary heat treatment to 1450°C. Magnification: a) 16 x, b 16x, c) 600 x, d) 1600 x.

3.4. *Mechanical Properties*

A three-point bending arrangement (support span 50 mm, height to span ratio 1:20 – 1:25, cross-head speed 0.5 mm/min) was used for flexural loading of the specimens (length 60 mm) and an INSTRON 4301 testing machine. Flexural strength σ_F [MPa] and modulus $\boldsymbol{E}_F$ [GPa] are listed in Table 2. Selected load-displacement characteristics are presented in Fig. 6.

Table 2. Mechanical Properties of the Investigated (Initial and Secondary Heat Treated) Specimens

Series	Specimen	Secondary HTT (°C)	Flexural Strength (MPa)	FST (%)	Flexural Modulus (GPa)
A	75	1450	217	6	83
	76	-	159	5	89
	77	-	174	5	87
B	29	1450	290	13	130
	30	1450	230	10	108
	32	-	230	10	112
	33	-	177	8	98
C	64	-	185	13	80
	65	-	205	16	72
	66	2800	216	15	61
	67	2800	323	20	63
D	1	1450	288	20	84
	2	1450	310	20	90
	4	-	181	13	74

To characterize the fiber strength utilization in the composite flexural strength, the value of "fiber strength translation" (FST) can be calculated according to:

$$FST\ (\%) = \frac{\sigma_F}{V_f \sigma_{fib}} \times 100$$

where σ_F, V_f, and σ_{fib} denote flexural strength of the composite, fiber volume fraction, and tensile strength of the employed fibers, respectively. As the exact values of σ_{fib} are not known to us, we have assumed the values 5 GPa and 3.5 GPa for the HT and HM fibers, respectively. The estimation is given in Table 2 and it allows us to compare different composites made of similar fibers, i.e., Series B and C (HM) and A and D (HT).

3.5. *Failure Behavior, Fracture Surfaces*

Fracture patterns of the investigated specimens (some shown in Fig. 6) can be discussed in terms of the corresponding fracture surface morphology. For this purpose, flexural

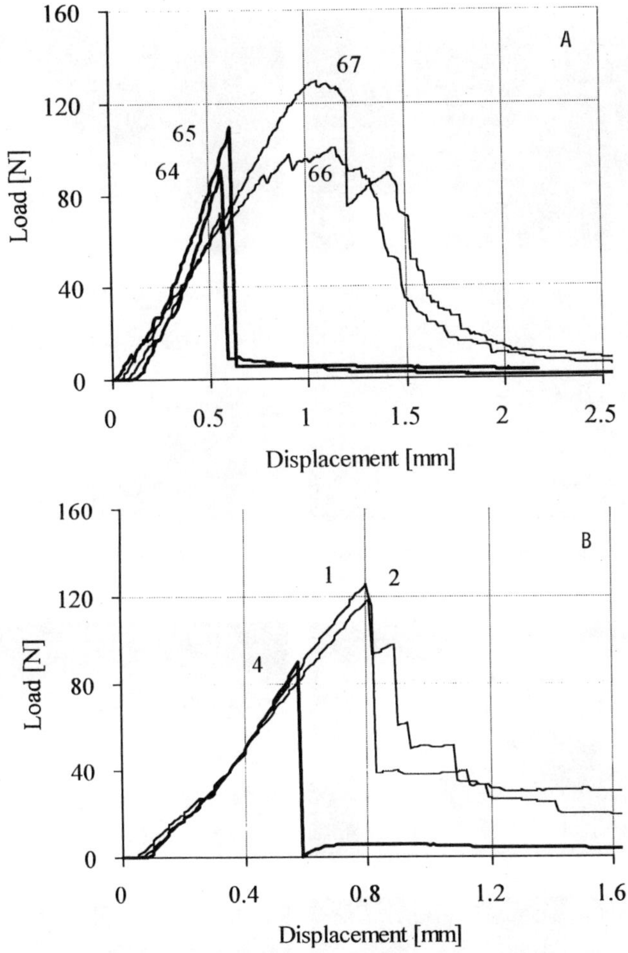

Fig. 6. Load-displacement characteristics of initial and secondary heat treated specimens: a) Series C (64, 65 - initial, 66, 67 - SHT at 2800°C), b) Series D (4 - initial, 1, 2 - SHT at 1450°C).

fracture surfaces (predominantly their tensile loaded regions) were examined and photographed using SEM.

The composites using phenolic resin as a matrix precursor and made of HT and HM fibers (Series A and B, respectively) behave similarly. The fracture is brittle both before and after SHT to 1450°C.

The most striking influence of the SHT (2800°C in this case) was observed with the composites made of polyfurfuryl alcohol-based matrix (Series C, Fig. 7). Here, even before

Fig. 7. SEM micrographs of the fractured composites of Series C before (a, c) and after (b, d) SHT to 2800°C.

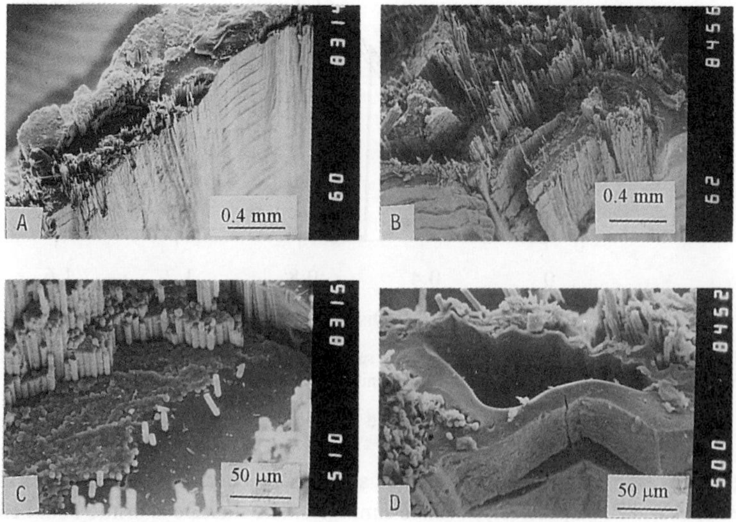

Fig. 8. SEM micrographs of the fractured composites of Series D before (a, c) and after (b, d) SHT to 1450°C.

SHT some limited fiber pull-out exists (Figs. 7a and 7c) but it becomes much more pronounced after the SHT (Fig. 7b). Moreover, the heat treated matrix itself undergoes a substantial reconstruction resulting in its (partial) breakdown to thin "leaves" which may support dissipation processes during fracture and the predominantly shear failure (Fig. 6a).

The behavior of the composites made with the use of furfuryl-siloxane matrix precursor (Series D, Fig. 8) is also strongly affected by the SHT. Before SHT the fracture is brittle (Figs. 8a and 8c) but after the heat treatment the fibers separate from the matrix more easily (Figs. 8b and 8d) which supports the observed mixed-mode failure (Fig. 6b).

3.6. Oxidation Behavior of the Investigated Composites

Oxidation studies were made thermogravimetrically using a Derivatograph Q 1500 D apparatus. Cube-shaped samples (approximately 2 x 2 x 2 mm) were heated at a rate of 10°C/hr in flowing dry air until a weight loss of 20–25% when the reaction was extinguished by flowing pure nitrogen. Reactivity of the specimens was characterized by plotting the weight loss vs. temperature (Fig. 9). The oxidized surfaces (perpendicular to the fiber direction) were photographed by SEM.

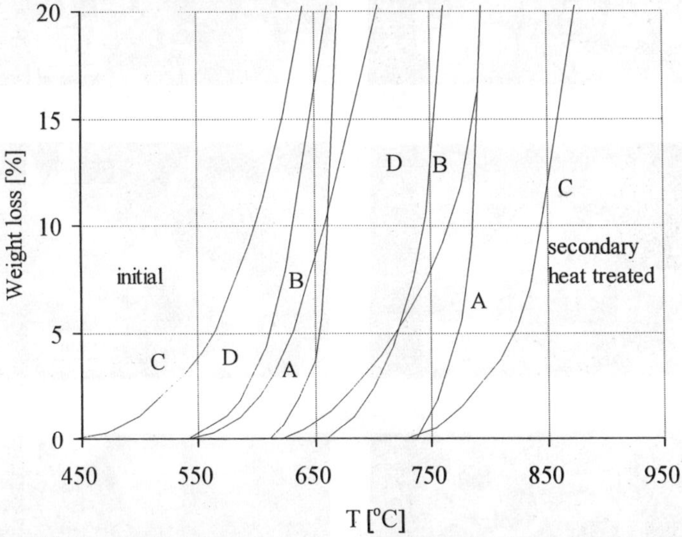

Fig. 9. Weight loss of initial and secondary heat treated composites (heating rate 10K/hr in dry air) to the weight loss 20%.

Obviously, SHT significantly improves the resistance to oxidative corrosion. In most cases the TG curves of the specimens heat-treated to 1450°C are shifted by 100–150°C towards higher temperature. With the Series C specimens (treated to 2800°C) a shift by 250–300°C with respect to the highly reactive initial specimen was detected.

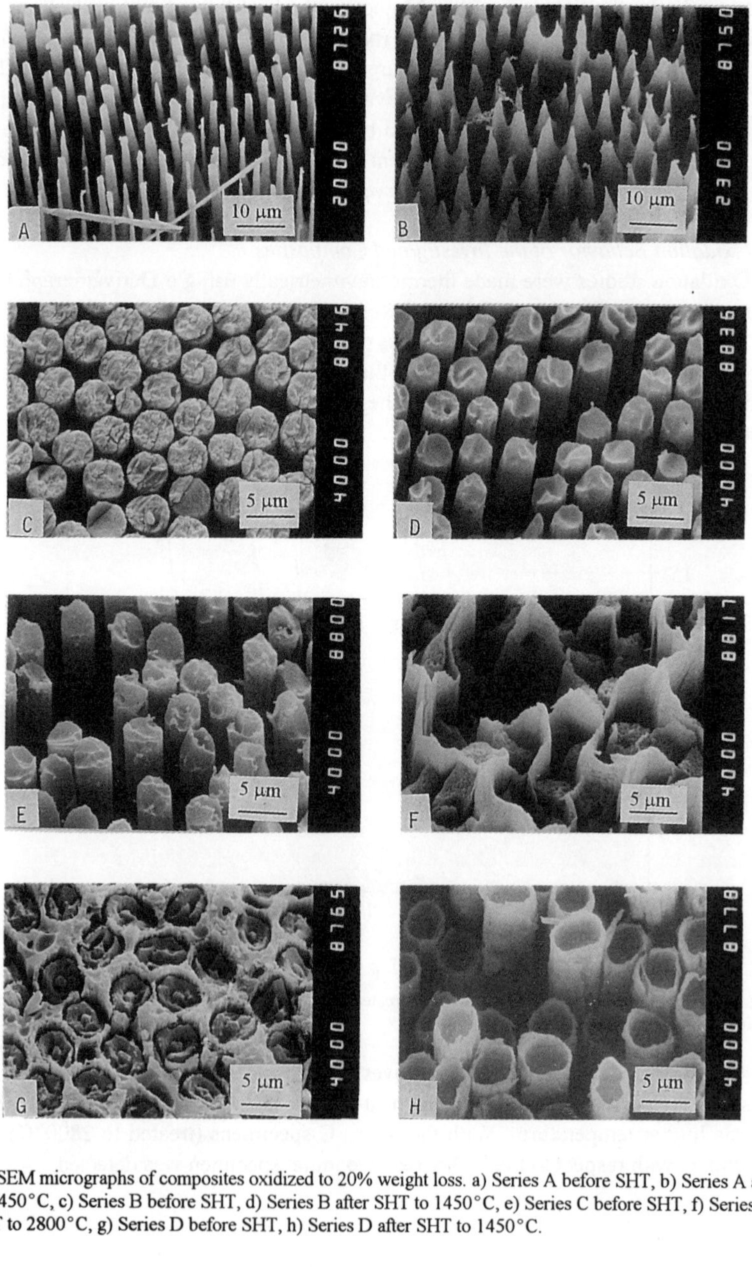

Fig. 10. SEM micrographs of composites oxidized to 20% weight loss. a) Series A before SHT, b) Series A after SHT to 1450°C, c) Series B before SHT, d) Series B after SHT to 1450°C, e) Series C before SHT, f) Series C after SHT to 2800°C, g) Series D before SHT, h) Series D after SHT to 1450°C.

The SHT affects predominantly the matrix reactivity as the HTT of 1450°C does not (except Series C) exceed the fiber manufacture temperature. The observed decrease of matrix reactivity can be explained as a consequence of: a) a diminished (after the SHT) specific surface area due to the supposed closing of the microporous system, and/or b) a lowered chemical activity of the surface stabilized by the SHT.

The HM fibers are obviously less reactive than the HT ones. Almost no oxidative loss of fibers was observed with the initial HM-reinforced Series B and C specimens (Figs. 10c and 10e) as well as with the secondary heat treated Series B specimens (Fig. 10d). On the other hand, with the secondary heat treated HT-reinforced Series A specimens (Fig. 10b) both matrix and fibers lose roughly comparable amounts and even with the initial specimens some degree of fiber tip sharpening was observed (Fig. 10a). For the secondary heat treated Series C specimens the exceptionally high oxidation loss of HM fibers (Fig. 10f) can be attributed to the high oxidation temperature of this material (Fig. 9).

Reduced reactivity can be expected for the carbonized matrix containing siloxane in the Series D composites. However, it seems to be counterbalanced by an easy access to the inner reactive surface due to the high porosity P_1 containing also transport pores. The burning mechanism of this matrix differs from those of other investigated composites in which the matrix gradually vanishes down to the burn-out. In the Series D specimens, the matrix burns from inside by pore expansion and/or opening while preserving its outer volume. After SHT, P_1 increases but the specific surface reactivity simultaneously diminishes. The inter-phase boundary seems to be the least reactive (Fig. 10h).

The polyfurfuryl alcohol based matrix seems to be more reactive than the phenolic based one unless the different reactivity of the secondary untreated Series A and C specimens is due to different fiber density. (The matrix reactivity comparison for SHT specimens A and C is impossible due to the involved HTT difference).

4. Conclusions

- Development of the pore structure of the investigated composites after SHT reflects differences between the polymer derived pyrolyzed matrix in the composites A, B, C and the mixed ceramic matrix prepared by sol-gel technique in the composite D.
- For all of the investigated composites both flexural strength and fiber strength translation increase after SHT. The FST reaches its maximum value, ≈20%, for secondary heat treated Series D composites.
- The initially brittle failure behavior of the investigated composites changes after SHT to other failure types for Series C (shear with a fiber pull-out) and Series D (mixed mode failure) but persists even after SHT with Series A and B.
- The increase in oxidation resistance of the Series A, B, and D composites after SHT is mutually comparable though a higher resistivity of silicon containing Series D composite was expected. The high porosity of the latter probably diminishes the advantage gained by its higher chemical stability.

5. Aknowledgement

This study was partially supported by the Grant Agency of the Czech Academy of Sciences within the project No. 246405.

6. References

1. L. M. Manocha, *Carbon* **32(2)** (1994) 213.
2. S. Appleyard and B. Rand, *Proc. 5th Int. Carbon Conference*, Essen (1992) 718.
3. J. E. Mark et al., *Inorganic Polymers*, Prentice Hall (1992).
4. J. Jortner, *Carbon* **24** (1986) 603.

INFLUENCE OF THE THERMAL TREATMENT ON MICROSTRUCTURE AND MECHANICAL PROPERTIES OF UNIDIRECTIONAL C/C FIBER COMPOSITES

E. RUDNAYOVÁ and J. DUSZA
Institute of Materials Research, Slovak Academy of Sciences, Watsonova 47, 04353 Košice, Slovakia

P. GLOGAR
Institute of Rock Structure and Mechanics, Academy of Sciences of the Czech Republic V Holešovičkách 41, 18209 Prague 8, Czech Republic

ABSTRACT

Room temperature flexural strength and modulus of carbonized and 2200 °C treated unidirectional carbon/carbon composites were evaluated and discussed in terms of findings of fractographic and ceramographic analyses. Methodical aspects of quantitative estimation of microstructural parameters (volume fraction of pores and fibers) are discussed.

1. Introduction

The mechanical properties of materials are determined in a decisive way by their microstructure, which is formed in the manufacturing process and can be positively or negatively influenced by changing the conditions of any step in the process. Continuous fiber carbon/carbon composites have a quite complex microstructure. Their architecture is determined principally by the way in which carbon fiber bundles are arranged with respect to each other. The distribution of the matrix carbon, both between and within the fiber bundles, is also relevant but it is of secondary importance to mechanical properties. In these composites there exists, besides the solid components, a wide scale of internal voids ranging in shape from nearly equiaxial pores to high aspect ratio cracks, in size from the nanometric to the millimetric scale. These voids usually have a profound influence on the mechanical properties. It has been shown that the thermal treatment also influences the microstructure and thus the properties of carbon/carbon composites[1]. And so, the properties of carbon-carbons are the result of sometimes very complex interactions between the different features present in the microstructure.

The present contribution deals with the room temperature flexural strength, fractographic investigation and microstructure characterization of unidirectional continuous fiber carbon/carbon composites (two series differing slightly in their porosity and fiber content) and with the influence of the high temperature treatment (graphitization) on the measured and statistically evaluated values.

2. Material and Experimental Methods

The unidirectional carbon/carbon composites were prepared from high strength carbon fibers T-800 soaked with phenolic resin (UMAFORM) by a prepreg technique. The stacked prepregs were cured at 120°C. The obtained 1.3 mm thick carbon-polymer composites were cut to ≈ 5 x 1.3 x 45 mm specimens. Two batches (labeled 1 and 2) of composites with a slightly different volume fraction of microstructure constituents were prepared by carbonization at 1000°C in nitrogen[2,3]. After this procedure, some of the specimens were heat treated at 2200°C in argon resulting in specimen series G1 and G2 while the remaining carbonized specimens were denoted as series Cl and C2. The materials were initially characterized by measurement of bulk density, open porosity and the estimation of the volume fraction of fibers.

The specimens were broken in three-point bending mode with a span of 40 mm and a cross-head velocity of 0.5 mm/min. Their fracture surfaces were investigated by SEM. Five specimens of each series were fractured. The work to failure was calculated as the area under the ascendant branch of the load-displacement curve up to the first load fall on the curve. The flexural modulus was calculated as the gradient of the linear section of the rising part on the load-displacement curve.

The samples for ceramography were cut from one end of the bars, transversely to the fiber alignment, by electrospark erosion in order to prevent possible damage and delamination of the individual layers. The embedded samples were carefully ground on abrasive papers beginning with the finest possible one up to 4/0 at very low forces and speeds. The polishing of specimens was made on satin with a 3 μm diamond paste for about 15 min followed by end polishing on velvet with 1 μm diamond paste.

The qualitative observation of microstructures was followed by quantitative estimation of their parameters. The volume fraction of each microstructure constituent was estimated by image analysis. The volume fraction of pores was checked, and the fiber diameters and fiber spacings were estimated by point analysis and by counting the fibers along a gauge line, respectively. The measurement and calculation of the point analysis and counting of intercepts were made according to references 4 and 5. The image analysis and point and/or intercept analyses were performed on micrographs taken from the polished cross sections of samples. These were scanned by scanner and processed by the image analyzer (DIPS).

3. Results and Discussion

The qualitative observation revealed no remarkable differences in the microstructure of samples from batches 1 and 2. Therefore the batches were distinguished only when describing the results of quantitative analysis. On the other hand, the microstructure appearance before and after the high temperature treatment (Series C and G, respectively) was quite different already at low magnifications (Figs. 1a and 1b). The samples possess different forms of voids (cracks and pores). These become more crack-like in the Series G

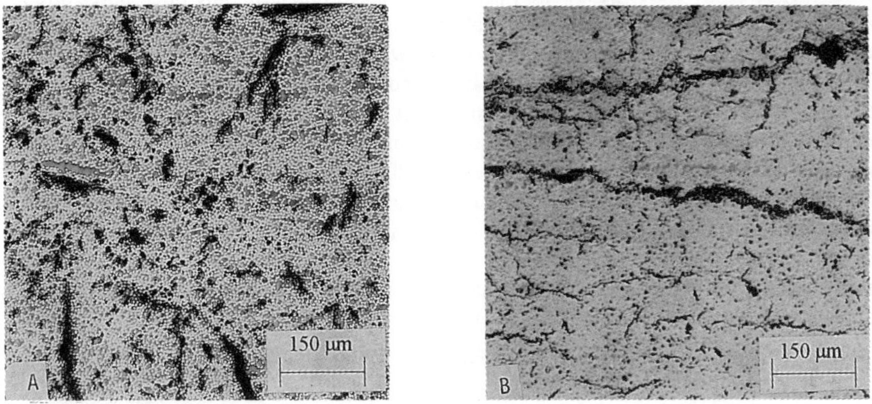

Fig. 1. Overall distribution of voids and fibers in the studied carbon/carbon composites before (a) and after (b) heat treatment at 2200°C.

specimens, and while the cross-section of fibers and the non-uniformity of their distribution within the matrix can be very easily distinguished in the untreated samples (Series C), the opposite is true for the samples of the G series. This is more evident from Figs. 2a and 2b at higher magnification (Fig. 2b is taken in polarized light, which makes the darker matrix luminous, chiefly between the fibers, less in the matrix "islands").

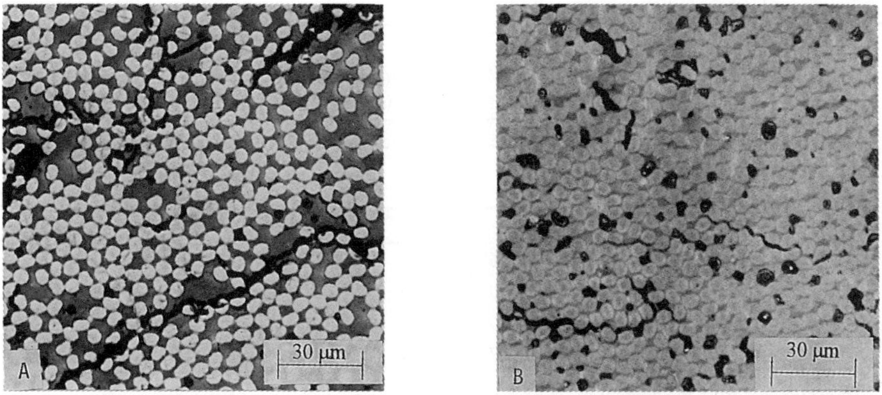

Fig. 2. Microstructure of the studied composites before (a) and after (b) heat treatment at 2200°C (Fig. 2b taken in polarized light).

Results of the quantitative analyses are summarized in Tables 1 and 2. The low porosity values measured by water penetration (*a*) reflect that in this case only open pores directly connected with the sample surface were taken into account. The volume fractions of voids determined by point[4] and image analyses (*c* and *d*, respectively) are mutually comparable. The plane fractions of fibers determined by methods *b*, *c*, and *d* differ remarkably and an attempt to assess their validity was desired. It was achieved by comparing the fiber diameters estimated using the three mentioned values of the plane fractions of fibers. First, the mean values of fiber diameters were estimated as precisely as possible by counting the fibers along a gauge line[5] on three selected micrographs (taken at possibly perfect specimen portions) of each series at high magnification (1000 x). The volume fraction of each microstructure constituent (fibers, matrix and pores) was estimated by point analysis on the same micrographs. From the obtained data, the mean fiber diameter was calculated and its value was the same for all series (4.8 μm ± 0.05 μm).

Table 1. Bulk Density, Volume Fraction of Voids, and Plane Fraction of Fibers Measured by Various Methods

Series	Bulk density ($g.cm^{-3}$)	Volume fraction of voids (%)			Plane fraction of fibers (%)		
		a	*c*	*d*	*b*	*c*	*d*
C1	1.615	4.3	9.8	9.2	64.8	55.0	45.4
C2	1.616	3.5	9.2	9.1	73.0	61.0	49.0
G1	1.646	6.1	9.2	8.4	68.0	62.3	70.0
G2	1.669	3.4	8.4	7.5	77.0	66.2	73.0

a - water penetration (ASTM C-20)
b - estimated from the amount of fibers inserted into the mold and the specimen cross-section
c - point analysis
d - image analysis

Table 2. Results of Calculation of Microstructural Parameters from Plane Fractions of Fibers Obtained by Three Methods (Table 1)

Series	Mean fiber diameter (μm)			Mean linear fiber spacing (μm)			Mean thickness of inter-fiber matrix layer (μm)
	b	*c*	*d*	*b*	*c*	*d*	*c*
C1	4.8	4.9	3.4	2.6	3.4	4.0	2.6
C2	5.3	4.6	3.5	2.0	2.9	3.8	2.4
G1	5.0	4.8	5.2	3.0	2.9	2.4	2.2
G2	5.4	4.8	5.2	1.7	2.4	1.9	1.8

b - estimated from the amount of fibers inserted into the mold and the specimen cross-section
c - point analysis
d - image analysis

Counting the fibers was repeatedly performed on a larger number of more representative micrographs (magnification 500 x) and from these data the fiber diameter and mean linear fiber spacing were estimated as given in Table 2. In doing that, values of plane fraction of fibers estimated by the methods *b, c* and *d* were used.

By comparing the fiber diameters (Table 2) and assuming as the most correct the value 4.8 μm obtained by the above mentioned special measurement, the point analysis[4] method seems to be the most reliable method. The method *b* based on a known number of fiber tows inserted into the mold, on their specific weight (0.231 $g.m^{-1}$) and on the fiber density (1.81 $g.cm^{-3}$ for the T-800 fiber as given in the supplier's data sheet) is negatively influenced by the uncertainty of actual number of filaments in the tow.

The image analysis results, *d*, cannot be taken as reliable either. In the case of Series C specimens, they are underestimated; in Series G, overestimated. The results can be improved by further tuning of the image processing procedure (thresholding of the images of the Series G specimens taken in a polarized light). Therefore, they should be considered as preliminary results only.

The mean linear fiber spacing was also calculated from the data obtained by counting the fibers along a gauge line. The obtained differences in its values (Table 2) reflect the differences in determination of the volume fraction of fibers. From these results it follows that the graphitized specimens are more densely packed. To find out whether the differences between carbonized and graphitized specimens are not caused by changes of volume fraction of voids only, the mean thickness of matrix layer between adjacent fibers was calculated by subtracting the volume fraction of voids as determined from the volume fraction of fibers obtained by point analysis (last column in Table 2). This calculation, together with the previous determination of fiber diameter, revealed that the matrix itself shrinks during the high temperature treatment.

Typical load vs. displacement plots taken during three-point strength tests for both types (Series C and G) are schematically presented in Figs. 3a and 3b. The macroscopic views of the corresponding fracture surfaces are given in Figs. 3c and 3d. The course of the plot and the flat morphology of the fracture surface in the case of Series C (Figs. 3 and 3c) give evidence on the brittle character of the failure. Their loading curve terminates on the X-axis, which means that no laminar fracture occurred in this state. The high temperature treated samples (Series G), on the other hand, show at the same magnification a much rougher fracture morphology with large amounts of fibers protruding out of the major fracture plane (Figs. 3b and 3d). The failure of individual fiber bundles and/or layers and the fiber pullout are evident here. The descendent branch of the loading curve proves extensive delamination in the compressed part of the bent specimen.

At higher magnifications, Figs. 4 and 5, the extent of the fiber pullout in untreated (Fig. 4) and treated (Fig. 5) specimens, on Fig. 4 also the delamination and intra-bundle pore form are visible. Fig. 6 documents the longitudinally grooved surface of fibers and the micro-mechanisms of the fiber matrix debonding as well as the matrix fracture. On Figs. 7 and 8 the micropattern on the fiber fracture surfaces in untreated (Fig. 7) and heat treated (Fig. 8) states are visible together with the micropatterns of the matrix fracture.

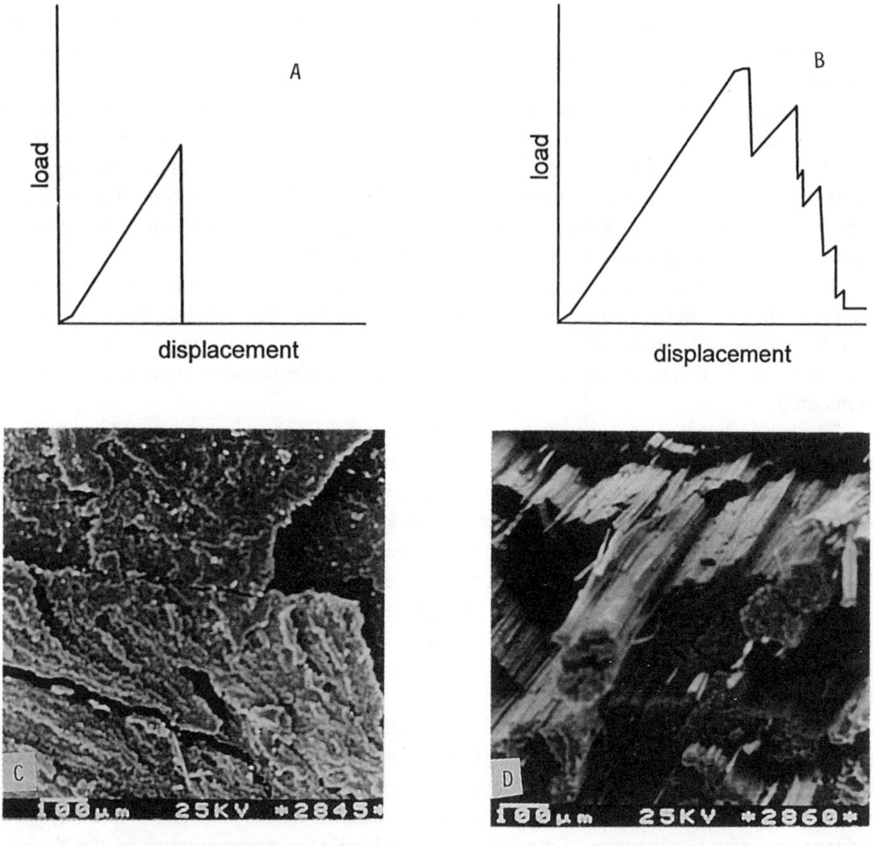

Fig. 3. Typical load vs. displacement plots for Series C (a) and G (b) and macroscopic views of the corresponding fracture surfaces of Series C (c) and Series G (d).

Table 3. Results of Flexural Tests

Series	Flexural strength (MPa)		Work to failure (N.mm)		Flexural modulus (GPa)	
	Mean value	St. Dev.	Mean value	St. Dev.	Mean value	St. Dev.
C1	456	28.1	18.0	2.3	164	12.6
C2	445	47.4	17.4	4.6	150	12.5
G1	683	31.6	31.4	1.6	170	12.0
G2	787	56.3	37.4	4.3	177	12.9

The results of the bending tests, Table 3, confirm the fractographic findings. It follows that the high temperature treated (G) specimens have much higher flexural strength and work to failure values than the untreated ones (C). The decision whether the differences in the flexural strength for batches 1 and 2 are significant or are caused by a natural scatter of values (due to the presence of voids) would be possible only on the basis of Weibull analysis,

Fig. 4. Fracture surface of specimen of Series C.

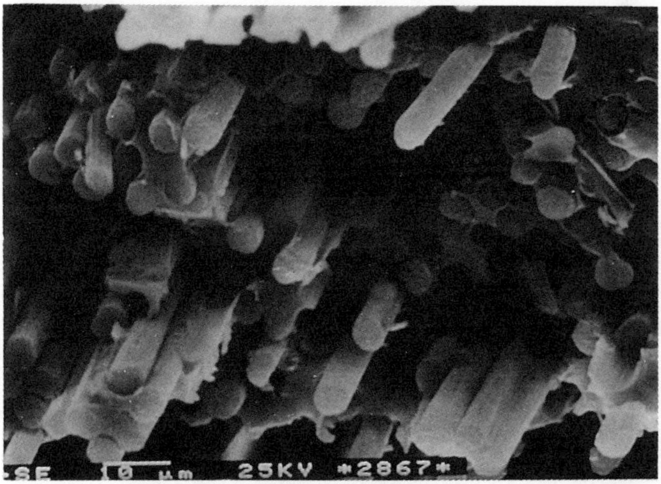

Fig. 5. Fracture surface of specimen of Series G.

but the number of specimens at our disposal did not enable us to carry it out. The values of flexural modulus are as expected; the differences between them probably represent normal scatter. The fiber/matrix debonding is facilitated by graphitization of intra-bundle matrix carbon and development of shrinkage cracks.

Fig. 6. Longitudinally grooved surface of fibers, micromechanisms of fiber/matrix debonding of fractured specimen of Series G.

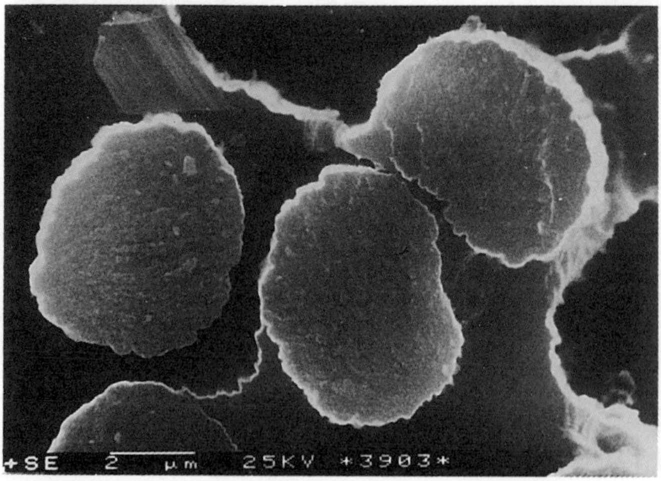

Fig. 7. Micropattern of the fiber and matrix fracture surface in specimen of Series C.

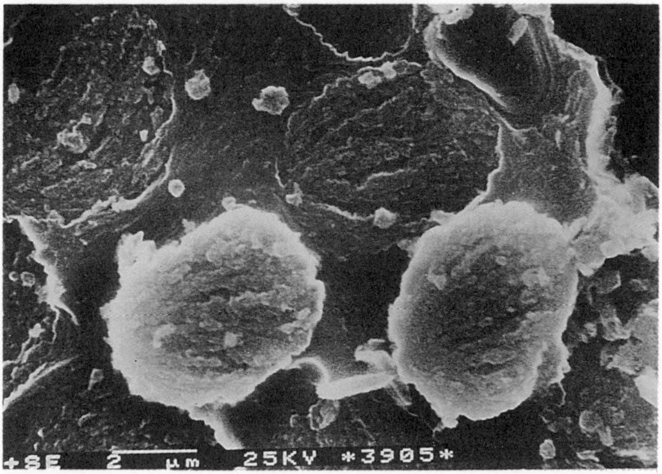

Fig. 8. Micropattern of the fiber and matrix fracture surface in specimen of Series G.

4. Conclusions

A quantitative analysis of microstructural parameters of the studied composite specimens enabled us to compare the reliability of three employed methods. The image analysis method needs further tuning to reach the reliability of the point analysis. After the secondary heat treatment ("graphitization") the volume fraction of voids decreased, the plane fraction of fibers increased, and both mean linear fiber spacing and mean thickness of the inter-fiber matrix layer decreased. The fiber/matrix debonding was facilitated by the heat treatment induced microstructural changes (graphitization of the matrix carbon and development of shrinkage cracks) which lead to an increase of strength and of work to failure in a three-point flexural test.

5. Acknowledgment

The study was partly supported by the Grant Agency of the Czech Republic within the project No. 246405 and partly by the Grant Agency of the Slovak Academy of Sciences within the project No. 2/1314/94.

6. References

1. B. Rand, in *Essentials of Carbon-Carbon Composites,* ed. J. C. R. Thomas (Royal Society of Chemistry, 1993), p. 67.
2. K. Balík, P. Glogar, and A. Tomanová, *CFI-Beihefte* **9** (1994) 200.
3. K. Balík, P. Glogar, and I. Sýkorová, *Proceedings of CARBON'94,* Granada, (1994) 746.
4. H. E. Exner et al., *Practical Metallography* **30** (1993) 216.
5. H. E. Exner et al., *Practical Metallography* **30** (1993) 322.

WEAR MECHANISMS IN CARBON-CARBON COMPOSITE AIRCRAFT BRAKES

T. J. HUTTON and B. MCENANEY
School of Materials Science,
University of Bath, Bath, BA2 7AY, UK

ABSTRACT

A laboratory scale dynamometer was used to produce low and high energy wear surfaces on a PAN/CVI carbon-carbon composite brake material. A larger dynamometer was used to produce wear debris under simulated taxiing and landing conditions and a full scale dynamometer was used to produce wear debris under simulated rejected take off (RTO) conditions. The wear surfaces and wear debris were examined using OM, SEM, and XRD. Abraded material, (type I surface debris) was found on low energy wear surfaces and film material (type II surface debris) was found on high energy surfaces. Wear debris produced under simulated taxi conditions consisted of small equiaxed abraded particles < 5 μm in size; under RTO conditions, wear debris consisted of flake-like particles 5–10 μm thick; under simulated landing conditions, the debris consisted of both abraded and flake-like material. XRD of the debris produced under taxi conditions revealed that much of the graphitic order in the parent composite had been destroyed. However, some order had been restored in the debris produced under RTO conditions as a result of stress graphitization. XRD of the debris produced under landing conditions showed the presence of an ordered and a disordered component. The XRD spectrum of the ordered component was similar to that of the base composite.

1. Introduction

Carbon-carbon composite aircraft brakes were initially developed in response to demand from the aircraft industry for light brakes for supersonic and vertical take off aircraft[1]. Although such brakes have been used increasingly for over twenty years, the mechanisms of wear are still not firmly established. A regenerative film wear mechanism has been proposed[2] in which abraded surface debris is deformed into a friction film that adheres to the wear surface. The friction film eventually delaminates, and then is crushed and recycled into another friction film. This is an imperfect process and some surface debris escapes from the brake. Very little is known either about the microstructural changes to the carbon friction materials during abrasion, friction film formation, delamination and repair, or about the influence of the various aircraft braking operations upon these processes. This paper reports a study of wear surfaces and wear debris produced after testing a PAN/CVI carbon-carbon composite aircraft brake material under high and low energy conditions. Scanning electron microscopy was used to identify the salient features of the wear surfaces and the wear debris. An X-ray diffraction study of the wear debris produced from various simulated aircraft braking operations was also carried out. In this paper a distinction is made between 'surface debris' which is examined on a wear surface and 'wear debris' which is examined after removal from the wear surface.

2. Experimental

2.1. Laboratory Scale Dynamometer Testing

A laboratory scale dynamometer was used to produce the wear surfaces. Testing was carried out using two circular samples (one rotor and one stator) with a raised annulus on each which created the wear surfaces. The annulus had an internal and external diameter of 25.4 mm and 50.8 mm respectively, creating a surface with a nominal contact area of 1520 mm^2 per sample. Experiments were carried out at 1800 rpm and 8000 rpm, corresponding approximately to speeds of rotation of friction surfaces in taxiing and landing operations. The samples were weighed and measured before testing under a constant torque of 3.4 Nm. The duration of experiments was 40 minutes and consisted of 20 cycles of 1 minute 'on' followed by 1 minute 'off'. The wear surfaces produced were examined in a JEOL 6310 SEM. There was not enough wear debris produced for XRD and SEM in these experiments, due to the small size of the samples used in this laboratory scale dynamometer.

2.2. Intermediate and Full Scale Dynamometer Testing

An inertial dynamometer connected to a three-disc aircraft brake (one rotor and two stators) was used to produce samples of wear debris under simulated aircraft braking conditions. The dynamometer test programs were designed to simulate 50 cold taxi cycles and 50 landing cycles. A full size dynamometer connected to a full-size aircraft brake was used to produce wear debris from a simulated rejected take off (RTO). The taxiing and landing wear debris were collected from beneath the brake, however, the RTO debris was collected by gently brushing the wear surfaces after dismantling the brake. The three different types of wear debris were examined in a JEOL 6310 scanning electron microscope and analyzed without any pre-treatment using a Phillips PW1710 X-ray diffractometer. The X-ray diffraction patterns were corrected for machine line broadening as well as Lorentz, polarization and dispersion effects. The curves were then analyzed to obtain values of $d_{(002)}$, L_a, and L_c. Values of L_a and L_c are sensitive to lattice imperfections which decrease the observed crystallite size. However, these parameters provide a semi-quantitative base on which to compare the wear debris samples.

3. Results and Discussion

3.1. SEM of Wear Surfaces

The wear surfaces produced after testing at 1800 rpm on the laboratory scale dynamometer appeared dull and unpolished to the eye, whereas those tested at 8000 rpm appeared smooth and highly polished. SEM of the wear surfaces produced at 1800 rpm revealed a roughened surface characterized by loosely attached debris, less than 5 μm in size, Fig. 1. This is a typical abraded, or type I surface debris produced under low energy conditions and it is similar to type I surface debris obtained by Murdie et al.[3] from a carbon-carbon composite with mesophase pitch-based fibers and a phenolic resin/CVI matrix. Low energy wear of carbon-carbon composites that produces type I surface debris also results in high wear rates[2–4]. Under low energy conditions there is insufficient energy to deform the abraded surface debris into friction film, or type II surface debris. Friction film surface debris can be seen on wear surfaces tested at 8000 rpm, Fig. 2. This is similar in morphology to the type II surface debris found by Murdie et al.[3] Fig. 2 shows a fragment of friction film material that has broken away from the main body of the film on the surface. If testing had continued, it is likely that the fragment would have been crushed to form smaller particles of surface debris, as expected from the regenerative friction film model for wear. The wear rates for these higher energy braking operations, when corrected for sliding distance, are much lower than those for low energy braking operations[4], which is thought to be due to recycling of wear debris at the friction interface.

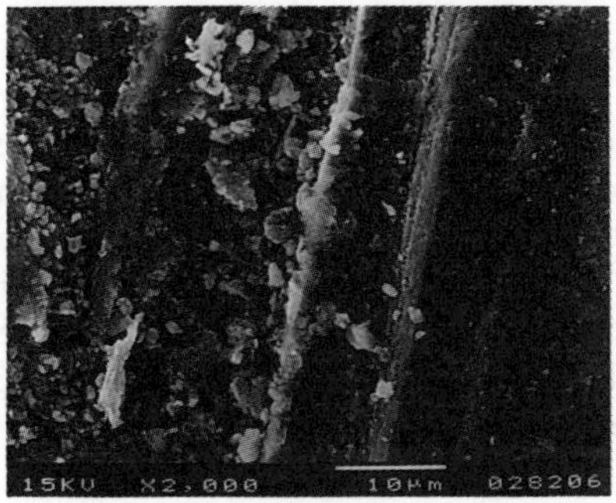

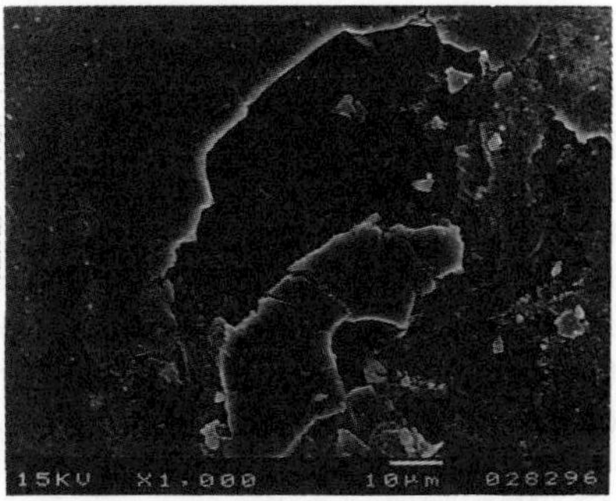

Scanning electron micrographs of wear surfaces: Fig. 1, produced at 1800 rpm; Fig. 2, produced at 8000 rpm.

3.2. *SEM of Wear Debris*

Scanning electron micrographs of the taxiing debris, landing debris, and RTO debris are shown in Figs. 3, 4, and 5, respectively. Taxi wear debris, Fig. 3, has a very similar morphology to the type I surface debris identified on wear surfaces produced under low energy conditions, Fig. 1, consisting mainly of equiaxed particles less than 5 µm in size. The small

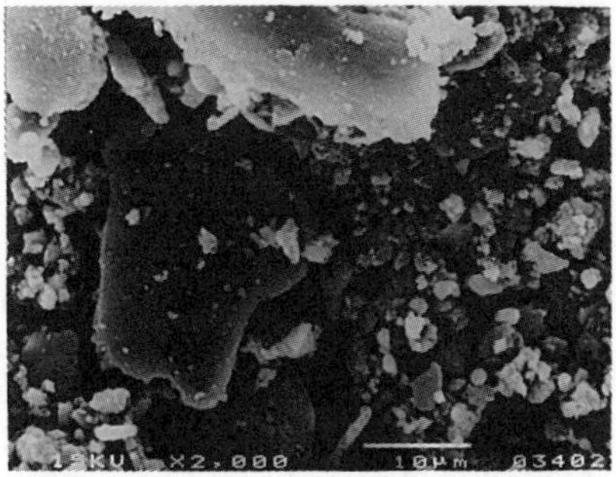

Scanning electron micrographs of wear debris: Fig. 3, produced under simulated taxiing conditions; Fig. 4, produced under simulated landing conditions.

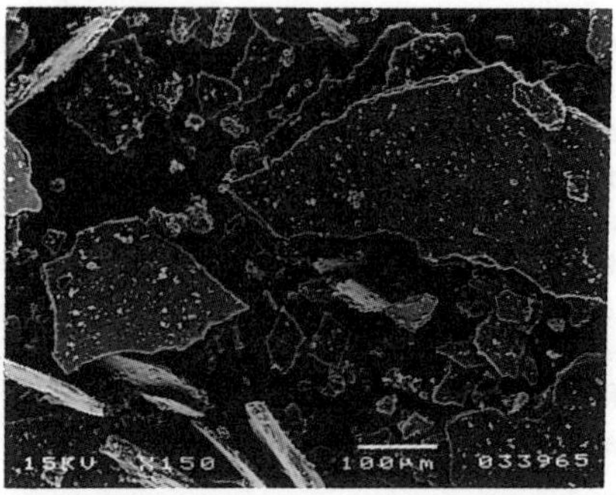

Fig. 5. Scanning electron micrograph of wear debris produced under simulated RTO conditions.

particle size and the correspondingly high surface energy lead to agglomeration of the particles. In contrast, RTO wear debris, Fig. 5, consists of flat, plate-like particles, 5–10 μm thick and sometimes over 1 mm wide. These large flakes are similar in morphology to type II surface debris formed on wear surfaces, Fig. 2. RTO debris was formed at the wear surface under extreme conditions, involving a power density of 1.5–2 MWm^{-2} and interface temperatures in excess of 2000°C[2]. The wear debris produced under simulated landing conditions, Fig. 4, shows evidence for both particulate and flake-like morphologies.

3.3. *XRD of Wear Debris*

XRD spectra for the base composite and three types of wear debris are shown in Fig. 6; values of interlayer spacing and crystallite size are shown in Table 1. The parent material has an average interlayer spacing of 3.38 Å and average crystallite sizes (L_a and L_c) of approximately 200 Å. This suggests that the X-ray spectrum of the composite is dominated by the graphitic character of the CVD matrix, rather than the less graphitic structure of the PAN-based fibers.

Comparison of Figs. 6(a) and 6(b) shows qualitatively that the wear debris obtained after simulated cold taxi conditions is much more disordered than the base composite with an interlayer spacing of 3.43 Å and values of L_a and L_c of 34 Å and 64 Å, respectively, Table 1. This indicates that as a result of the taxi cycles the graphitic order that was present in the base composite is almost completely destroyed in the wear debris. The extent of disruption of the graphitic order in the base composite was of a similar order to that reported in a study of the effects of grinding upon the structure of a natural graphite[5]. Cold taxi conditions are thought to cause abrasion at the wear surface which disrupts the graphitic structure of the parent composite and produces wear debris on a very fine scale (extending to less than 0.1 μm). The low power density and wear surface temperature associated with the cold taxiing (180 kWm^{-2} of interface and ~ 100°C[2]) limit the ability of the wear debris to deform plastically and to produce a friction film, resulting in high rates of wear[3,4].

Table 1. Interlayer Spacings and Crystallite Sizes for the Base Composite and Wear Debris

Material	$d_{(002)}$ (Å)	L_a (Å)	L_c (Å)
Base Composite	3.38	200	200
Cold Taxi Wear Debris	3.43	34	64
Landing Wear Debris [a]	3.38	-	200
Landing Wear Debris [b]	-	37	10
Rejected Take Off Debris	3.39	150	65

a: ordered component; b: disordered component

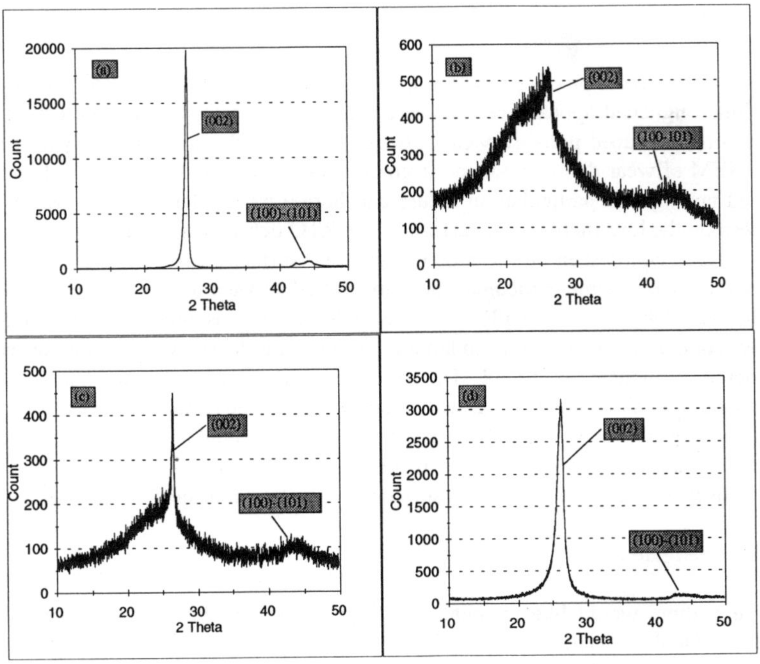

Fig. 6. X-ray spectra for (a) the base composite, (b) cold taxi wear debris, (c) landing debris, (d) RTO debris.

The X-ray diffraction pattern for wear debris obtained after 50 simulated landings contains a sharp graphitic peak and an amorphous halo at the 2θ value associated with the (002) interlayer spacing, Fig. 6(c), indicating the presence of two components in this material. Deconvolution of the peaks enabled the L_c value of the ordered component to be determined; the L_a value could not be determined as it is obscured by the broad peak at $2\theta \sim 42^\circ$. The L_c value suggests that the ordered component has a graphitic character that is similar to that of the parent composite. This may indicate that the wear debris obtained after simulated landing conditions contains fragments that retain elements of the microstructure of the base composite. The disordered component has a degree of graphitic disorder that is comparable to that produced after the simulated taxi cycles.

The X-ray diffraction pattern for the wear debris obtained after a simulated RTO, Fig. 6(d), indicates the presence of a single component in which the degree of graphitic order is greater than that for the wear debris obtained after simulated cold taxiing, but not as great as the parent material, Table 1. The high rate of energy conversion and high interface temperatures, together with the high shear forces produce conditions favorable for shear-stress assisted graphitization. In a recent study, Bustin et al.[6] have shown that shear-stress assisted graphitization of an anthracite coal can occur at temperatures in the range 600–900°C.

4. Conclusions

The wear surfaces produced by low speed testing on the laboratory scale dynamometer had a dull roughened appearance and were characterized by type I surface debris. High speed testing produced polished wear surfaces that were characterized by type II surface debris. SEM of wear debris produced from carbon-carbon composite aircraft brakes after simulated taxiing was particulate in nature and bore a close relation to the type I surface debris seen on the low energy wear surfaces. The RTO debris was flake-like and resembled type II surface debris seen on the high energy wear surfaces. SEM of the landing debris showed it to contain some particulate and some flake-like wear debris.

The wear debris produced after simulated cold taxiing was much more disordered than the parent composite. After simulated landing cycles, wear debris containing an ordered and a disordered component was found. The structure of the disordered component was similar to that found after simulated taxiing, while that of the ordered component was similar to the structure of the parent composite. Wear debris found after a simulated rejected off was more ordered than that found after taxiing. This ordering was attributed to stress graphitization at the wear surface as a result of the high surface temperature and shear stresses.

5. Acknowledgments

The authors would like to thank the Engineering and Physical Sciences Research Council and Dunlop Aviation Division for their financial support, and Dr. R. Fisher and Mr. D. Johnson for useful discussions.

6. References

1. I. L. Stimson and R. Fisher, *Phil. Trans. Roy. Soc. Lond.* **A294**, 583 (1980).
2. S. Awasthi and J. L. Wood, *Advanced Ceramic Materials* **3**(5) 449 (1988).
3. N. Murdie, C. P. Ju, J. Don and F. A Fortunato, *Carbon* **29**(3) 335 (1991).
4. T. J. Hutton and B. McEnaney, unpublished work.
5. J. B. Aladekomo and R .H. Bragg, *Carbon* **28**(6) 897 (1990).
6. R. M. Bustin, J-N. Rouzaud and J. V. Ross, *Carbon* **33**(5) 679 (1995).

APPLICATION OF IODINE TREATMENT ON FABRICATION OF PITCH BASED C/C COMPOSITES

E. YASUDA, H. KAJIURA*, and Y. TANABE
Research Laboratory of Engineering Materials, Tokyo Institute of Technology
Nagatsuta, Midori, Yokohama, 226 Japan
**Research Fellow of the Japan Society for the Promotion of Science*

A. KAIHO and I. SHIOTA
Department of Chemical Engineering, Kogakuin University
Nakano, Hachioji, Tokyo, 192 Japan

ABSTRACT

The authors investigated the effect of iodine treatment on fabrication of pitch based C/C composites. Coal tar pitch having the softening point of 101°C was used as a matrix precursor in this research. The specimen, $6\phi x 20 mm^3$ in size and $V_f = 43\%$, was put into a test tube with iodine powder. After sealing the tube, the specimen was heated at 90°C for iodine treatment. The iodine treated specimens were heated at a rate of 2.5°C/min up to 800°C in Ar flow for carbonization. Carbon yield of the specimen was increased from 73% to 93% after 20 hours of iodine treatment. For the 20 hour treated specimen, the matrix pitch did not flow out from the fiber bundle after carbonization. Optical microstructure of the matrix in the carbonized specimens changed from a flow type texture to a mosaic one with the increasing of the iodine treatment period.

1. Introduction

Carbon/carbon composites (C/C composites) are expected to be used under critical conditions where other materials cannot serve. Pitch is widely used as a matrix precursor of C/C composites. In the process of C/C composite production, however, impregnation and carbonization processes are repeated to increase their density because the carbon yield is limited and the matrix can flow out from the fiber bundle during heat treatment. Carbonization under high pressure has also been used to increase the carbon yield. Some researchers[1–4] have attempted to change the properties of pitch to make it suitable for fabrication of either carbon fiber or pitch based C/C composites.

Preoxidation treatment was reported to be an effective process to change the properties of pitch such as softening point. J. B. Barr et al.[2] reported that air blowing oxidation made the softening point of pitch higher as a result of dehydrogenative polymerization supported by oxygen. J. L. White et al.[3] carried out the oxidation process to stabilize the pitch matrix of C/C composites. C. Yamaguchi et al.[4] proposed the mechanism of the air blowing reaction. They reported that the reaction occurred through the radical species.

Some papers indicate that iodine doping is an effective method for coal or pitch characterization[5–7] because iodine forms charge transfer complexes with components in pitch and creates radicals whose concentration reflects the structure of components in the pitch. The authors believed that the pitch was stabilized by iodine because it is an oxidizing agent. In a previous paper[8], the effect of iodine treatment on the carbonization behavior of coal tar pitch was reported. The authors applied this iodine treatment to fabricating pitch based C/C composite. Considering the results of the previous paper, it is expected that the treatment will increase the carbon yield and change the microstructure of the carbonized specimen. We studied the: (1) effect of iodine treatment, and (2) effect of a preoxidation treatment, which was reported to be an effective process to stabilize pitch, along with the iodine treatment.

2. Experimental Procedure

Coal tar pitch supplied by Nippon Steel Chemical Corporation was used as a matrix precursor. Characteristics of the pitch are shown in Table 1. PAN based M40B fiber, supplied by Toray Industries, Inc., was used as a reinforcement fiber. Before impregnation of pitch, the fiber was heated at 200°C for 5 hours in flowing air to remove the sizing agent. The approximate 1% weight reduction via this treatment indicates that the sizing agent was almost entirely removed. After removing the sizing agent, the fiber was put in a glass tube 6mm in diameter and 20mm in length. The fiber orientation was parallel to the longitudinal orientation of the glass tube. The fiber was immersed in fused pitch for 1 hour, and the specimen was subsequently subjected to an autoclave treatment under 1MPa at 250°C for 1 hour. After impregnation, the specimen was removed from the glass tube.

Table 1. Characteristics of Coal Tar Pitch Used

Softening Point	101°C
Fixed Carbon	53.3%
Toluene Insoluble	14.6%
Quinoline Insoluble	0.59%
Ash	0.02%
Sulfur	0.50%

Iodine treatment was carried out at 90°C to prevent melting of the specimen. One part of the CF/pitch specimen with two parts of iodine powder by weight were put into a test tube and separated from each other by carbon felt. After sealing, the test tube was put into an oven and heated at 90°C for times varying between 1 and 20 hours. Preoxidation treatment to stabilize the pitch in the iodine treated specimen was carried out in an electric furnace at 170°C for 3 hours in an air flow. Since the exhaust gas seemed to contain a gas phase of iodine, it was passed through xylene to remove the iodine. Then the specimen was heated at a rate of 2.5°C/min up to 800°C in flowing Ar for carbonization. The duration at 800°C was 1 hour. The carbon yield was calculated from the weight of the original CF/pitch specimen.

Iodine distribution in the specimen was investigated with an electron probe X-ray microanalyzer (Shimadzu, EPMA-1400). The microstructure in the carbonized specimens was examined in their polished section under a polarized light microscope. Gas mass spectrum analysis was carried out to study reactions occurring during heat treatment by using TG-DTA/MS (Rigaku Thermoflex TAS200, Hewlett Packard 5971A Mass Selective Detector).

3. Results and Discussion

Fig. 1 shows the change in weight of the specimen after iodine treatment, and carbon yield of the specimen carbonized at 800°C as a function of iodine treatment period. The weight increased in rough proportion to the treatment period. For the 20 hour iodine treated specimen, a weight increase of about 60% was obtained. The iodine treated specimen was cut into two parts through the center of the specimen. The cut face, normal to the longitudinal direction of the fiber axis, was examined with EPMA to determine the distribution of iodine. The results are shown in Fig. 2. For the 3 hours iodine treated specimen, iodine was detected primarily in the outer region. Meanwhile, in the 20 hour iodine treated specimen, iodine was found throughout the specimen. From this result, we concluded that iodine diffuses from the surface to the inside with time. Although there were no pores before the iodine treatment, some pores were observed in the iodine treated specimen. There appeared to be some micropores in the specimen. The pores might expand during iodine treatment and some pores were observed after the treatment.

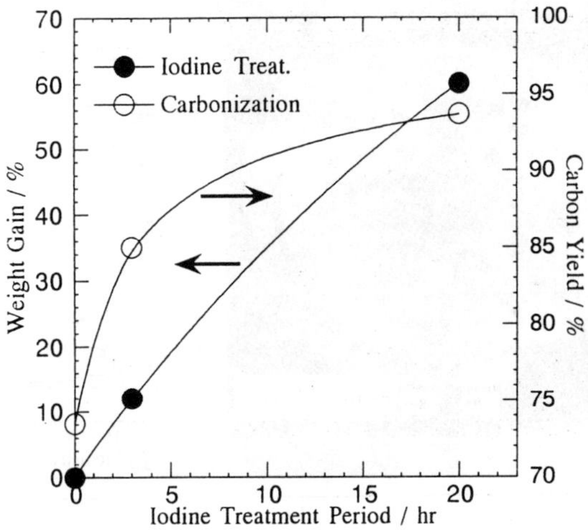

Fig. 1. Weight change of specimens after iodine treatment and the carbon yield as a function of the iodine treatment period.

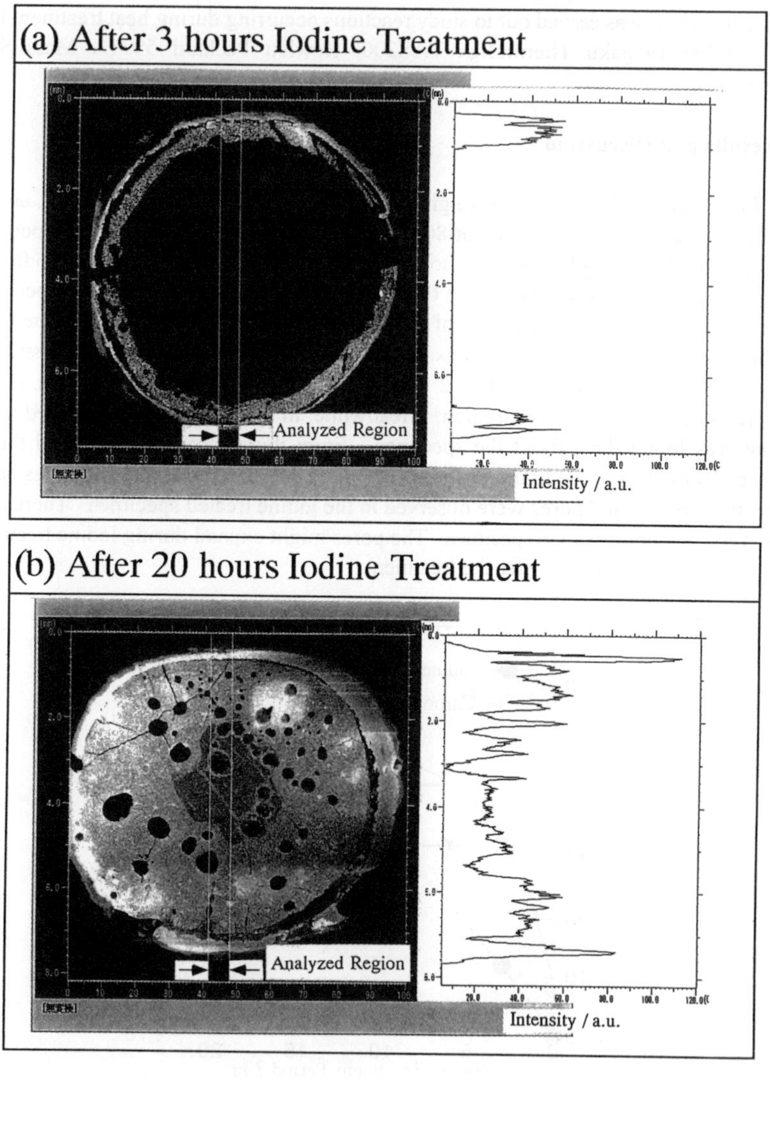

Fig. 2. Microstructure and iodine distribution of the iodine treated specimen.

As shown in Fig. 1, the carbon yield of the specimen increased with an increase in the iodine treatment period; 93% carbon yield was obtained for the 20 hour iodine treatment. From this result, the iodine treatment is considered to be an effective process to increase the carbon yield of the pitch.

Fig. 3 indicates the specimen's shape at each treatment. As shown in this figure, a little expansion could be seen in the specimens after iodine treatment. This expansion was considered to be due to pores that were discussed previously. After carbonization, it was observed that the matrix flowed out from the fiber bundle for non iodine treated specimen and for the 3 hour iodine treated specimen. For the 20 hour iodine treated specimen, the matrix didn't flow out and the shape of the specimen was maintained during carbonization.

Fig. 4 indicates the microstructures of the carbonized specimens. Through observation, microstructural change in the carbon fiber was not observed. However, the matrix microstructure was obviously changed by iodine treatment. As shown in the photographs, the flow type texture was observed for the non iodine treated specimen, whereas the mosaic type texture was observed in the iodine treated specimens and it seemed to be finer with increasing iodine treatment period.

Fig. 5 shows the results of gas mass spectrum analysis and weight change of the specimen as a function of temperature. (a), (b) and (c) indicate total detected gas, m/e=254 and m/e=128 intensities of which m/e corresponded to the molecular weight of I_2 and HI, respectively. Here, the analysis was carried out on the 20 hour iodine treated specimen. m/e = 254 was detected over the entire measured temperature range. m/e = 128 was detected only from 220 to 450°C. This indicates that some part of the introduced iodine volatilized as the HI molecule from 220 to 450°C. The hydrogen atom in the HI molecule comes from the components in the pitch. It was conceived that the hydrogen loss from the carbon precursor produces radical species. The polymerization reaction that occurred between components in the pitch is considered to be accelerated by the radical species and results in an increase of molecular weight of the components. Here, solvent fractionation indicated the increase of molecular weight of the component in the pitch[8]. The viscosity of liquid phase which appeared during heat treatment was increased due to the increased molecular weight. Molecular movement was suppressed in the high viscosity liquid phase. The mosaic texture is considered to be caused by the suppression of component stacking in the high viscosity liquid.

In Fig. 6, weight change of the specimen during iodine treatment and preoxidation treatments after the iodine treatment is shown as a function of iodine treatment period, together with the carbon yield. From this figure, the weight of the preoxidized specimen was less than that of the iodine treated specimen. Considering the results of gas analysis shown in Fig. 5, the weight reduction of the specimen during preoxidation treatment was mainly due to volatilization of I_2 from the specimen. However, about one fourth of introduced iodine remained after the treatment. The carbon yield of the specimen increased with the increased iodine treatment period; a 95% carbon yield was obtained after 20 hours of iodine treatment. Comparing Figs. 6 and 1, the carbon yield of the specimens carbonized with and without preoxidation treatment were almost the same. From this result, the iodine treatment is considered to be a remarkable process to increase the carbon yield of this pitch.

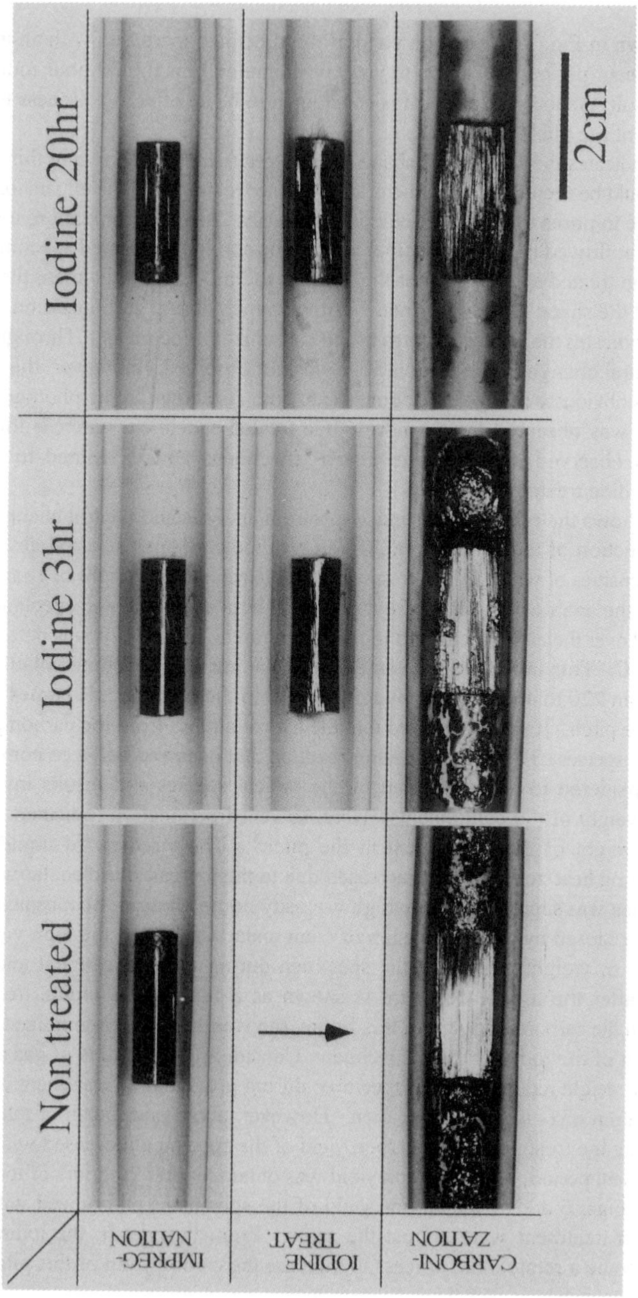

Fig. 3. Specimen's shape at each treatment.

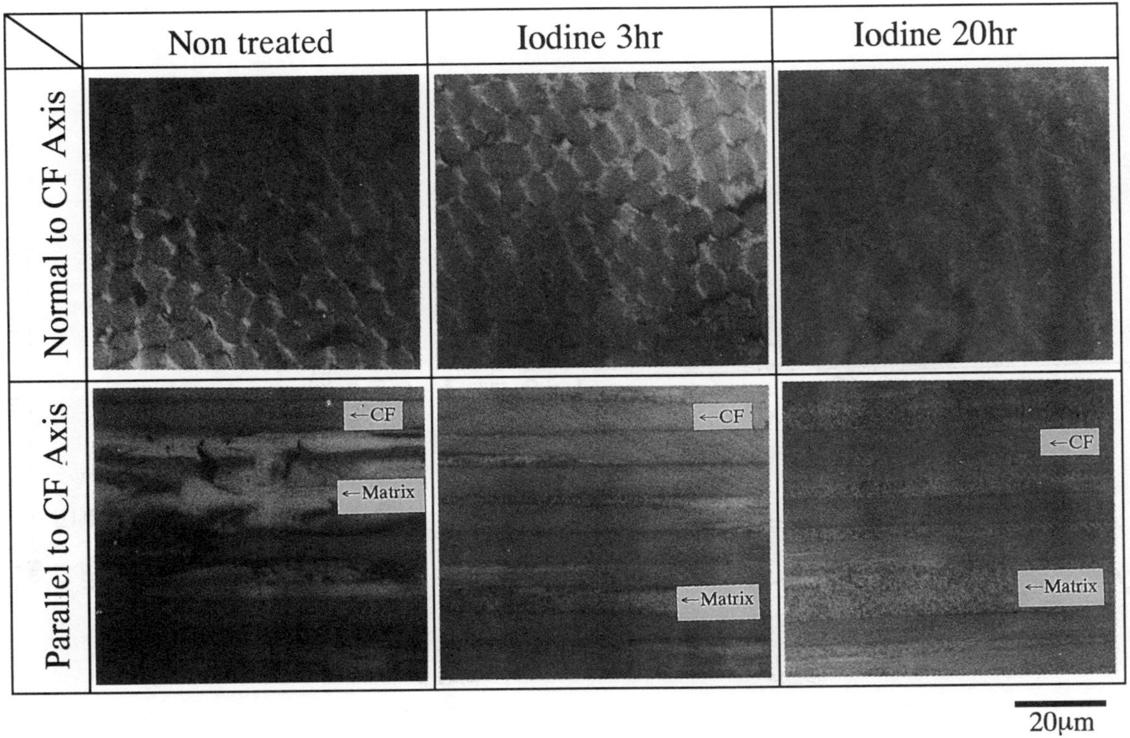

Fig. 4. Microstructure of carbonized specimens.

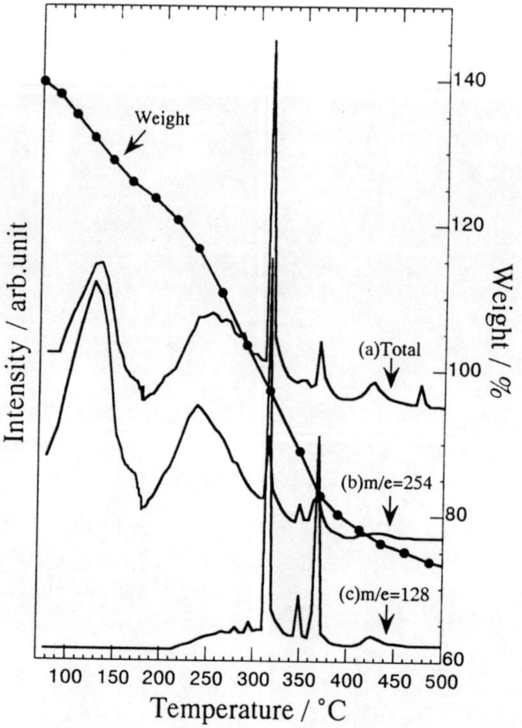

Fig. 5. The gas mass spectrum and weight change of the 20 hours iodine treated specimen as a function of temperature.

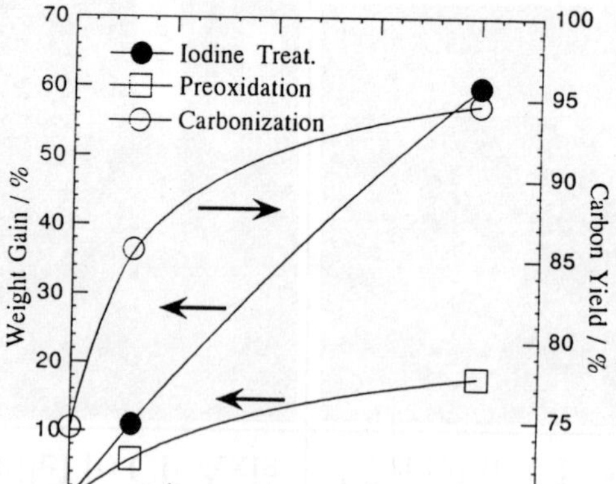

Fig. 6. Weight change of specimens after iodine and preoxidation treatment, together with the carbon yield as a function of iodine treatment period.

4. Conclusions

The authors studied the effect of the iodine treatment and the preoxidation treatment on the fabrication of pitch based C/C composite. Through experiments, the iodine treatment was considered to be a remarkable process to control the properties of this coal tar pitch.

Our conclusions are as follows:

- The carbon yield of the coal tar pitch was increased by the iodine treatment.
- For the 20 hour iodine treated specimen, the matrix didn't flow out from the fiber bundle, and the specimen's shape was maintained after carbonization.
- The microstructure of matrix was changed from flow type texture to mosaic one with increasing iodine treatment period.
- These changes noted above are considered to be mainly caused by dehydrogenative polymerization promoted by iodine.

5. Acknowledgments

One of the authors, H. Kajiura, is indebted to the Ministry of Education, Science and Culture of Japan for financial support by Grant-in-Aid for JSPS Fellows No. 2339. The authors are grateful to Mr. T. Senda (Rigaku Corporation) for gas mass spectrum analysis. Several helpful discussions with Prof. S. Yamada (Nishi Tokyo University) are gratefully acknowledged.

6. References

1. S. Otani, *Carbon* **3** (1965) 31.
2. J. B.Barr and I. C. Lewis, *Carbon* **16** (1978) 439.
3. J. L. White and P. M.S heaffer, *Carbon* **27** (1989) 697.
4. C. Yamaguchi, J. Mondori, A. Matsumoto, H. Honma, H. Kumagai and Y. Sanada, *Carbon* **33** (1995) 193.
5. T. Yokono, T. Imamura and Y. Sanada, *Sekiyu Gakkaislti* **30** (1987) 59, in Japanese.
6. H. de Lapes, T. Tokono, K. Murakami, Y. Sanada and H. Marsh, *FUEL* **66** (1987) 866.
7. M. Sasaki and Y. Sanada, *Sekiyu Gakkaislti* **34** (1991) 218.
8. E. Yasuda, H. Kajiura and Y. Tanabe, *TANSO*, **170** (1995) 286.

Round Table Discussion

of

Worldwide Carbon Research

CARBON RESEARCH IN THE CZECH REPUBLIC[1]

P. GLOGAR
Institute of Rock Structure and Mechanics
Academy of Sciences of the Czech Republic,
V Holešovičkách 41, 182 09 Prague 8, Czech Republic

1. Introduction

At present, there is no industry producing advanced carbon materials in the Czech Republic. Carbon research is not stimulated by a need to improve current products and is therefore carried out only at the academic level (at universities and in research institutes). Carbon research is sponsored almost exclusively by one of two existing grant agencies, the Grant Agency of the Czech Republic (GACR) and the Grant Agency of the Academy of Sciences of the Czech Republic (GAASCR).

The main research areas at present include thin layers, composites, biomaterials, carbon separation membranes, electrochemical carbon, carbon electrodes, fullerenes and coal tar pitch.

2. Thin Layers

2.1. *Study of Ultra-thin Films and Their Systems Prepared by rf Magnetron Sputtering*

Principal investigator: J. Hrdina, Institute of Physics, ASCR[1], Prague, phone (+422)66052959, grant project No. 2010536 of the GAASCR[2]. In this project, nitrogenated carbon is prepared by reactive rf magnetron sputtering. Attempts are being made to prepare a hypothetical β-C_3N_4 phase whose superhardness is predicted to be comparable to that of diamond. The goal is to prepare wear resistant coatings.

2.2. *Study of Laser Deposition of Thin Layers*

Principal investigator: M. Jelínek, Institute of Physics, ASCR[1], Prague, phone (+422)66052733, grant project No. 02/93/0464 of the GACR[3]. This project involves thin films of amorphous or diamond-like carbon and doped amorphous C:N films that are created by laser ablation on fused silica, silicon, and steel substrates. Graphite or glass-like carbon targets are used as carbon sources. Findings include that mechanical properties, adhesion, electrical resistivity and optical gap depend considerably on deposition temperature and that room temperature yields the best diamond-like carbon films.

---

[1]This review reflects the author's personal opinion and presents data provided by individual investigators.

2.3. Diamond-like Carbon (DLC) Films

Principal investigator: P. Louda, Dept. of Material Science, Technical University Liberec, phone (+4248)26441, ext. 113. Low-pressure radio frequency plasma discharge is employed for synthesis of DLC films in this study. The films' mechanical properties (hardness, adhesion, elasticity and friction) are evaluated as is their biological tolerance (for medical applications). Thermal properties of the DLC films are also measured.

3. Composites

3.1. Manufacture and Properties of Carbon-carbon Composites

Principal investigator: K. Balík, Inst. of Rock Structure and Mechanics, ASCR[1], Prague, phone (+422)66009212, grant project No. 104/94/1789 of the GACR[3]. In this project, mechanical properties and structure of carbon composites reinforced by carbon fibers and/or fabrics are pursued. The influence of matrix precursor type and processing parameters (including those of impregnation by pitch or resin) is investigated. The effort aims to produce materials for medical applications such as bone replacements.

3.2. Mechanical Properties and Structure of CFRC Materials with Resin-based and Pitch-based Matrix

Principal investigator: P. Glogar, Inst. of Rock Structure and Mechanics, ASCR[1], Prague, phone (+422)66009339, grant project No. 246405 of the GAASCR[2]. In this project, unidirectional CFRC differing in their matrix precursor and processing parameters are investigated in terms of their void structure, strength, and failure behavior. Processing conditions leading to relatively strong materials which do not fail in a brittle mode are preferred.

3.3. The Theoretical Basis of a New Process for Carbon-carbon Composite Manufacturing

Principal investigator: M. Hájek, Inst. of Rock Structure and Mechanics, ASCR[1], Prague, grant project No. 106/93/0839 of the GACR[3]. This project aims to establish the principles of a single step process to manufacture CFRC with a coal tar pitch based carbon matrix. Studies focus on the solidification stage of the pitch pyrolysis from the point of view of its rheology, coking value and degree of anisotropy.

4. Biomaterials

4.1. New Technologies for Advanced Biomaterials

Principal investigator: P. Filip, Faculty of Metallurgical Engineering, Technical University Ostrava, phone (+4269)6993299, grant project No. 106/95/0480 of the GACR[3]. The goal of this project is to develop a fibrous composite (thin Ti/Ni fibers in a carbon matrix) for use as an implant material exploiting the shape memory of Ti/Ni for fixation. Mechanical properties of the carbon matrix are used to achieve the required toughness and modulus for the implant.

4.2. Prospective Biocompatible Structural Materials

Principal investigator: V. Starý, Dept. of Material Science, Faculty of Mechanical Engineering, Czech Technical University, Prague, phone (+422)24357226, grant project No. 106/95/0359 of the GACR [3]. This study involves a search for a biomaterial that is suitable for implants. Biocompatibility, suitability of mechanical properties and their change during *in vitro* treatment under conditions similar to those of the biological environment are considerations in this work. CFRC and CFRP (PEEK) composites are considered prospective candidates.

5. Carbon Separation Membranes

5.1. Catalytic and Separation Processes on the Microporous Carbon Membranes

Principal investigator: B. Bernauer, Dept. of Inorganic Chemistry, University of Chemical Technology, Prague, phone (+422)24354018, grant project No. 104/94/0596 of the GACR[3]. The project involves the development of a suitable carbon membrane with a high separation capacity prepared by carbonization of selected polymer precursors. A microporous structure exhibiting the properties of molecular sieves may be, up to a certain extent, controlled by the choice of the polymer precursors and of the preparation conditions.

6. Electrochemical Carbon

6.1. New Adsorbents for Chromatography Based on Electrochemical Carbon and its Derivatives

Principal investigator: L. Kavan, J. Heyrovský Institute of Phys. Chemistry, ASCR[1], Prague, phone (+422)66053975, grant project No. 203/93/0251 of the GACR[3]. Electrochemical carbonization of fluoropolymers, chemical modification of carbons, gas chromatography and HPLC are all topics of this study.

6.2. Optical and Electrical Properties of Electrochemical Carbon

Principal investigator: L. Kavan, J. Heyrovský Institute of Phys. Chemistry, ASCR[1], Prague, phone (+422)66053975, grant project No. D4/0004/94 EC-COST. Topics of this study include Raman spectroscopy, polyyne formation and degradation, electronic conductivity, doping and quantum chemical calculations.

7. Carbon Electrodes

7.1. Alkali Metal Insertion into the Carbon Electrodes

P. Krtil, J. Heyrovský Institute of Phys. Chemistry, ASCR[1], Prague, phone (+422) 66053826, has submitted an NSF grant application. Topics addressed in the application are carbon electrodes, aprotic electrolyte solutions and electrochemical quartz microbalance.

8. Fullerenes

8.1. Uptake of Hydrocarbon Vapors by Fullerene-60 Particles

Principal investigator: A. Zukal, J. Heyrovský Institute of Phys. Chemistry, ASCR[1], Prague, phone (+422)66053865, grant project No. 203/95/1228 of the GAASCR[2]. Topics of this project include adsorption, structural studies using X-ray diffraction and the mechanism of sorption.

9. Coal Tar Pitch

9.1. Molecular Transformations of Polyaromatics at the Solidification Stage in Carbonization

Principal investigator: L. Brož, Institute of Rock Structure and Mechanics, ASCR[1], Prague, grant project No. 203/94/1790 of the GACR[3]. As a basis for this study, a hypothetical path of polyaromatic mesophase formation from alternant polyaromatic hydrocarbons was postulated. Some real phenomena such as the free-spin concentration profile in the course of pitch carbonization, and the pregraphitic structure of the solidified mesophase, are predicted by the model.

9.2. A Study on Binder Pitch Properties, Namely Restriction of Benzpyrene Formation during Heat Treatment with Attention to Rheological Properties

Principal investigator: M. Kaloč, Faculty of Metallurgy and Material Engineering, Technical University Ostrava, phone (+4269)6995133.

10. Industrial Plants

There are two industrial plants in the Czech Republic connected with production and utilization of carbon, DEZA Co. and TESLA Vršovice.

10.1. DEZA Co.

DEZA Co., Valašské Meziříčí, is a specialized coal-tar processing plant founded in 1968. Approximately 200 kt/yr of raw coal tar is produced in the domestic metallurgical coke manufacture, yielding pure anthracene-based compounds and industrial mixtures such as impregnants, lacquers and binders. The carbon black production facilities now belong to Cabot Co. The basic line of the pitch-based production is a range of electrode binder pitches (for aluminum and steel making industry), refractory binders, soft carbon binders and road tars. The DEZA pitches contain a negligible amount of detrimental heavy metals (vanadium less than 0.1 ppm, total alkali content less than 250 ppm).

10.2. TESLA Vršovice

TESLA Vršovice is a production plant for radio-frequency power (transmitting) tubes with grids made of self-supporting pyrolytic carbon. The production procedure consists in the deposition of these layers (thickness several hundreds of micrometers) from butane onto a shaped graphite substrate and their safe lifting. The "hot-wall" method (resistance heating of the substrate) is employed. The semi-products are then laser machined to reach the desired grid structure.

The only full-scale plant producing standard carbon materials (mainly carbons for electric engines, brushes, etc.) in the former Czechoslovakia is "Elektrokarbon Topol' čany." It is situated in the Slovak Republic and is therefore not covered in this review.

11. Abbreviations and Addresses

1. ASCR - Academy of Sciences of the Czech Republic, Národní 3, CZ-111 42 Praha 1, phone (+422)24220384, fax (+422)24240572.
2. GAASCR - Grant Agency of the Academy of Sciences of the Czech Republic, Národní 3, CZ-111 42 Praha 1, phone (+4122)24240533, fax (+422)24240515.
3. GACR - Grant Agency of the Czech Republic, Národní 3, CZ-111 42 Praha 1, phone (+422)24240588, fax (+422)24240565

- Institute of Physics, ASCR,
  Na Slovance 2, CZ-180 40 Praha 8,
  phone (+422)66051111, fax (+422)821227
- Institute of Rock Structure and Mechanics, ASCR,
  V Holešovičkách 41, CZ-182 09 Praha 8,
  phone (+422)66009111, fax (+422)6880105, 6880649
- J. Heyrovský Institute of Physical Chemistry, ASCR,
  Dolejškova 3, CZ-182 23 Praha 8,
  phone (+422)66051111, fax (+422)8582307
- Faculty of Metallurgy and Material Engineering, Technical University Ostrava,
  Třída 17. listopadu, CZ-708 33 Ostrava - Poruba,
  phone (+4269)6993299, fax (+4269)6994401
- Dept. of Chemical Technology, University of Chemical Technology,
  Technická 5, CZ-166 28, Praha 6,
  phone (+422)24351111, fax (+422)24311082
- Dept. of Material Science, Faculty of Mechanical Engineering,
  Czech Technical University,
  Karlovo nám. 13, CZ-121 35, Praha 2,
  phone (+422)24351111, fax (+422)290302
- Dept. of Material Science, Technical University Liberec,
  Hálkova 6, CZ-461 17 Liberec 1,
  phone (+4248)26441, fax(+4248)23317

DEVELOPMENT AND APPLICATION OF CARBON FIBER AND POLYMER COMPOSITES IN CHINA

Z. SHEN, H. WANG, and Y. LU
Institute of Carbon Fiber Composites, Beijing University of Chemical Technology, Beijing, 100029 China

ABSTRACT

In this paper, development and application of carbon fibers and resin based composites are introduced in China, especially in Beijing University of Chemical Technology (BUCT). There are four parts in this paper. They are: (1) the history and present situation of carbon fiber and resin composites, (2) the application of carbon fiber and resin composites, (3) brief introduction of organizations and institutes, and (4) outlook of carbon fiber and resin composites.

1. Introduction

The relationship between the development of material and technical science is very close. The importance of carbon fiber and its composites has been widely known for over thirty years. The development and application of carbon fiber has become one symbol of a nation's general power. The low density, high tensile strength and high modulus of carbon fiber reinforcement results in its high specific strength and specific modulus. Usually the values of reinforcement specific strength and modulus are $\geq 6.5 \times 10^6$ cm and $\geq 6.5 \times 10^8$ cm respectively, for high performance composites. Advanced composites materials (ACM) are defined as specific strength $\geq 4.0 \times 10^6$ cm and specific modulus $\geq 4.0 \times 10^8$ cm. Therefore, carbon fibers are one of the best reinforcements and their composites are ACM because they have the desirable properties of this material. They have been used in aviation and aerospace for many years. In recent years, the market of ACM has been transferring from the aerospace industry to the transportation, machinery, chemical and sports industries. The ratio of applied field will occupy 60% in civilian market in end of century have been predicted by specialists promoting the ACM industry[1,2].

In this paper, development and application of carbon fibers and resin based composites in China, especially in Beijing University of Chemical Technology (BUCT) are introduced[3-5].

2. Carbon Fiber

2.1. PAN Based Carbon Fibers

2.1.1. Precursor

When carbon fibers were just studied in 1960s polyacrylonitrile fiber was used as precursor. Since 1970s to 1990s a few factories have produced special acrylic precursors, as shown in Table 1.

Table 1. Specification and Production Capacity of Precursor

Manufacturer	Capacity t/y	Process	Tensile Strength CN/dtex	Fineness dtex
Lanzhou Factory Chem. Fibers	60	NaSCN	5	1.11
Jilin Factory of Fiber Materials	90 + 200*	HNO3	5	1.21
Yuci Factory of Chem. Fibers	380	DMSO	4–5	1.23
Shanghai Institute of Synthetic Fibers	5	DMSO	> 4.5	1.10
*being built				

2.1.2. Carbon Fibers

In the 1960s general purpose carbon fibers were produced by batch process. In 1973 the air method of continuous carbonization process was developed by Shanxi Institute of Coal Chemistry, which built the carbon fiber industry in China. A pilot line with capacity of 1 t/y based on air method was built in 1976 and carbon fibers were produced with tensile strength 2.6 GPA, modulus 200 GPa and elongation 1.3%. In the 1990s a 100 t/y factory was opened at Jilin Factory of Fiber Material and carbon fibers were produced with tensile strength ≥ 3.0 GPa, modulus ≥ 220 GPa and elongation ≥ 1.4%. In BUCT a 6 t/y pilot line was also built in the 1990s. Now BUCT has a device for producing high modulus carbon fibers. In the 1990s carbon fibers with high tensile strength and intermediate modulus are developing at BUCT. The total production capacity in China is about 150 t/y.

2.2. Pitch Based Carbon Fibers

General purpose pitch based carbon fibers with tensile strength ≥ 1.0 GPa, modulus ≥ 40 GPa and elongation ≥ 2.0% were developed in 1978 and a 10 t/y pilot line has been built in 1993. Recently an isotropic pitch based carbon fibers factory with capacity of 200 t/y was imported from Ashland, USA, and built in Liaoning province. High performance mesophase pitch based carbon fibers were developed in BUCT with tensile strength 3.0–3.4 GPa, modulus 760–830 GPa and elongation 0.4% which is nearly the level of the advanced countries.

2.3. Rayon Based Carbon Fibers

There are three factories producing rayon based fibers, as listed in Table 2.

Table 2. Rayon Based Fibers Production Factories

Manufacturer	Capacity (t/y)
Nantong Company of Activated Carbon Fibers	10
Liaoyuan Factory of Special Fibers	2
Liaoyuan Factory of Fibers Materials	3

3. Polymer Composites

3.1. Thermosets

Most thermosets are used as the matrix in carbon fiber composites. The application of epoxy resin and polyimides have notably been successful because of their excellent adhesive characterization. The toughness of epoxy resin can be increased by adding thermoplastics such as TPI, PES, PEEK, and PEK-C into the epoxy resin. It is then called modified epoxy resin. Tables 3 and 4 show the properties of modified epoxy resin which are the research results of BUCT.

Table 3. Properties of PEK-C and PES Modified Epoxy Resin

Systems	Tensile strength (MPa)	Modulus (GPa)	Elongation (%)	Tg (°C)
TGDDM / F-51 / DDS	86.1	3.9	2.5	238
TGDDM / F-51 / DDS / PEK-C	86.5	3.6	3.1	240
TGDDM / E-51 / DDS	83.4	3.9	2.8	235
TGDDM / E-51 / DDS / PEK-C	96.5	3.8	3.6	233
TGDDM / E-51 / DDS / PES	82.5	3.8	3.1	225

Table 4. Properties of PAI Modified Epoxy Resin

Systems	Flexture strength (MPa)	Modulus (GPa)	Fracture energy (GIC)	Tg (°C)
TGDDM / DDS	385	58.6	812	—
TDE-85 / DDS	466	58.6	988	280
TDE-85 / DDS / PAI	531	49.0	1612	290

3.2. Thermoplastic Composites

All types of thermoplastic composites used in the world are produced or developed in China. In the 1960s, PA was produced. In the 1970s, PET and PC were produced. In the 1980s, PPO, PES, PPS, POM and PBT were developed. In the early 1990s, PEEK and PEK-C were developed. Thermoplastics have the advantages of unlimited shelf life, low scrap or recyclability, rapid fabrication cycle and toughness or impact resistance. During 1986–1995, BUCT has researched short CF reinforced PEEK composites used to magnetic pulp with severe environment. The properties of PEEK/CF composites are shown in Table 5. In the 1990s, BUCT developed PA and POM/CF composites which were used for rapier heads and gears in textile machinery. Some products of PA and POM composites are shown in Fig. 1. BUCT also researched continuous CF/thermoplastic composites. Their properties are shown in Table 6.

Table 5. Properties of Short Carbon Fibers/PEEK Composites

Content (%)	Tensile strength (MPa)	Modulus (GPa)	Elongation (%)	Impact strength (KJ/m)
0	98	4.3	11.0	8.5
10	171	10.9	3.7	9.7
20	181	13.1	2.0	10.4
30	231	24.5	1.7	11.2

Table 6. Continuous Carbon Fibers/Thermoplastics Composites

Samples		Flexture strength (MPa)	Impact strength (KJ/m)	ILSS (MPa)
PEK-C,	30%	877	114	41
	38%	1048	98	46
	44%	1042	83	53
PES,	30%	950	70	50
	40%	971	55	51
	50%	980	56	66
PDC-2,	30%	493	108	59
	40%	699	109	66
	50%	1999	55	66

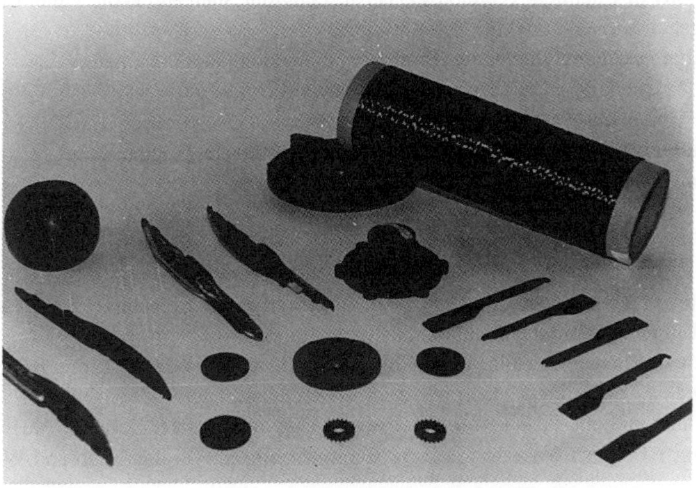

Fig 1. CFRTP samples (rapier head, pump and gears).

4. Applications

The properties of our CF are a little inferior to Japan's T-300, but our application is very active in China. The products include:

- space and aviation products
- sealing products: e.g., high temperature-resistant and corrosion-resistant packing
- sporting goods: rod of badminton rackets and tennis rackets, ice hockey sticks, cars and racing boats, etc.
- textile machines: rigid arrow shaft, arrow head, pulley, handle and static resistant brush.
- medical appliances: bed plate of X-ray machine, head holder, negative box and artificial limbs
- chemical industry: magnetic pumps and the valves
- machinery, traffic and other: parts of robot, gear, tabular spring of vehicle, bike, brake of aircraft and vehicle, cement reinforced products and acoustics material

5. Brief Introduction of Organization Institutes

5.1. Scientific Institutes (Main Researcher About 300 to 400)

5.1.1. Universities

- China National Engineering Research Center for Carbon Fiber (belongs to Beijing University of Chemical Technology) (Prof. Zengmin Shen and Prof. YaFei Lu)
- Zhongshan University (Prof. Hanmin Zeng)
- Shanghai Jiao Tong University (Prof. Renjie Wu and Guoding Zhang)
- China Textile University (Prof. Baojun Qian and Ding Pan)
- Anhui University (Prof. Rengyuan Li)
- Harbin University of Technology (Prof. Yuezen Wei and Zhiqian Zang)
- Hu'nan University (Prof. Zhongyu Xu)

5.1.2. Institutes

- Institute of Shanxi Coal Chemistry of Chinese Academy of Sciences (Prof. Shuan Qian, Lang Liu and Maozhang Wan)
- Institute of Shenyan Metallurgy of Chinese Academy of Sciences (Prof. H. M. Chen)
- Institute of Aeronautical Materials (Senior Specialist Fengfan Zhang)
- Shanghai Synthetic Fibers Research Institute (Senior Engineer Dengbao Xu)
- Nanjing Fiberglass Research Design (Senior Engineer Weijuan Shao and Wenguang Yang)
- Harbin FRP Institute (Senior Engineer Qingbao Kong)

- Beijing Materials Industry FRP Research Institute (Prof. Xuegui Zhang and Youying Wu)
- Shanghai FRP Institute (Senior Engineer Huaili Lu)

5.2. Societies (There is not a Carbon Society in China, but we have established many specialized commissions of carbon graphite and composite materials)

- Specialized Commission of Carbon Materials under the Chinese Metal Society
- Specialized Commission of Carbon-graphite Materials under the Chinese Electro-technical Society
- Specialized Commission of Polymer Composite Material and Technology under Chinese Society of Composite Material
- Specialized Commission of Composite Materials under Aeronautics Society

5.3. Journal

- ATCA MATERIAE COMPOSITAE SINICA (Edited and published by Chinese Society of Composite Materials, sponsored by Beijing University of Aeronautics and Aeronautics).
- CARBON (sponsored by the Socialized Commission of Carbon-graphite Materials under the Chinese Electrotechnical Society), issued in China.
- CARBON TECHNIQUES (sponsored by Jilin Carbon Plant), issued in China.

6. Outlook

- According to theoretical calculation, the real tensile strength of present CF in China is only 3% of its theoretical strength. So there is great potential to improve our CF's mechanical properties. We should put more attention into raw materials to eliminate flaws and microcracks. Strain should be exerted in various technological range (spinning of precursor, pre-oxidation and carbonization). It is also important to arrange carbon layers in order. From above points, we can design many methods to improve CF's properties.
- While improving properties, we should consider methods to lower costs and increase utilization.
- There are too many people in our country. Woods and steels are in serious short supply. The environmental pollution from asbestos is very serious. We should actively develop low cost CF and its composites products to meet the request of renewable industry materials and saving energy.
- Our industrial foundation is weak. The intensity of investment is not enough. For present CF and its composites, research projects are more, but industrialization of economic scale is less. I hope we can strengthen the kinds of cooperation to promote the practical development.

7. References

1. Shigeru, Matsuda (1987). *Reinforced Plastics*, **vol. 33**.
2. Chengjun Liu, Wangrou Fu (1991). *Proceedings Conference on New Development of Composites Material.* China, Ji'nan, pp. 22-26.
3. Zengmin Shen, YaFei Lu, Huiqiong Wang and Reishan Li (1989). *Glass Fiber/ Composite* (Chinese), **no. 5**, pp. 39-46.
4. Zengmin Shen, Huiqiong Wang, Zhengping Wang and YaFei Lu (1989). *Advance in Chem. Ind* (Chinese), **no. 2**, pp. 30-37.
5. YaFei Lu, Zengmin Shen (1990). *1st Symp on New Chem. Mater of China* (Chengdu, China).